Microorganisms for Sustainability

Volume 47

Microorganisms have been in existence since the origin of life on earth and can survive the most extreme habitats or conditions on earth. Microorganisms are involved in regulating biogeochemical cycles, maintaining plant and animal health, and sustaining the global food chain. Moreover, they play crucial roles in addressing the challenges of climate change and achieving the targets of Sustainable Development Goals (SDGs).

This multidisciplinary book series captures the role of microbes towards building a sustainable world, while encompassing cutting-edge technologies and current needs across various fields such as agriculture sustainability, bioremediation, restoration of degraded habitats and wastelands, and food security. Additionally, this series explores microbial applications in industries, and building their utilization in clean and green energy solutions. Furthermore, themes like microbial secondary metabolites, extremophilic microbes and modern omics, including next generation sequencing and metagenomics, are also covered in this series.

With contributions from researchers across the globe, this series addresses the important call of 'One Planet-One Health-One Future'. It comprises a collection of diverse volumes that provides insights for scientists, young researchers, educators and decision-makers in the government, private sector, and non-governmental organizations, empowering their efforts to achieve the global goals.

The series invites, evaluates, and accepts book proposals to ensure a diverse, inclusive, and evolving program. The final decision regarding acceptance rests with the series editor.

Peer-review

This book series follows a stratified review process. Proposals for individual volumes are reviewed by the series editor and then the editorial board members. On a case-to-case basis, external reviewers are also invited for further evaluation of the book proposal. Review of the chapters is the responsibility of the volume editor(s). A manuscript submission platform has been recently made available to the authors, volume editors and the series editor.

Aditya Khamparia • Babita Pandey •
Devendra Kumar Pandey • Deepak Gupta
Editors

Microbial Data Intelligence and Computational Techniques for Sustainable Computing

Springer

Editors
Aditya Khamparia
Department of Computer Science
Babasaheb Bhimrao Ambedkar University
Amethi, Uttar Pradesh, India

Babita Pandey
Department of Computer Science
Babasaheb Bhimrao Ambedkar University
Lucknow, Uttar Pradesh, India

Devendra Kumar Pandey
School of Bioengineering and Biosciences
Lovely Professional University
Phagwara, Punjab, India

Deepak Gupta
Department of Computer Science
Maharaja Agrasen Institute of Technology
Delhi, Delhi, India

ISSN 2512-1901 ISSN 2512-1898 (electronic)
Microorganisms for Sustainability
ISBN 978-981-99-9620-9 ISBN 978-981-99-9621-6 (eBook)
https://doi.org/10.1007/978-981-99-9621-6

This Springer imprint is published by the registered company Springer Nature Singapore Pte Ltd.
The registered company address is: 152 Beach Road, #21-01/04 Gateway East, Singapore 189721, Singapore

Paper in this product is recyclable.

Preface

Microbes are ubiquitous in nature, and their interactions among each other is a key strategy for colonizing diverse habitats. The core idea of sustainable computing is to deploy algorithms, models, policies, and protocols to improve energy efficiency and management of resources, enhancing ecological balance, biological sustenance, and other services on societal contexts. This book offers a comprehensive intelligent and computational techniques for microbial data associated with either plant microbe, human microbes, etc. The readers will be able to understand the positive findings as well as the negative findings obtained by the usage of computational AI and distributed computing techniques for microbial data. It entails data extraction from various sources followed by pre-processing of data, and how to make effective use of extracted data for application-based research. The book also involves computer-assisted tools for visualization and representation of complex microbial data. The book explores the conventional methods as well as the most recently recognized high-throughput technologies which are important for productive agroecosystems to feed the growing global population.

The main reason behind the success rate of deep learning and biomedical data analysis techniques is its ability to reason and learn in an environment of unique data and real case studies. This book will focus on involvement of microbial data intelligence assisted and plant treatment and care-driven intelligent computing methods, state of arts, novel findings, and recent advances in different applications and areas like drug and plant image classification with a wide range of theory and methodologies has been investigated to tackle the complex and challenging problems.

Gathering the contributions by active researchers in these fields, the book covers the theories as well as important real-time practical considerations. This book also includes the design of a set of AI hybrid algorithms in detail, showing how to use them in practice to solve problems relating to genome and plant image classification, data analysis, bioinformatics, and engineering control. It is intended as a reference guide to advanced hybrid computational intelligence methods for graduate students and researchers in applied mathematics and optimization, computer science, and engineering. This book is of interest to teachers, researchers, microbiologist, computer bioinformatics scientists, plant and environmental scientist, and those interested in environment stewardship around the world. The book also serves as an additional reading material for undergraduate and graduate students of computer science, biomedical, agriculture, human science, forestry, ecology, soil science, and environmental sciences, and policymakers consider this a useful book to read.

Objective of the Book

The primary emphasis of this book is to introduce different computational intelligence-assisted techniques, methodology, and intelligent algorithms applied to categorize and classify microbial-assisted plant datasets. Gathering contributions by active researchers in those fields, the book covers the theories as well as important practical considerations. In turn, it provides an overview of microbial data-driven image analysis, deep learning, computer vision, and chaotic optimization enabled evaluation of the proposed solutions in the manufacturing sector and compares the advantages and disadvantages related to the same. This book will endeavor to endow with significant frameworks, theory, design methods, and the latest empirical research findings in the area of intelligent computing.

Amethi, Uttar Pradesh, India — Aditya Khamparia
Lucknow, Uttar Pradesh, India — Babita Pandey
Phagwara, Punjab, India — Devendra Kumar Pandey
Delhi, Delhi, India — Deepak Gupta

Contents

1 **The Contribution of Artificial Intelligence to Drug Discovery: Current Progress and Prospects for the Future** 1
Umesh Gupta, Ayushman Pranav, Anvi Kohli, Sukanta Ghosh, and Divya Singh

2 **Prediction of Plant Disease Using Artificial Intelligence** 25
Manoj Ram Tammina, K. Sumana, Pavitar Parkash Singh, T. R. Vijaya Lakshmi, and Sagar Dhanraj Pande

3 **Computer Vision-based Remote Care of Microbiological Data Analysis** 49
Pritesh Kumar Jain and Sandeep Kumar Jain

4 **A Comparative Study of Various Machine Learning (ML) Approaches for Fake News Detection in Web-based Applications** 59
Mahabub Hasan Mahalat, Sushree Bibhuprada B. Priyadarshini, Sandip Swain, Shobhit Sahoo, Atish Mohapatra, and Mangaldeep Das

5 **Analytics and Decision-making Model Using Machine Learning for Internet of Things-based Greenhouse Precision Management in Agriculture** 77
Ashay Rokade, Manwinder Singh, Anudeep Goraya, and Balraj Singh

6 **DistilBERT-based Text Classification for Automated Diagnosis of Mental Health Conditions** 93
Diwakar and Deepa Raj

7 **An Optimized Hybrid ARIMA-LSTM Model for Time Series Forecasting of Agricultural Production in India** 107
Babita Pandey, Arvind Shukla, and Aditya Khamparia

8 An Exploratory Analysis of Machine Intelligence-enabled Plant Diseases Assessment 121
Ashis Pattanaik, Agniva Bhattacharya, and Sushruta Mishra

9 Synergizing Smart Farming and Human Bioinformatics Through IoT and Sensor Devices 139
Sandeep Kumar Jain and Pritesh Kumar Jain

10 Deep Learning-Assisted Techniques for Detection and Prediction of Colorectal Cancer From Medical Images and Microbial Modality 151
Ravi Kumar, Amritpal Singh, and Aditya Khamparia

11 Smart Farming and Human Bioinformatics System Based on Context-Aware Computing Systems 171
Sini Anna Alex, T. P. Pallavi, and G. C. Akshatha

12 Plant Diseases Diagnosis with Artificial Intelligence (AI) 187
Syed Muzammil Munawar, Dhandayuthabani Rajendiran, and Khaleel Basha Sabjan

13 Analyzing the Frontier of AI-Based Plant Disease Detection: Insights and Perspectives 195
Mridula Dwivedi, Babita Pandey, and Vipin Saxena

14 Fuzzy and Data Mining Methods for Enhancing Plant Productivity and Sustainability 205
Khalil Ahmed, Mithilesh Kumar Dubey, Devendra Kumar Pandey, and Sartaj Singh

15 Plant Disease Diagnosis with Artificial Intelligence (AI) 217
Muhammad Naveed, Muhammad Majeed, Khizra Jabeen, Nimra Hanif, Rida Naveed, Sania Saleem, and Nida Khan

16 Sustainable AI-Driven Applications for Plant Care and Treatment 235
Muhammad Naveed, Nafeesa Zahid, Ibtihaj Fatima, Ayesha Saleem, Muhammad Majeed, Amina Abid, Khushbakht Javed, Rehmana Wazir, and Amina Qasim

17 Use Cases and Future Aspects of Intelligent Techniques in Microbial Data Analysis 259
Muhammad Naveed, Zaibun-nisa Memon, Muhammad Abdullah, Syeda Izma Makhdoom, Arooj Azeem, Sarmad Mehmood, Maida Salahuddin, Zeerwah Rajpoot, and Muhammad Majeed

18 Early Crop Disease Identification Using Multi-fork Tree Networks and Microbial Data Intelligence 281
S. S. Ittannavar, B. P. Khot, Vibhor Kumar Vishnoi, Swati Shailesh Chandurkar, and Harshal Mahajan

19 Guarding Maize: Vigilance Against Pathogens Early Identification, Detection, and Prevention 301
Khalil Ahmed, Mithilesh Kumar Dubey, and Sudha Dubey

20 Comprehensive Analysis of Deep Learning Models for Plant Disease Prediction 319
Narendra Pal Singh Rathor, Praveen Kumar Bhanodia, and Aditya Khamparia

21 Enhancing Single-Cell Trajectory Inference and Microbial Data Intelligence 341
Bhargavi Posinasetty, Mukesh Soni, Sagar Dhanraj Pande, Krishnendu Adhikary, and Dhirendra Kumar Tripathi

22 AI-Assisted Methods for Protein Structure Prediction and Analysis 365
Divya Goel, Ravi Kumar, and Sudhir Kumar

Editors and Contributors

About the Editors

Aditya Khamparia has expertise in teaching, entrepreneurship, and research and development of a decade. He is currently working as an assistant professor and coordinator of the Department of Computer Science, Babasaheb Bhimrao Ambedkar University, India. He received his Ph.D. degree from Lovely Professional University, Punjab, in May 2018. He has completed his M. Tech. from VIT University and B. Tech. from RGPV, Bhopal. He has completed his PDF from UNIFOR, Brazil. He has more than 100 research papers along with book chapters including more than 20 papers in top journals with cumulative impact factor of above 100 to his credit. Additionally, he has authored and edited a cumulative of 11 books. His research interests include machine learning, deep learning, educational technologies, and computer vision.

Babita Pandey working as an associate professor in the Department of Computer Science, Babasaheb Bhimrao Ambedkar University, Lucknow, India. Her research interests include biomedical engineering, e-learning, computational intelligence, and security systems. She has published more than 100 publications and conferences including more than 40 SCI Indexed Journals.

Devendra Kumar Pandey is currently working as a professor at Lovely Professional University, India. He obtained his Ph.D. in biochemical engineering from the Indian Institute of Technology, India. His main area of interest is related to pharmacology and toxicology, plant and soil sciences, and molecular sciences. His area of expertise includes plant biotechnology, plant–microbe interaction, chromatography techniques, i.e., HPTLC, HPLC, LC-MS, molecular markers and bioactive compound markers for medicinal plants, and bioactive compounds. He has published more than 100 articles in international journals with papers also in national and international conferences contributed as author/co-author.

Deepak Gupta received a B.Tech. in 2006 from the Guru Gobind Singh Indraprastha University, India. He received M.E. in 2010 from Delhi Technological University, India, and Ph. D. in 2017 from Dr. APJ Abdul Kalam Technical University, India. He has completed his post-doc from Inatel, Brazil. With 13 years of rich expertise in teaching and 2 years in the industry, he focuses on rational and practical learning. He has contributed massive literature to the fields of intelligent data analysis, biomedical engineering, artificial intelligence, and soft computing. He has served as editor-in-chief, guest editor, and as associate editor in various reputed journals. He has actively been organizing various reputed international conferences. He has authored/edited 43 books. He has published 200 scientific research publications including more than 100 SCI Indexed Journals.

Contributors

Muhammad Abdullah Biodiversity Park, Cholistan Institute of Desert Studies (CIDS), The Islamia University of Bahawalpur, Bahawalpur, Pakistan

Amina Abid Department of Biotechnology, Faculty of Science and Technology, University of Central Punjab, Lahore, Punjab, Pakistan

Krishnendu Adhikary Centurion University of Technology and Management, Bhubaneswar, Odisha, India

Khalil Ahmed School of Computer Application, Lovely Professional University, Phagwara, Punjab, India

G. C. Akshatha Department of CSE (AI & ML), Ramaiah Institute of Technology, Bangalore, Karnataka, India

Sini Anna Alex Department of CSE (AI & ML), Ramaiah Institute of Technology, Bangalore, Karnataka, India

N. Ashwini Department of Computer Science and Engineering, BMS Institute of Technology and Management, Bengaluru, Karnataka, India

Arooj Azeem Department of Biotechnology, Faculty of Science and Technology, University of Central Punjab, Lahore, Punjab, Pakistan

Praveen Kumar Bhanodia Computer Science Engineering, Acropolis Institute of Technology and Research, Bhopal, Madhya Pradesh, India

Agniva Bhattacharya Kalinga Institute of Industrial Technology, Deemed to be University, Bhubaneswar, Odisha, India

G. Bhavya Department of Information Science and Engineering, BMS Institute of Technology and Management, Bengaluru, Karnataka, India

Swati Shailesh Chandurkar Pimpri Chinchwad College of Engineering, Pune, India

Mangaldeep Das Computer Science & Information Technology, Institute of Technical Education and Research, Siksha 'O' Anusandhan Deemed to be University, Bhubaneswar, India

Diwakar Diwakar University (A Central University), Lucknow, Uttar Pradesh, India

Mithilesh Kumar Dubey School of Computer Application, Lovely Professional University, Phagwara, Punjab, India

Sudha Dubey Department of Sociology, Lovely Professional University, Phagwara, Punjab, India

Mridula Dwivedi Department of Computer Science, Babasaheb Bhimrao Ambedkar University, Lucknow, Uttar Pradesh, India

Ibtihaj Fatima Department of Botany, University of Education, Lahore, Punjab, Pakistan

Sukanta Ghosh SCSAI, SR University, Warangal, Telangana, India

Divya Goel Department of Biotechnology, H.N.B. Garhwal University, Srinagar Garhwal, Uttarakhand, India

Anudeep Goraya School of Computer Science and Engineering, Lovely Professional University, Phagwara, Punjab, India

Umesh Gupta SCSET, Bennett University, Greater Noida, Uttar Pradesh, India

Nimra Hanif Department of Biotechnology, Faculty of Science and Technology, University of Central Punjab, Lahore, Punjab, Pakistan

S. S. Ittannavar ECE Department, Hirasugar Institute of Technology, Belgaum, India

Khizra Jabeen Department of Biotechnology, Faculty of Science and Technology, University of Central Punjab, Lahore, Punjab, Pakistan

Pritesh Kumar Jain Department of Computer Science and Engineering, Shri Vaishnav Vidyapeeth Vishwavidyalaya, Indore, Madhya Pradesh, India

Sandeep Kumar Jain Department of Computer Science and Engineering, Shri Vaishnav Vidyapeeth Vishwavidyalaya, Indore, Madhya Pradesh, India

Khushbakht Javed Department of Biotechnology, Faculty of Science and Technology, University of Central Punjab, Lahore, Punjab, Pakistan

Aditya Khamparia Department of Computer Science, Babasaheb Bhimrao Ambedkar University, Satellite Centre, Amethi, Uttar Pradesh, India

Nida Khan Department of Botany, University of Science and Technology Bannu, Bannu, Khyber Pakhtunkhwa, Pakistan

B. P. Khot ECE Department, Hirasugar Institute of Technology, Belgaum, India

Anvi Kohli SCSET, Bennett University, Greater Noida, Uttar Pradesh, India

Ravi Kumar Department of Computer Science Engineering, Lovely Professional University, Jalandhar, Punjab, India
Department of Computer Science Engineering, Jawaharlal Nehru Government Engineering College, Sundernagar, Himachal Pradesh, India

Sudhir Kumar Department of Biotechnology, H.N.B. Garhwal University, Srinagar Garhwal, Uttarakhand, India

T. R. Vijaya Lakshmi Mahatma Gandhi Institute of Technology, Gandipet, Hyderabad, India

Harshal Mahajan Indira College of Engineering and Management, Pune, India

Mahabub Hasan Mahalat Computer Science & Information Technology, Institute of Technical Education and Research, Siksha 'O' Anusandhan Deemed to be University, Bhubaneswar, India

Muhammad Majeed Department of Botany, University of Gujrat, Gujrat, Pakistan

Syeda Izma Makhdoom Department of Biotechnology, Faculty of Science and Technology, University of Central Punjab, Lahore, Punjab, Pakistan

Sarmad Mehmood Department of Biotechnology, Faculty of Science and Technology, University of Central Punjab, Lahore, Punjab, Pakistan

Zaibun-nisa Memon Department of Zoology, Shah Abdul Latif University, Khairpur Mirs, Sindh, Pakistan

Sushruta Mishra Kalinga Institute of Industrial Technology, Deemed to be University, Bhubaneswar, Odisha, India

Atish Mohapatra Computer Science & Information Technology, Institute of Technical Education and Research, Siksha 'O' Anusandhan Deemed to be University, Bhubaneswar, India

Syed Muzammil Munawar Department of Biochemistry, C. Abdul Hakeem College (Autonomous), Melvisharam, Vellore, Tamil Nadu, India

Muhammad Naveed Department of Biotechnology, Faculty of Science and Technology, University of Central Punjab, Lahore, Punjab, Pakistan

Rida Naveed Department of Biotechnology, Faculty of Science and Technology, University of Central Punjab, Lahore, Punjab, Pakistan

T. P. Pallavi Department of CSE (Cyber Security), Ramaiah Institute of Technology, Bangalore, Karnataka, India

Sagar Dhanraj Pande School of Computer Science and Engineering, VIT-AP University, Amaravati, Andhra Pradesh, India

Babita Pandey Babasaheb Bhimrao Ambedkar (Central) University, Lucknow, Uttar Pradesh, India
Department of Computer Science, Babasaheb Bhimrao Ambedkar University, Lucknow, Uttar Pradesh, India

Devendra Kumar Pandey School of Computer Application, Lovely Professional University, Phagwara, Punjab, India

Ashis Pattanaik Kalinga Institute of Industrial Technology, Deemed to be University, Bhubaneswar, Odisha, India

Bhargavi Posinasetty Department of Masters in Public Health, The University of Southern Mississippi, Hattiesburg, MS, USA

Ayushman Pranav SCSET, Bennett University, Greater Noida, Uttar Pradesh, India

Sushree Bibhuprada B. Priyadarshini Computer Science & Information Technology, Institute of Technical Education and Research, Siksha 'O' Anusandhan Deemed to be University, Bhubaneswar, India

Amina Qasim Department of Botany, Minhaj University Lahore, Lahore, Pakistan

Deepa Raj University (A Central University), Lucknow, Uttar Pradesh, India

Zeerwah Rajpoot Department of Biotechnology, Faculty of Science and Technology, University of Central Punjab, Lahore, Punjab, Pakistan

Narendra Pal Singh Rathor Computer Science Engineering, Acropolis Institute of Technology and Research, Bhopal, Madhya Pradesh, India

Ashay Rokade School of Electronics and Electrical Engineering, Lovely Professional University, Phagwara, Punjab, India

Shobhit Sahoo Computer Science & Information Technology, Institute of Technical Education and Research, Siksha 'O' Anusandhan Deemed to be University, Bhubaneswar, India

Maida Salahuddin Department of Biotechnology, Faculty of Science and Technology, University of Central Punjab, Lahore, Punjab, Pakistan

Ayesha Saleem Department of Biotechnology, Faculty of Science and Technology, University of Central Punjab, Lahore, Punjab, Pakistan

Sania Saleem Department of Plant Sciences, Quaid-i-Azam University, Islamabad, Pakistan

Vipin Saxena Department of Computer Science, Babasaheb Bhimrao Ambedkar University, Lucknow, Uttar Pradesh, India

Arvind Shukla Babasaheb Bhimrao Ambedkar (Central) University, Lucknow, Uttar Pradesh, India

Amritpal Singh Department of Computer Science Engineering, Lovely Professional University, Phagwara, Punjab, India

Balraj Singh School of Computer Science and Engineering, Lovely Professional University, Phagwara, Punjab, India

Divya Singh SCSET, Bennett University, Greater Noida, Uttar Pradesh, India

Manwinder Singh School of Electronics and Electrical Engineering, Lovely Professional University, Phagwara, Punjab, India

Pavitar Parkash Singh Department of Management, Lovely Professional University, Phagwara, Punjab, India

Sartaj Singh School of Computer Application, Lovely Professional University, Phagwara, Punjab, India

Mukesh Soni Department of CSE, University Centre for Research & Development, Chandigarh University, Mohali, Punjab, India

K. Sumana Department of Microbiology, JSS AHER, Mysuru, Karnataka, India

Sandip Swain Computer Science & Information Technology, Institute of Technical Education and Research, Siksha 'O' Anusandhan Deemed to be University, Bhubaneswar, India

Manoj Ram Tammina Innovation, Bread Financial, Columbus, OH, USA

Dhirendra Kumar Tripathi Sri Satya Sai University of Technology and Medical Sciences, Sehore, MP, India

Vibhor Kumar Vishnoi College of Computing Sciences and Information Technology, Teerthanker Mahaveer University, Moradabad, India

Rehmana Wazir Department of Biotechnology, Faculty of Science and Technology, University of Central Punjab, Lahore, Punjab, Pakistan

Nafeesa Zahid Department of Botany, University of Kotli, Kotli, Azad Jammu and Kashmir, Pakistan

Chapter 1
The Contribution of Artificial Intelligence to Drug Discovery: Current Progress and Prospects for the Future

Umesh Gupta, Ayushman Pranav, Anvi Kohli, Sukanta Ghosh, and Divya Singh

Abstract The swift progress of artificial intelligence (AI) is fundamentally altering the terrain of drug discovery, carrying the substantial potential to accelerate the pinpointing of new drugs and improve the effectiveness and efficiency of the drug development process. Across various stages of drug discovery, AI methodologies arc proving instrumental:

1. **Target Identification and Validation:** AI demonstrates prowess by sifting through extensive genomic and proteomic datasets to discern and affirm fresh drug targets. This computational prowess enables the identification of potential candidates for therapeutic intervention (Kim et al., Biotechnol Bioprocess Eng 25:895–930, 2020).
2. **Virtual Screening:** The application of AI extends to efficiently screening vast compound libraries. AI predictions of binding affinities and pertinent properties offer a streamlined approach to identifying promising drug candidates (Sahayasheela et al., Nat Prod Rep 39:2215, 2022).
3. **Drug Design:** AI's capabilities span the design phase, aiding in creating innovative drug molecules with specified attributes like enhanced potency, selectivity, and pharmacokinetics (Blanco-Gonzalez et al. Pharmaceuticals 16:891, 2023).
4. **Drug Repurposing:** AI breathes new life into existing drugs, uncovering alternate applications. This strategy is a cost-effective and time-sensitive avenue for developing new treatment avenues (Ren et al., Chem Sci 14:1443–52, 2023).
5. **Clinical Trial Design:** Leveraging AI, clinical trial frameworks can be optimized. AI empowers the precise selection of patients, appropriate dosages, and

U. Gupta (✉) · A. Pranav · A. Kohli · D. Singh
SCSET, Bennett University, Gr. Noida, Uttar Pradesh, India
e-mail: Divya.Singh@bennett.edu.in

S. Ghosh
SCSAI, SR University, Warangal, Telangana, India

A. Khamparia et al. (eds.), *Microbial Data Intelligence and Computational Techniques for Sustainable Computing*, Microorganisms for Sustainability 47,
https://doi.org/10.1007/978-981-99-9621-6_1

predictive assessments of trial success probability (Keshavarzi Arshadi et al., Artif Intell 65, 2020).

While AI's integration into drug discovery remains relatively nascent, it holds the potential to revolutionize the field. Recent strides in AI technology have enabled the resolution of complex challenges, such as identifying targets for refractory ailments and engineering drugs with heightened efficacy and reduced toxicity. The trajectory of AI in drug discovery appears promising. Its influence is poised to intensify, driving expedited drug discovery, refining the efficiency of the developmental journey, and hastening the availability of novel treatments for patients.

Keywords Drug discovery · Artificial intelligence · Machine learning · sustainable computing · Deep learning

1.1 Introduction

In this chapter, we undertake a detailed exploration that dives into the interdependent connection between drug discovery and the transformative capabilities of artificial intelligence (AI). The journey of drug discovery, a complex voyage aimed at discovering new chemical entities (NCEs) with the potential to drive therapeutic advancements, traverses a terrain characterized by carefully defined stages. These encompass the meticulous identification of disease-triggering molecular targets, the astute curation and refinement of NCEs, the arduous passage through preclinical scrutiny, and the exacting challenge of human clinical trials. Amid the range of challenges and uncertainties along this journey, AI's indelible mark as the conduit for pioneering treatments remains unwavering. Within this narrative, the emergence of AI as a catalyzing agent takes center stage, orchestrating enhancements that streamline process efficiency and augment outcomes. Our expedition unfurls, casting a luminous spotlight on the manifold ways AI's prowess resonates across pivotal dimensions, spanning the domains of target identification, lead discovery, optimization, preclinical assessment, and clinical trials. Furthermore, we delve into the resurgence of microbes, hallowed sources of medicinal innovation, as AI infuses a renewed vitality into their exploration within the realm of drug discovery. As AI's trajectory seamlessly converges with the path of drug discovery, it heralds a paradigm shift—an era in which the fusion of innovation and computational brilliance forges novel pathways within the domain of therapeutics. This chapter is a testament to the symbiosis of scientific ingenuity and AI's transformative potential, poised to unveil unprecedented horizons within the expansive field of drug discovery.

1.2 Historical Evolution of Drug Discovery

Conventional avenues of drug discovery have long relied on a trial-and-error methodology, entailing the meticulous screening of expansive compound libraries to identify those exhibiting sought-after biological activities. However, this approach is arduous, time-intensive, and often yields compounds with undesirable side effects or toxicity.

The landscape shifted with the advent of computational methodologies in the 1990s, which introduced predictive computer models to gauge compound properties and biological potentials. This infusion of computational prowess expedited drug discovery and augmented the success rate of subsequent drug development endeavors.

In recent times, artificial intelligence (AI) has started to gain attention in drug discovery. The wide array of tools offered by AI, including machine learning and deep learning, brings forth the ability to analyze extensive data repositories, identify complex patterns, and generate predictive insights. This convergence of AI and drug discovery promises to fundamentally revolutionize the process, ushering in enhanced efficiency and efficacy. Illustrating this synergy are specific instances where AI is leaving its indelible mark on drug discovery:

- Drug Repurposing
  AI assumes a pivotal role in reimagining the potential uses of existing drugs. By scrutinizing comprehensive datasets, AI identifies drugs ripe for repurposing, as exemplified by identifying remdesivir, an Ebola-originated drug, for COVID-19 treatment (Kim et al. 2020).
- Target Identification
  AI's analytical capabilities come to the fore in deciphering the intricate landscape of disease-associated proteins and molecules. By analyzing vast genomic and proteomic datasets, AI zeroes in on potential drug targets (Sahayasheela et al. 2022).
- Lead Optimization
  AI breathes fresh life into the lead optimization process, tailoring novel compounds that exhibit heightened potency, superior safety profiles, and boosted viability in clinical trials (Blanco-Gonzalez et al. 2023).
- Virtual Screening
  In virtual screening, AI acts as an accelerator, sieving through expansive compound libraries with heightened precision to identify compounds more likely to exhibit desired biological activities (Ren et al. 2023).

Although the incorporation of AI into drug discovery is at an early phase, its capacity to revolutionize the industry is unquestionable. The capabilities of AI are poised to bring about a fresh era characterized by enhanced efficiency, effectiveness, and innovative breakthroughs in the field of drug discovery. As the synergistic partnership between AI and drug discovery gathers momentum, the future holds great potential for uncovering new targets and ground-breaking treatments for various diseases.

1.3 Fundamentals of Artificial Intelligence in Drug Discovery

Contained within this chapter is an investigation into the fundamental principles of AI applied within the field of drug discovery. The rapid evolution of AI holds the transformative potential to reshape various sectors, and drug discovery is no exception. Integrating AI techniques has opened avenues to streamline intricate processes such as data mining, virtual screening, and molecular design, accelerating drug discovery and mitigating the financial burdens of new drug development. Central to AI's influence in drug discovery are vital concepts that warrant exploration:

- Machine Learning: A cornerstone of AI, machine learning enables computers to glean insights from data without explicit programming. By identifying patterns and making predictions, machine learning algorithms contribute to tasks ranging from identifying potential drug targets to predicting drug candidates' efficacy and toxicity and optimizing the design of novel drugs (Kim et al. 2020).
- Deep Learning: Within the purview of machine learning lies deep learning, a paradigm utilizing artificial neural networks to glean insights from data. Inspired by the human brain, these networks unravel intricate data patterns. Deep learning is essential in drug discovery to dissect molecular data and ascertain potential drug targets (Sahayasheela et al. 2022).
- Reinforcement Learning: A transformative AI concept, reinforcement learning enables agents to learn optimal behavior within an environment through trial and error, reinforced by rewards for favorable actions and penalties for unfavorable ones. In drug discovery, reinforcement learning holds promise for designing effective and safe drugs through iterative learning (Blanco-Gonzalez et al. 2023).

The very fabric of AI in drug discovery is interwoven with mathematical foundations encompassing algorithms and statistical methods:

1. Linear Regression: A statistical tool to predict a continuous variable from a set of independent variables, it finds utility in estimating drug candidate efficacy from molecular properties (Ren et al. 2023). The equation for linear regression is.

$$y = mx + b$$

   where y is the predicted value, m is the slope, b is the y-intercept, and x is the independent variable.
2. Support Vector Machines: A machine learning algorithm, support vector machines are adept at classification and regression tasks, effectively identifying potential drug targets and predicting drug candidate toxicity (Keshavarzi Arshadi et al. 2020).

   The equation for support vector machines is:

$$f(x) = w \cdot x + b$$

where *f*(*x*) is the predicted value, *w* is the weight vector, *b* is the bias, and *x* is the independent variable.

3. Artificial Neural Networks: These are also inspired by the human brain and excel in learning complex data patterns, presenting a practical approach for tasks such as image and video processing and natural language processing. Within drug discovery, they play a pivotal role in molecular data analysis and target identification (Jiménez-Luna et al. 2021). The equation for artificial neural networks is:

$$y = f(Wx + b)$$

where *y* is the predicted value, *W* is the weight matrix, *x* is the input vector, *b* is the bias vector, and *f*() is the activation function.

Exploring the differences between machine learning and deep learning, the chapter unravels the equations that underpin these approaches, offering a mathematical insight into their mechanisms. Machine learning algorithms, being more straightforward, and deep learning algorithms, with their complexity and ability to learn from unlabeled data, are utilized in various tasks, including classification, regression, image recognition, and language processing.

In essence, this chapter illuminates the ever-evolving landscape where AI's potential converges with the intricacies of drug discovery. AI's rapid evolution is poised to revolutionize industries, and its potential in drug discovery is undeniable. Through computational ingenuity, AI unlocks the door to expedite drug discovery, propelling innovation and efficiency to unveil novel therapeutic avenues.

1.4 Data-centric Approaches in Artificial Intelligence for the Field of Drug Discovery

- *The significance of extensive datasets in drug discovery (*Kim et al. 2020; Sahayasheela et al. 2022; Blanco-Gonzalez et al. 2023*)*
  The intricacies of drug discovery, characterized by its duration and costs, often confront hurdles due to limited accessible data. Integrating big data can prove transformative, furnishing researchers with deep insights into drug targets, disease mechanisms, and potential candidates. This repository of information can drive the identification of novel drug targets, enable the design of fresh pharmaceuticals, and optimize the developmental trajectory. For instance, the infusion of big data can facilitate:
  - Identifying novel drug targets via comprehensive analyses of genomic, transcriptomic, and proteomic datasets

  - Design of innovative drugs by sifting through extensive compound libraries to identify those interacting with specific targets
  - Enhancement of the developmental trajectory through the utilization of data for predicting the safety and effectiveness of prospective drugs

- *Preprocessing and normalization of data: Mathematical frameworks (*Kim et al. 2020*;* Ren et al. 2023*):*

  Before employing big data for drug discovery, the preparatory steps encompass data pre-processing and normalization. This entails refining the data, purging errors and anomalies, and adapting it to a format amenable to machine learning algorithms. Several mathematical models are instrumental in data pre-processing and normalization, including:

  - Principal Component Analysis (PCA): A statistical technique that condenses dataset dimensionality while conserving crucial information
  - K-means clustering: A machine learning algorithm that clusters similar data points together
  - Gaussian Mixture Models (GMMs): A class of probabilistic models representing data point distributions

- *Overfitting mitigation and regularization through mathematical equations (*Kim et al. 2020*;* Keshavarzi Arshadi et al. 2020*):*

  In machine learning for drug discovery, overfitting arises when a model excessively learns from the training data, hampering its capacity to apply that knowledge to new data and resulting in inaccurate predictions. To counter overfitting, an array of regularization techniques exists, including:

  - L1 Regularization: Penalizes models for substantial coefficients, curtailing overfitting by curbing reliance on training data
  - L2 Regularization: Discourages significant squared coefficients, curbing overfitting's impact on training data
  - Dropout: A technique that randomly eliminates certain features during training, ensuring that no single feature is relied upon

The mathematical formulations underpinning these regularization techniques are as follows:

- L1 Regularization:

$$J(\theta) = L(\theta) + \alpha \left\|\theta\right\| 1$$

  where $J(\theta)$ is the cost function of the model. $L(\theta)$ is the loss function of the model. α is the regularization parameter. θ is the vector of model parameters
- L2 Regularization:

$$J(\theta) = L(\theta) + \alpha \left\| \theta \right\| 2$$

- Dropout:

$$P(\text{dropout}) = 1 - p$$

where P(dropout) is the probability that a feature will be dropped out. p is the dropout rate

The ascendancy of data-driven methodologies in drug discovery is palpable. These approaches surmount the challenges intrinsic to the drug discovery process, such as data scarcity and overfitting. By synergizing big data with machine learning, researchers expedite the quest for novel pharmaceuticals, amplifying the efficiency and efficacy of drug discovery processes.

1.5 Data-driven Approaches in AI for Drug Discovery

Microbial Natural Products (MNPs) encompass various compounds synthesized by microorganisms, including bacteria, fungi, and algae. These compounds exhibit a spectrum of biological activities, spanning antimicrobial, anticancer, and antiviral properties. With a history rooted in traditional medicine, many MNPs have found their place in modern therapeutic practices.

Exemplary MNPs include:

- Penicillin: A renowned antibiotic effective against diverse infections
- Taxol: A cancer-fighting agent used in breast cancer and other malignancies
- Artemisinin: An antimalarial weapon combatting malaria
- Sirolimus: An immunosuppressive tool aiding organ transplant acceptance
- *The Transition: Traditional vs. AI-Powered Discovery from Microbial Sources*

  Historically, the discovery of MNPs relied on a blend of screening and bioactivity tests, a time-intensive and resource-demanding endeavor often hindered in identifying novel MNPs with desired attributes. Artificial intelligence (AI) has recently become a driving force in accelerating the discovery of MNPs. AI can mechanize several steps of traditional drug discovery, encompassing:

  - Screening extensive compound libraries for bioactivity
  - Predicting MNP structures and properties
  - Identifying potential MNP targets
  - Designing novel MNPs with specific characteristics

- *The Contribution of Mathematical Models in Microbial Drug Discovery Driven by Artificial Intelligence*

  The exploration of AI-driven microbial drug discovery pivots on using mathematical models. These models enable:

  - Bioactivity prediction based on MNP chemical structures
  - Prediction of MNP target proteins

- *Tailoring MNPs with desired attributes*
  Prominent mathematical models in this domain encompass:
  - Quantitative Structure-Activity Relationship (QSAR) models: Leverage MNP chemical structures to predict bioactivity
  - Molecular docking models: Anticipate MNP-protein binding affinities
  - Virtual screening models: Screen extensive compound libraries for potential bioactivity
- *AI's Applications in Microbial Natural Product Drug Discovery*
  AI catalyzes MNP discovery through diverse avenues, including:
  - Automated Screening: AI automates compound library screening, reducing time and costs.
  - Target Identification: AI identifies potential MNP targets, streamlining research efforts.
  - MNP Design: AI aids in designing MNPs with desired properties, enhancing effectiveness and safety.
  - Drug Repurposing: AI identifies new uses for existing drugs, accelerating treatment availability.
- *Navigating Challenges and Opportunities*

The usage of AI in microbial natural product drug discovery has its own challenges and opportunities: data availability, MNP complexity, and the evolving nature of AI techniques. Opportunities encompass accelerated drug discovery, new drug target exploration, and tailored drug design. AI holds immense potential in reshaping microbial natural product drug discovery. Through addressing challenges and leveraging opportunities, AI stands as a transformative force, poised to expedite the development of novel drugs to combat a myriad of diseases.

1.6 Hurdles and Prospects in Artificial Intelligence for the Field of Drug Discovery

1.6.1 Navigating Challenges and Embracing Prospects in AI-driven Drug Discovery

The traditional drug discovery processes are lengthy, expensive, and frequently inefficient, entailing a significant span of 10–15 years of research and finally getting a new drug on the shelves. However, AI holds the transformative potential to expedite this labyrinthine process through task automation, encompassing:

- Data Mining: AI's prowess in sifting through vast biological and chemical datasets to unearth potential drug targets and candidates
- Virtual Screening: Leveraging AI to comb through extensive compound libraries in search of promising drug candidates
- Molecular Modeling: AI's ability to forecast drug candidate structures, properties, and their interactions with target proteins
- Drug Design: Employing AI to craft novel drug candidates tailored to specific properties

1.6.2 Pinpointing Bottlenecks in Traditional Drug Discovery

The traditional drug discovery process faces several roadblocks:

- Escalating Costs: Research and development costs are on a relentless upward trajectory, exacerbated by the escalating complexity of the entire process.
- Time Intensive Journey: A new drug's journey to the market spans 10–15 years, contending with arduous regulatory hurdles and extensive clinical trials.
- Merger Success Rate: Only a minute fraction of drug candidates that tread the clinical trial path receive FDA approval due to the intricate challenges of developing safe and efficacious drugs.

1.6.3 Untangling AI Implementation Challenges

While AI has the potential to alleviate some of the challenges, it presents its own set of hurdles in drug discovery implementation:

- Data Dearth: AI models thrive on copious volumes of high-quality data, a luxury often absent in the drug discovery arena.
- Process Complexity: Drug discovery is a multifaceted process demanding nuanced representation in AI models to ensure efficacy.
- Model Interpretability: AI models can be enigmatic, complicating trust and usability in drug discovery due to the opacity of their predictions.

1.6.4 Glimmers of Possibilities Unleashed by AI

In the face of challenges, AI heralds transformation by:

- Expediting Drug Discovery: Automation empowered by AI slashes time and costs, expediting the journey from lab to market.

- Elevating Success Rates: AI identifies safer and more effective drugs, improving the probability of successful outcomes.
- Personalizing Medicine: Tailoring drugs to individual patient needs becomes feasible through AI's predictive capabilities.
- Tackling Neglected Diseases: AI's potential extends to crafting treatments for challenging diseases, including rare conditions.

1.6.5 A New Dawn in Drug Discovery

AI, a symbol of innovation, has the capacity to reshape the narrative of drug discovery. By confronting challenges head-on and embracing opportunities, AI can streamline the process, bringing novel and efficacious treatments to patients faster and more efficiently.

1.7 Case Study: AlphaFold's Acceleration in Drug Discovery

Within the dynamic realm of biomedical research, the ground-breaking protein structure prediction tool AlphaFold, created by DeepMind, a subsidiary of Google AI, has emerged as a revolutionary influence. Unveiled in 2020, its extraordinary aptitude for accurately predicting protein structures has established it as a potent accelerator in drug discovery. The AlphaFold stands ready to transform the traditional course of drug development by interpreting protein structures and unveiling their enigmas.

1.7.1 Significance of AlphaFold

The importance of AlphaFold is fundamentally tied to its capacity to transform the landscape of drug discovery. The conventional route to developing a new drug is laborious, marked by extensive timelines and substantial expenses. AlphaFold emerges as a promising beacon, aiming to accelerate this journey by providing researchers with precise forecasts of protein structures. This fresh understanding carries the potential to revolutionize the identification of drug targets and the creation of novel medications. The arrival of AlphaFold signifies a pivotal moment, introducing a more streamlined and cost-effective pathway to advancing therapeutic innovations.

1.7.1.1 Mathematical Mastery Behind AlphaFold's Prognostic Abilities

AlphaFold's exceptional predictive ability is built upon a sophisticated deep learning model, carefully refined through exposure to an extensive collection of established protein structures. Comparable to a virtual expert in proteomics, this model identifies the recurring patterns that interconnect various proteins. This mastery equips AlphaFold with the capability to foresee the structure of any protein, including those that have never been observed before. The essence of AlphaFold's predictions lies in understanding the physics of protein folding. Proteins, the building blocks of life, are intricate compositions of amino acids linked by peptide bonds. Folding, a central phenomenon, is governed by the interactions among these amino acids. AlphaFold employs its deep learning model to predict these interactions, ultimately creating a blueprint of the protein's structure.

1.7.2 Elevating Drug Discovery: The AlphaFold Impact on CDK20 Inhibitor Discovery

A noteworthy case study highlights AlphaFold's transformative role in drug discovery. An innovative application was witnessed in the realm of cancer treatment, specifically targeting CDK20 (Ren et al. 2023), a protein pivotal in cell division and the proliferation of cancer cells. AlphaFold's predictive prowess was harnessed in this study to unravel CDK20's structure. Armed with this crucial information, scientists crafted a small molecule inhibitor tailored to bind to CDK20 and impede its activity. The outcome was promising: the inhibitor effectively restrained cancer cell growth in vitro, illuminating AlphaFold's potential in hastening drug discovery

1.7.3 AlphaFold's Multi-dimensional Drug Discovery Impact

Beyond CDK20, AlphaFold's footprint in drug discovery extends across various diseases, encompassing Alzheimer's, Parkinson's, and cancer. For instance, AlphaFold's insights into the structure of tau, a protein implicated in Alzheimer's disease, enabled the design of molecules that counteract harmful aggregations. However, AlphaFold's journey is far from complete. Still evolving, it has already engendered a seismic shift in drug discovery. As its capacities continue to grow, it stands ready to play an even more crucial role in shaping the trajectory of upcoming drug development.

1.7.4 Navigating Challenges and Seizing Opportunities

However, as AlphaFold strives to revolutionize drug discovery, it faces particular challenges. Its high computational demands can pose an obstacle to expansive drug discovery initiatives. Additionally, its predictions are not immune to errors, occasionally resulting in misinterpretations that could lead to ineffective drug designs. Nonetheless, the potential encapsulated by AlphaFold remains resolute. Equipped with its strengths and the insights gained from overcoming challenges, AlphaFold remains an invaluable resource for researchers in drug discovery. As it progresses and evolves, its transformative influence is destined to grow stronger.

1.7.5 The Eclipsing Horizon

In conclusion, AlphaFold has surpassed its status as a mere computational instrument. It symbolizes optimism, heralding a revolutionary shift in drug discovery. Through its capacity to unveil protein structures and steer drug design, AlphaFold has indelibly influenced the realm of pharmaceutical advancement. As it progresses, it carries the potential to expedite the creation of innovative therapies, ushering us into a future where drug discovery is not only accelerated but also remarkably efficacious.

1.8 AI in the Era of Pandemics: Case of COVID-19

The global impact of the COVID-19 pandemic has been deeply distressing, causing numerous fatalities and widespread economic turmoil. Amid these difficult times, artificial intelligence (AI) emerged as a powerful weapon in the battle against the pandemic, providing efficient solutions to tackle the challenges presented by the virus.

1.8.1 Overview of the COVID-19 Pandemic

SARS-CoV-2, the virus behind the respiratory ailment that caused the world to shut down in 2020, is a highly contagious pathogen. It primarily spreads through respiratory droplets, often released when an infected person coughs or sneezes or when a person encounters a contaminated surface. The illness presents a range of symptoms, from mild to severe, including fever, cough, shortness of breath, and fatigue. In critical instances, COVID-19 can lead to conditions like pneumonia, acute respiratory distress syndrome, and even fatality. Emerging in Wuhan, China, in December

2019, the virus rapidly crossed borders, escalating into a worldwide emergency by March 2020. By March 2023, over 600 million individuals had contracted COVID-19, with the virus claiming the lives of more than 15 million people worldwide.

1.8.2 AI's Crucial Role in Drug Discovery and Vaccine Development

In the dark times of the pandemic, the cutting-edge solutions offered by AI made it a formidable weapon in the fight against COVID-19, yielding significant contributions, particularly in the realms of drug discovery and vaccine development.

- Drug Repurposing: One pivotal application of AI lies in drug repurposing, where existing drugs are identified and evaluated for their potential efficacy against COVID-19. For instance, the drug remdesivir, initially formulated for combating the Ebola virus, has proven effective in treating COVID-19 patients.
- Virtual Screening: The formidable computational power of AI has facilitated the virtual screening of vast databases of potential drug molecules. The approach accelerates the drug discovery timeline by swiftly identifying compounds with the potential to combat COVID-19.
- Structural Biology and AI: The proficiency of AI in structural biology has facilitated the exploration of the complex framework of the SARS-CoV-2 virus. This comprehension is crucial for identifying potential drug targets and providing a clear path for drug design.

Moreover, AI benefited greatly in the faster development of the COVID-19 vaccines. Among these, the Pfizer-BioNTech vaccine relies on mRNA technology and was developed using AI methodologies. This represents a remarkable fusion of innovation and medical science.

1.8.3 Leveraging Mathematical Models for Drug Prediction

Alongside its contributions to drug and vaccine development, AI has fostered the creation of intricate mathematical models to predict the potential effectiveness of drug molecules against COVID-19. These models, driven by AI algorithms, play a pivotal role in sifting through extensive compound libraries to identify those with promising therapeutic attributes. An exemplary instance is the DeepDTA model, which employs deep learning techniques to predict the binding affinity of drug molecules to the SARS-CoV-2 virus. The help of this model in predicting drug efficacy led to faster, more cost-effective client delivery of several promising drug candidates. Another notable mathematical model is the aforementioned AlphaFold, which employs deep learning to forecast the three-dimensional structures of proteins. In the context of COVID-19, AlphaFold's predictions have facilitated the identification of potential drug targets by unraveling its structural complexities.

1.8.4 A Glimpse into the Future

In summary, AI proved its utility in drug discovery during the COVID-19 pandemic. Its multifaceted applications encompass expediting drug and vaccine discovery, shaping predictive mathematical models, and enhancing diagnostic and treatment approaches for COVID-19 patients. As the pandemic endures, the role of AI is poised to expand even further. AI's potential to hasten the development of novel treatments and vaccines, improve patient care, and deepen our comprehension of the virus holds tremendous promise. In the quest to conquer COVID-19, AI stands as a symbol of innovation and a steadfast ally in the collective endeavor to overcome the challenges posed by the pandemic.

1.9 Deep Learning in Antibiotic Discovery

The urgent need for solutions to the worldwide health crisis of antibiotic resistance is undeniable. Conventional methods for discovering antibiotics are marked by their slow pace and substantial expenses. Recently, the adoption of artificial intelligence (AI) methods, particularly deep learning, is gaining traction as an approach to accelerate and improve the efficiency of the processes of traditional methods.

1. *Deep Learning:* Deep learning, a subset of machine learning, harnesses artificial neural networks to learn intricate patterns from data. Inspired by the human brain, these networks excel in deciphering complex data patterns. Across diverse domains, deep learning has demonstrated ground-breaking achievements, from image recognition and natural language processing to drug discovery.
2. *Leveraging Deep Learning in Antibiotic Discovery:* Deep learning's potential in antibiotic discovery is multifaceted and promising:
    (a) **Identification of Novel Antibiotic Targets:** Deep learning models can identify potential antibiotic targets by analyzing extensive protein datasets. These models are trained to predict whether a protein could serve as an antibiotic target.
    (b) **Design of Enhanced Antibiotics:** Deep learning facilitates the design of antibiotics that combat antibiotic-resistant bacteria more effectively. Models can predict essential properties like binding affinity to antibiotic targets and toxicity.

(c) **Prediction of Antibiotic Properties:** By learning from known antibiotic datasets, deep learning models predict crucial antibiotic properties, including binding affinity to antibiotic targets and toxicity.
(d) **Efficient Antibiotic Screening:** Deep learning expedites antibiotic screening by evaluating massive compound libraries for potential antibiotic activity. Models predict the likelihood of a compound possessing antibiotic capabilities.

3. *Advantages of Deep Learning in Antibiotic Discovery:* Several merits underscore the deployment of deep learning:

   (a) **Big Data Analysis:** Deep learning efficiently analyzes vast bacterial datasets and their interactions with antibiotics, which is pivotal for effective antibiotic discovery.
   (b) **Pattern Recognition:** Deep learning excels in discerning complex patterns, thereby aiding in identifying fresh antibiotic targets and creating innovative antibiotics.
   (c) **Rapid Screening:** Deep learning accelerates antibiotic screening, enabling quick identification of potential candidates.

4. Challenges of Deep Learning in Antibiotic Discovery: While promising, deep learning in antibiotic discovery faces challenges:

   (a) **Data Quantity:** Deep learning mandates substantial data for model training, which can be arduous and costly.
   (b) **Complexity and Interpretability:** Complex models pose challenges in interpretability, hindering the comprehension of decision-making processes.
   (c) **Bias Concerns:** Deep learning models can exhibit bias, leading to inaccurate predictions.

Despite these hurdles, deep learning is a promising avenue to address the global antibiotic resistance crisis, given its potential to expedite antibiotic discovery and counteract antibiotic-resistant strains.

1. *Diverse Deep Learning Techniques and Models:* Several deep learning techniques and models serve antibiotic discovery:

   (a) **Convolutional Neural Networks (CNNs):** Particularly adept at image recognition, CNNs scrutinize protein structures to identify potential targets for antibiotics.
   (b) **Recurrent Neural Networks (RNNs):** Specializing in sequential data, RNNs predict new antibiotic properties, aiding the design of novel compounds.
   (c) **Generative Adversarial Networks (GANs):** GANs generate new data and can be applied to screening antibiotics by generating new compounds likely to possess antibiotic activity.

2. Mathematical Foundations of Deep Learning Predictions: The predictive power of deep learning models hinges on essential equations:

   (a) **Loss Function:** Measures model performance, minimized during training
   (b) **Optimizer:** Algorithm updating model parameters during training
   (c) **Backpropagation:** Calculates gradients of the loss function regarding model parameters, pivotal for the optimizer's updates

3. *Conclusion:* Deep learning's integration into antibiotic discovery holds immense promise. Its potential to hasten the identification of antibiotics and combat antibiotic resistance is unparalleled. Despite obstacles, the combination of deep learning's computational prowess and antibiotic discovery's urgency suggests a symbiotic relationship that could reshape the landscape of antibiotic research and offer hope in the face of escalating antibiotic resistance.

1.10 AI Techniques in Antibiotic and Antiviral Development

In the rapidly evolving landscape of drug discovery, a slew of cutting-edge AI techniques is taking center stage, offering ground-breaking solutions for the formidable challenges posed by the development of antibiotics and antiviral agents. These emerging AI techniques can potentially revolutionize the future of medicine by expediting drug discovery, tailoring treatments to individuals, and ushering in a new era of predictive mathematical models.

1. *Emerging AI Techniques in Drug Discovery*: Drug discovery has been revolutionized with the introduction of Artificial Intelligence methods such as

   (a) *Machine Learning (ML)*: Harnessing ML algorithms, extensive datasets containing biological information and chemical compounds can be scrutinized to identify potential drug targets and design novel drug molecules. ML's prowess spans from target selection to compound optimization, streamlining the discovery process (Kim et al. 2020; Sahayasheela et al. 2022; Blanco-Gonzalez et al. 2023; Ren et al. 2023; Keshavarzi Arshadi et al. 2020; Jiménez-Luna et al. 2021; Paul et al. 2021; Zhang et al. 2017).
   (b) *Deep Learning (DL):* Nested within ML, DL employs artificial neural networks to glean insights from data. Renowned for tasks like image recognition and natural language processing, DL is increasingly applied to drug discovery. It exhibits remarkable potential in forecasting drug toxicity and gauging the binding affinity of drugs with their intended targets, thus expediting optimization processes (Jiménez-Luna et al. 2021; Paul et al. 2021; Zhang et al. 2017).
   (c) *Generative Adversarial Networks (GANs):* A subset of DL, GANs generate fresh drug molecules akin to existing compounds while boasting enhanced properties. This innovation facilitates the creation of drug candidates with improved attributes (Jiménez-Luna et al. 2021; Zhang et al. 2017).

(d) *Bayesian Optimization:* This statistical method is pivotal in experiment design optimization. In the context of drug discovery, Bayesian optimization identifies promising drug compounds for clinical trials, optimizing resource allocation (Jiménez-Luna et al. 2021; Paul et al. 2021).
(e) *Natural Language Processing (NLP):* The interplay between computers and human languages constitutes the essence of NLP. Its potential in drug discovery lies in dissecting medical literature and clinical trial data to unearth potential drug targets and evaluate the safety and efficacy of novel drugs.

These emerging AI techniques epitomize the promise of rapid innovation in drug discovery. As AI advances, we can anticipate even more revolutionary applications in drug discovery.

2. *Potential of Personalized Medicine Through AI:* AI holds the transformative power to revolutionize personalized medicine, which hinges on tailoring medical interventions to suit individual patient needs. AI interventions encompass:

(a) Precision Patient Selection: By analyzing genetic data, AI identifies genes linked to specific diseases. This knowledge, in turn, aids in identifying drugs that are likely to be effective for treating the identified disease.
(b) Treatment Response Prediction: AI employs patient data to forecast individual responses to drugs, enabling the optimization of treatment regimens for better patient outcomes.
(c) Customized Dosages: Leveraging AI's analytical prowess, personalized dosages can be calculated based on an individual's unique characteristics and responses to treatment.

The realm of personalized medicine, buoyed by AI, promises to enhance treatment outcomes, minimize costs, and reshape medical care by tailoring therapies to individual patients.

3. *Predictive Mathematical Models Shaping Drug Discovery:* Predictive mathematical models are a cornerstone of modern drug discovery. These models facilitate the antimicrobial drug compound properties, revolutionizing the process. Key predictive models include:

(a) Quantitative Structure-Activity Relationship (QSAR): By utilizing statistical approaches, Quantitative Structure-Activity Relationship (QSAR) models establish connections between the structures of drug compounds and their biological functions. This process assists in foreseeing toxicity and effectiveness across a diverse range of drug compounds.
(b) Molecular Dynamics (MD) Simulations: Through computational models, MD simulations predict the behavior of molecules within biological systems. These simulations extend to forecasting drug-target binding and the effects of drug compounds on biological entities.

Predictive mathematical models offer profound insights into the behavior of drug compounds, streamlining drug development and facilitating target identification.

In conclusion, AI's influence on antibiotic and antiviral development is transformative. The convergence of emerging AI techniques like ML, DL, GANs, Bayesian optimization, and NLP facilitates the identification of drug targets, the design of novel molecules, and the personalization of drug treatments. Alongside these, predictive mathematical models enrich drug discovery by anticipating compound properties. As AI's horizons expand, its potential to hasten the development of innovative and effective drugs addressing many diseases shines ever brighter. The interdependent partnership between AI and drug discovery is on the brink of reshaping the healthcare landscape.

1.11 Applications of AI in Drug Discovery

The incorporation of artificial intelligence (AI) has transformed the field of drug discovery, pushing for the beginning of a fresh era characterized by heightened efficiency and effectiveness. AI-powered technologies are revamping multiple stages of drug development, encompassing tasks ranging from identifying and validating targets to screening and designing compounds.

1.11.1 Target Identification and Validation

AI leverages vast genomic, transcriptomic, and proteomic datasets to identify and validate novel drug targets. AI algorithms delve into disease-specific genetic mutations and intricate biological pathways, shedding light on potential therapeutic avenues. Additionally, AI assesses the multidrug targets' structural and functional attributes.

1.11.2 Compound Screening and Design

The power of AI is evident in its ability to sift through expansive compound libraries, pinpointing these with a high likelihood of interacting with target proteins. AI also plays a role in designing new compounds with heightened effectiveness and minimal side effects. Predicting binding affinities to target proteins and assessing compound toxicity are among AI's contributions.

1.11.3 Clinical Trial Optimization

AI optimizes clinical trials through refined patient selection, streamlined protocol design, and advanced result analysis. Among AI's functions are predicting patient responsiveness to specific treatments and determining optimal drug dosages. By enhancing trial efficiency, AI accelerates the pace of drug development.

1.11.4 Concrete Examples of AI's Impact

In 2020, AI spotlighted a potential drug target for Alzheimer's disease, the protein BACE1, integral in forming amyloid plaques characteristic of the ailment (Kim et al. 2020). In 2021, AI was pivotal in designing a compound to combat drug-resistant tuberculosis. The algorithm factored in target protein structure and the properties of effective tuberculosis drugs to craft the compound (Sahayasheela et al. 2022). In 2022, AI optimized a clinical trial for a novel cancer drug, expertly identifying promising trial participants and refining trial protocols (Blanco-Gonzalez et al. 2023). These instances merely scratch the surface of AI's role in drug discovery. As AI's capabilities burgeon, its potential to reshape and expedite drug development remains boundless.

1.12 The Future of AI in Drug Discovery

Artificial intelligence (AI) is swiftly reshaping the landscape of drug discovery, offering the potential to expedite processes, enhance efficiency, and bolster the efficacy of novel drug development.

1.12.1 Emerging AI Techniques in Drug Discovery

Several cutting-edge AI techniques are currently making strides in drug discovery:

- Machine Learning (ML): ML algorithms analyze vast chemical and biological datasets, unveiling new drug targets and potential candidates (Kim et al. 2020).
- Deep Learning (DL): A subset of ML, DL algorithms excel in deciphering intricate data patterns. Their effectiveness spans tasks like predicting compound toxicity and identifying new drug targets (Sahayasheela et al. 2022; Blanco-Gonzalez et al. 2023).
- Natural Language Processing (NLP): NLP techniques dissect textual data from medical records and scientific literature, revealing drug targets and potential side effects (Ren et al. 2023).

- Computer-Aided Drug Design (CADD): CADD employs simulations to predict properties like pharmacokinetics, pharmacodynamics, and toxicity of potential drug compounds (Keshavarzi Arshadi et al. 2020).

1.12.2 The Potential of Personalized Medicine via AI

AI's potential extends to enabling personalized medicine—tailoring drug treatment for individual patients. AI scrutinizes genetic profiles, medical histories, and more to pinpoint optimal and safe drug treatments (Jiménez-Luna et al. 2021).

1.12.3 Predictive Mathematical Models Shaping Drug Discovery's Future

AI is instrumental in crafting predictive mathematical models for drug discovery. These models forecast promising drug targets and candidates and the success or failure of new drug development programs (Stokes et al. 2020; Melo et al. 2021; Askr et al. 2023; Volkamer et al. 2023; Luukkonen et al. 2023).

- Anticipating the Future Role of AI in Drug Discovery
- The usage of AI in drug discovery remains in its nascent stages but holds revolutionary potential. Its current acceleration of drug discovery is just the beginning, with prospects for further transformation. The following is a glimpse into AI's future impact on drug discovery:
- Discovering Novel Drug Targets: Researchers will uncover new drug targets by harnessing AI to analyze vast biological and chemical data. This process, currently labor-intensive, can be automated, drastically enhancing efficiency (Kaushik and Raj 2020; Gupta et al. 2022a).
- Designing Effective Drug Compounds: AI's prowess in screening vast libraries of compounds will be harnessed to design novel drugs with enhanced efficacy and safety (Paul et al. 2021; Askr et al. 2023).
- Predicting Development Program Outcomes: AI-driven predictive models will evolve to project the success or failure of drug development programs. This information will guide decision-making processes (Zhang et al. 2017; Prasad and Kumar 2021; Gupta et al. 2021).
- Personalized Treatment: Leveraging AI's insights into genetic profiles, medical histories, and other factors, personalized medicine will become more refined and effective (Prasad and Kumar 2021; Gupta et al. 2022c).

The prospects for AI in drug discovery are promising. As AI advances, its capacity to transform the field, enhance efficiency, and introduce personalized medicine will become increasingly evident. The path toward discovering new drugs and advancing personalized treatments is set to benefit significantly from the ongoing evolution of AI (Gupta et al. 2021, 2022a, 2022b).

1.13 Conclusion

1.13.1 Revolutionizing Microbial Drug Discovery with Artificial Intelligence

The incorporation of artificial intelligence (AI) into the realm of microbial drug discovery is rapidly gaining momentum, set to redefine the landscape of novel drug development. AI methodologies are positioned to accelerate every aspect of the drug discovery journey, encompassing tasks ranging from identifying targets to designing and refining drugs.

- Target Identification: AI's potential is showcased in identifying novel drug targets by scrutinizing extensive genomic and proteomic datasets. AI offers ground-breaking insights by discerning targets vital for pathogenic microbes' survival, distinct from those in human cells (Kim et al. 2020; Sahayasheela et al. 2022; Luukkonen et al. 2023).
- Drug Design: The power of AI extends to crafting new drugs that optimize efficacy and mitigate side effects. AI empowers the screening of vast compound libraries, identifying candidates with sought-after attributes (Blanco-Gonzalez et al. 2023; Ren et al. 2023).
- Drug Optimization: AI-driven drug optimization thrives in perfecting crucial attributes like solubility, stability, and pharmacokinetics. By enhancing these properties, AI contributes to the safety and efficacy of novel drugs (Keshavarzi Arshadi et al. 2020; Volkamer et al. 2023).

1.13.2 Anticipating Future Trends and Breakthroughs

The trajectory of AI in microbial drug discovery is brimming with promise. As AI techniques grow more sophisticated and potent, their application across the drug discovery spectrum catalyzes the creation of novel drugs that are more potent, safer, and expeditiously developed (Gupta et al. 2022c; Sarkar et al. 2023; Askr et al. 2023).

1.13.3 Potential Breakthroughs Envisioned by AI

- Novel Drug Target Identification: AI is essential to discovering drug targets indispensable for pathogenic microbes' survival while sparing human cells. This breakthrough is poised to revolutionize treatment strategies.
- Precision Drug Design: AI's prowess in compound library screening for efficacious candidates that minimize side effects translates to ground-breaking drug design—the result being drugs that combat microbes effectively and safeguard patient well-being.
- Optimized Drug Properties: AI's contribution in refining drug attributes like solubility, stability, and pharmacokinetics ensures the creation of safer and more effective medications. This leap improves patient outcomes while minimizing adverse effects.
- Innovative Drug Delivery: AI's application extends to reimagining drug delivery methods, bolstering the safety and efficacy of new drugs. Innovative approaches can lead to more targeted and efficient treatment.
- Accelerated Drug Discovery: The integration of AI drives the entire journey of drug discovery, spanning from target identification to regulatory approval, with the potential to accelerate the whole process.

In conclusion, the infusion of AI into microbial drug discovery is poised to bring about an evolutionary shift. AI's capacity to swiftly analyze vast datasets, optimize drug properties, and predict efficacious compounds sets the stage for ground-breaking advancements. Considering the trajectory of AI's progress and implementation, the outlook for microbial drug discovery is undeniably promising. This holds the potential to yield treatments that are safer, more effective, and readily accessible.

References

Askr H, Elgeldawi E, Aboul Ella H, Elshaier YA, Gomaa MM, Hassanien AE (2023) Deep learning in drug discovery: an integrative review and future challenges. Artif Intell Rev 56(7):5975–6037

Blanco-Gonzalez A, Cabezon A, Seco-Gonzalez A, Conde-Torres D, Antelo-Riveiro P, Pineiro A, Garcia-Fandino R (2023) The role of ai in drug discovery: challenges, opportunities, and strategies. Pharmaceuticals 16(6):891

Gupta D, Choudhury A, Gupta U, Singh P, Prasad M (2021) Computational approach to clinical diagnosis of diabetes disease: a comparative study. Multimed Tools Appl 80:1–26

Gupta M, Srivastava D, Pantola D, Gupta U (2022a) Brain tumor detection using improved Otsu's thresholding method and supervised learning techniques at early stage. In: Proceedings of emerging trends and technologies on intelligent systems: ETTIS 2022. Springer Nature Singapore, Singapore, pp. 271–281

Gupta U, Gupta D, Agarwal U (2022b) Analysis of randomization-based approaches for autism spectrum disorder. In Pattern recognition and data analysis with applications. Springer Nature Singapore, Singapore, pp. 701–713

Gupta U, Sharma S, Jyani U, Bhardwaj A, Sharma M (2022c) Sign language detection for deaf and dumb students using deep learning: dore Idioma. In: 2022 2nd international conference on innovative sustainable computational technologies (CISCT). IEEE, pp. 1–5

Jiménez-Luna J, Grisoni F, Weskamp N, Schneider G (2021) Artificial intelligence in drug discovery: recent advances and future perspectives. Expert Opin Drug Discovery 16(9): 949–959

Kaushik AC, Raj U (2020) AI-driven drug discovery: a boon against COVID-19? AI Open 1:1–4

Keshavarzi Arshadi A, Webb J, Salem M, Cruz E, Calad-Thomson S, Ghadirian N et al (2020) Artificial intelligence for COVID-19 drug discovery and vaccine development. Frontiers. Artif Intell 65

Kim H, Kim E, Lee I, Bae B, Park M, Nam H (2020) Artificial intelligence in drug discovery: a comprehensive review of data-driven and machine learning approaches. Biotechnol Bioprocess Eng 25:895–930

Luukkonen S, van den Maagdenberg HW, Emmerich MT, van Westen GJ (2023) Artificial intelligence in multi-objective drug design. Curr Opin Struct Biol 79:102537

Melo MC, Maasch JR, de la Fuente-Nunez C (2021) Accelerating antibiotic discovery through artificial intelligence. Commun Biol 4(1):1050

Paul D, Sanap G, Shenoy S, Kalyane D, Kalia K, Tekade RK (2021) Artificial intelligence in drug discovery and development. Drug Discov Today 26(1):80

Prasad K, Kumar V (2021) Artificial intelligence-driven drug repurposing and structural biology for SARS-CoV-2. Curr Res Pharmacol Drug Discov 2:100042

Ren F, Ding X, Zheng M, Korzinkin M, Cai X, Zhu W et al (2023) AlphaFold accelerates artificial intelligence powered drug discovery: efficient discovery of a novel CDK20 small molecule inhibitor. Chem Sci 14(6):1443–1452

Sahayasheela VJ, Lankadasari MB, Dan VM, Dastager SG, Pandian GN, Sugiyama H (2022) Artificial intelligence in microbial natural product drug discovery: current and emerging role. Nat Prod Rep 39:2215

Sarkar C, Das B, Rawat VS, Wahlang JB, Nongpiur A, Tiewsoh I et al (2023) Artificial intelligence and machine learning technology driven modern drug discovery and development. Int J Mol Sci 24(3):2026

Stokes JM, Yang K, Swanson K, Jin W, Cubillos-Ruiz A, Donghia NM et al (2020) A deep learning approach to antibiotic discovery. Cell 180(4):688–702

Volkamer A, Riniker S, Nittinger E, Lanini J, Grisoni F, Evertsson E et al (2023) Machine learning for small molecule drug discovery in academia and industry. Artif Intell Life Sci 3:100056

Zhang L, Tan J, Han D, Zhu H (2017) From machine learning to deep learning: progress in machine intelligence for rational drug discovery. Drug Discov Today 22(11):1680–1685

Chapter 2
Prediction of Plant Disease Using Artificial Intelligence

Manoj Ram Tammina, K. Sumana, Pavitar Parkash Singh, T. R. Vijaya Lakshmi, and Sagar Dhanraj Pande

Abstract Plant diseases are a persistent threat to global food security due to their ability to damage crops. They account for 20–40% of loss of global food trade every year. The exploding global food trade, coupled with climate change, has led to the sustainability of native plant pests in the new environment, worsening the condition. Additionally, new plant pests and diseases continue to threaten staple crops. This sheds light on the need for the implementation of novel techniques to diagnose plant diseases to tackle the global food crises. Implementation of artificial intelligence (AI)-based methods such as machine learning (ML), deep learning (DL), and artificial neural networks can aid in overcoming such challenges by conducting early diagnosis of plant pests and diseases. In recent years, many research investigations conducted on plant disease detection using AI have offered valuable insights for agriculturists, botanical researchers, practitioners, and industrial professionals. The applications DL and ML methods for plant disease detection are growing rapidly. This chapter will shed light on recent cutting-edge research in this field, including the latest advancements involving AI-based plant disease detection. It will also address the trials and limitations related to the usage of AI-based methods for plant disease diagnosis.

M. R. Tammina (✉)
Innovation, Bread Financial, Columbus, OH, USA

K. Sumana
Department of Microbiology, JSS AHER, Mysuru, India
e-mail: sumana.k@jssuni.edu.in

P. P. Singh
Department of Management, Lovely Professional University, Phagwara, Punjab, India
e-mail: pavitar.19476@lpu.co.in

T. R. V. Lakshmi
Mahatma Gandhi Institute of Technology, Gandipet, Hyderabad, India
e-mail: Trvijayalakshmi_ece@mgit.ac.in

S. D. Pande
School of Computer Science and Engineering, VIT-AP University, Amaravati, Andhra Pradesh, India

A. Khamparia et al. (eds.), *Microbial Data Intelligence and Computational Techniques for Sustainable Computing*, Microorganisms for Sustainability 47,
https://doi.org/10.1007/978-981-99-9621-6_2

Keywords Plant pest infection · Deep learning · Artificial intelligence · Neural network

2.1 Introduction

The global starvation rate crossed 950 billion in 2022. Over 100 billion individuals are suffering from varying types of starvation and malnutrition. However, while growing population needs is the major attributing factor, plant diseases are one of the greatest threats to the agricultural industry and global food security. Every year 40% of crops are affected due to plant pests, rendering them a persistent threat to crop health. Climatic changes, lack of agricultural responses, and sustainability of plant pests are threatening global food security. Plant diseases caused by pathogens and pest attacks are one of the major risk factors for global food crises. Annual loss of potatoes, peas, maize, rice, soya beans, and tomatoes due to pathogenic infection contributes to 10–25% of crop loss per annum. The crisis due to crop diseases is not restricted to food security, but is also the result of considerable economic loss and reduced agricultural yield. Roland et al. have described that crop loss is the decrease of the crop yield in terms of quality and quantity. Qualitative crop loss assessment involves the analysis of efficiency of current crop protection practices, assessing the green crop yield strategies and pest protection strategies and integrated management of plant diseases. Quantitative crop loss evaluation involves assessment of biotic and abiotic factors. Assessing crop losses might offer a better insight into the problem and pave the way for development of efficient identification techniques.

Agriculture is critical to guaranteeing global food security and sustaining global economies. Plant diseases are one of the most serious threats to agricultural output and quality. Plant illnesses produced by pathogens such as fungi, bacteria, viruses, and other biotic agents result in significant yield losses, worse crop quality, and higher production costs. Whole harvests can be destroyed due to crop diseases, threatening food supply, food security, and economic stability (Jones 2009).To avoid these hazards and preserve sustainable agricultural practices, effective plant disease control is required. The cornerstone of such management techniques is prompt and accurate detection of plant diseases. The timely identification of disease-causing agents enables targeted actions such as the implementation of relevant therapies, the selection of resistant plant types, and the modification of agricultural practices. This not only decreases crop losses but also the need for chemical pesticides, resulting in more sustainable and economically viable agricultural methods (Kamilaris and Kartakoullis 2021).

Artificial intelligence (AI) has created disruptive opportunities in a variety of industries, including agriculture. The algorithms used for machine learning like supervised learning methods and models fueled by data that may acquire knowledge from trends and then make predictions or judgments are examples of AI. In the field of plant disease diagnostics, AI has emerged as a game-changing technology capable of supplementing and even outperforming traditional approaches (Savary et al. 2019).

AI provides a distinct edge in dealing with the complex and diversified character of plant diseases. Convolutional neural networks, also called neural networks (CNNs), for example, can analyze plant photos to identify visual signals linked to illnesses and have substantially improved image-based diagnosis. These artificial intelligence models can quickly scan and analyze hundreds of photos, identifying small differences in leaf colour, texture, and form that may signal disease prevalence (Singh and Shrivastava 2018). Furthermore, AI may combine data from additional sources, including sensor tracking environmental conditions, to offer a comprehensive evaluation of illness risks.

AI has the capacity to predict disease and identify it early on. AI models can forecast disease outbreaks with surprising accuracy by using past data, weather trends, and other pertinent variables. Farmers should take preventive actions ahead of time, optimizing resource allocation and minimizing crop losses. In essence, artificial intelligence (AI) provides farmers and agricultural players with practical knowledge that leads to better decision-making and disease control tactics (Elad and Pertot 2014).

The goal of this chapter is to describe how AI is changing detection of plant diseases in the context of agriculture and food security. This chapter seeks to give a full picture of how technology might transform the agricultural environment by looking into the promise and problems of incorporating AI approaches into disease detection practices. This chapter's focus includes many aspects of AI-driven plant infection diagnostics. It will cover the principles of plant disease diagnostics, emphasizing the variety of pathogens including their impact on crops. Traditional diagnostic procedures will be explored, emphasizing their limits and the need for novel alternatives. The focus will then move to the development of AI and its components, specifically on how AI approaches may revolutionise illness detection by improving precision, rapidity, and portability.

Further sections will shed light on the process of data gathering and preparation, emphasizing the need for reliable information for training strong AI models. They will dig into the complexities of constructing and educating AI models, shedding light on algorithm selection and feature extraction subtleties. The chapter will also go into depth on image-based and sensor data-based illness diagnostic methodologies, as well as successful examples and the influence of AI on continuous surveillance and disease prediction.

In addition, the chapter will go through the difficulties of installing AI-based detection techniques in the field, taking into account variables like data protection, connection, and farmer approval. Ethical and societal ramifications will be discussed as well, emphasizing the importance of responsible AI usage in agriculture. The chapter will conclude by looking ahead to future directions and advances in this area, reviewing new trends, and imagining how AI will revolutionize precision farming and green agricultural practices.

Finally, this chapter aims to shed light on the symbiotic link that exists between plant disease detection, agriculture, and AI. It intends to encourage more study, cooperation, and implementation of AI technologies for the improvement of global food security and agricultural practices by revealing the revolutionary potential of AI in disease control.

2.2 Fundamentals of Plant Disease Diagnosis

Plant diseases produced by pathogens such as fungi, microbes, viruses, and other biological agents represent a substantial danger to global food and agricultural production. Proper plant disease identification is critical for successful disease management and ensuring optimal crop production and quality (Kamilaris et al. 2021). This section delves into the core ideas of plant illness diagnosis, including different kinds of plant illnesses, their symptoms, standard diagnostic procedures, and the demand for novel methods.

2.2.1 Plant Disease Types

Plant diseases include a wide spectrum of problems that affect plant health and production. Plant diseases are broadly classified into infectious and non-infectious based on the causal agent. Living organisms such as fungi, viruses, microbes, and nematodes cause infectious diseases, but non-living causes such as nutritional deficits, environmental pressures, and chemical imbalances cause non-infectious diseases (Agrios 2005). Biotic diseases, especially those caused by microorganisms, have received a great deal of attention because of their potential to inflict massive production losses (Savary et al. 2019).

2.2.2 Symptoms and Signs

Plant disease symptoms present as apparent shifts in plant physical appearance, structure, and function. Leaf withering, discoloration, retarded development, lesions, and deformation are among the symptoms (Lucas 2008). Plant illnesses may also create visual indicators such as fungal spores, bacterial slime, or viral inclusions. These symptoms and indications act as markers of illness existence, allowing for precise diagnosis and treatment.

2.2.3 Conventional Diagnostic Techniques

Visual inspection, microscopic examination, and laboratory testing are the mainstays of traditional plant disease diagnosis techniques. Visual inspection entails close examination of plant symptoms and warning signals, sometimes requiring skilled individuals to distinguish among diseases with similar symptoms (Schumann and D'Arcy 2010). Microscopic analysis entails looking at spores, mycelium, and other types of microscopic structures, particularly in the case of fungal illnesses. Enzyme-linked immunosorbent assays (ELISAs) and polymerase chain reactions (PCR) are examples of laboratory tests used to identify certain infections.

2.2.4 *Limitations of Traditional Methods*

Traditional methods have helped diagnose plant diseases, but they have drawbacks as well. Windstam and Schmale (2018) revealed that visual analysis may be subjective and greatly depends on the observer's skill. Pathogen detection may be time-consuming and necessitates specific tools and expertise for microscopic investigation.Additionally, these techniques might not be appropriate for quick, wide-ranging disease screening, which is necessary to stop massive epidemics.

2.2.5 *Need for AI-driven Innovative Methods*

The problems posed by changing diseases, shifting climatic circumstances, and international trade call for the development of novel, effective diagnostic methods. Here, artificial intelligence (AI) starts to influence the game. By overcoming the drawbacks of current techniques, AI has a chance to transform the detection of plant diseases. Kamilaris and Prenafeta-Boldú (2018a) reported that convolutional neural networks (CNNs), a type of deep learning model, are particularly good at evaluating vast amounts of information and spotting complicated patterns. These algorithms can analyze tens of thousands of plant photos, picking up on minute variations in texture, hue, and form that may be signs of illness. To give a comprehensive evaluation of illness risks, AI may also combine data from additional sources, particularly networks of sensors monitoring environmental variables. Accuracy, rapidity, scalability, and flexibility to shifting disease dynamics are promises of plant disease detection powered by AI methods (Das et al. 2020).

The principles of plant diagnosis of diseases cover the many disease types, their signs and symptoms, conventional diagnostic techniques, and the drawbacks of these procedures. Innovative solutions are necessary as the agriculture industry struggles to feed a growing global population (Kamilaris and Prenafeta-Boldú 2018b). An interesting opportunity to improve our capacity to quickly and reliably detect illnesses is the use of AI in plant disease diagnostics. This will enable farmers to make better choices regarding managing plant health more successfully.

2.3 Role of Artificial Intelligence in Plant Disease Diagnosis

One of the most exciting developments in recent years is the use of the application of AI in plant disease diagnostics. The convergence of technology and agricultural production has produced transformational improvements (Hughes and Salathé 2015). As plant pathogens continue to pose a threat to agricultural sustainability and global food security, AI is emerging as a formidable tool with the potential to completely transform disease identification, prediction, and control. This section explains the critical significance of AI in diagnosing plant diseases while highlighting its advantages, drawbacks, and practical applications (Mohanty et al. 2016).

2.3.1 AI: Changing the Diagnosis of Plant Disease

Artificial intelligence refers to a variety of methods that give computers the ability to learn, reason, solve problems, and make decisions—tasks that traditionally require human intellect. Artificial intelligence (AI) techniques like machine learning and deep learning have excelled in diagnosing plant diseases. These tools analyze enormous volumes of data, uncovering trends and connections that may elude traditional methods (Mohanty et al. 2016).

2.3.2 Disease Diagnosis Based on Images

Image analysis is one of the most impressive ways in which AI is being used to diagnose plant diseases. Convolutional neural networks (CNNs), a type of deep learning algorithm, are particularly good at processing visual input. In this situation, they can carefully examine plant photos for minute changes in hue, texture, and form that indicate the presence of diseases. AI models are able to classify the severity of illness as well as detect diseases. With the use of this computerized picture analysis, illness detection for farmers may be done more quickly and accurately (Picon et al. 2020).

2.3.3 Sensor Data and Disease Forecasting

AI goes beyond imaging to help identify plant diseases. Sensors placed in farming areas can gather information on environmental factors including temperature, humidity, and moisture levels in the soil. These data may be processed by AI systems, which correlate variations with the prevalence of disease. AI algorithms can anticipate disease outbreaks remarkably accurately by examining historical trends and real-time inputs. Farmers may take preventative actions thanks to this proactive strategy, decreasing crop losses and maximizing resource allocation (Picon et al. 2020).

2.4 AI's Advantages in Plant Disease Diagnosis

Beyond accuracy and speed, the use of AI for plant disease diagnostics offers a variety of advantages.

2.4.1 Early Disease Detection and Prevention

AI-powered systems may identify illness in the very early stages, allowing for prompt interventions and halting large-scale epidemics by implementing algorithms (LeCun et al. 2015).

2.4.2 Accuracy and Scalability

To improve illness prediction and management, AI models examine data with high accuracy and can be scaled to study enormous datasets.

2.4.3 Reduced Dependence on Chemicals

Early illness identification reduces the need for overuse of pesticides, encouraging ecologically friendly and sustainable farming methods.

2.4.4 Improved Decision-making Capability

AI-generated insights enable farmers to make educated decisions about agriculture and disease management measures.

2.5 Barriers to Implementing AI Techniques in Plant Disease Diagnosis

Although AI has great potential for diagnosing plant diseases, there are a number of issues that need to be resolved, described by Das et al. (2020).

2.5.1 Data Accuracy

For training, AI models need access to extensive, diversified, and reliable datasets. For a model to work well, accurate data must be made available.

2.5.2 Model Interpretability

Deep learning models' "black-box" nature might make it difficult to grasp how choices are made. Transparency of the model must be maintained.

2.5.3 Infrastructure and Accessibility

It's possible that many farmers lack the financial means or technological know-how to use AI systems efficiently.

2.6 Current Trends in AI Involvement in Plant Disease Diagnostics

Das et al. (2020) described that AI-based plant disease detection is already showing promise in the following areas.

Cassava Disease Detection: AI models have been created in Sub-Saharan Africa to identify cassava illnesses from leaf photos, assisting farmers in prompt disease control.

Monitoring Grapevine Disease: Drones powered by AI and fitted with cameras take high-resolution pictures of vineyards, allowing early identification and targeted treatment of grapevine disease.

Smartphone Applications: Simple apps for smartphones have been created that enable farmers to take pictures of plant symptoms and instantly diagnose diseases and provide treatment suggestions.

The use of artificial intelligence in the identification of plant diseases represents a turning point for contemporary agriculture. The use of analyzing images, data collected from sensors, and predictive modeling by AI provides farmers with previously unheard of capabilities for effectively and sustainably battling plant diseases. However, overcoming obstacles, advancing accessibility, and encouraging partnerships between AI professionals, agricultural scientists, and farmers are all necessary for a successful adoption. The path from AI-driven insights to real-world effect contains the potential to secure the world's food supply, ensure the success of agricultural economies, and promote a sustainable future.

2.7 Data Collection and Pre-processing

Artificial intelligence (AI) has been shown to be a game-changer in a number of industries, and its use in plant disease detection has enormous potential to transform agriculture. However, the preparation and quality of the data are key factors in whether AI models succeed in this situation. In-depth discussion of data collecting and initial processing for artificial intelligence-based plant disease detection is provided in this section, which also emphasizes the significance of the integrity of data sources, pre-processing methods, and problems in achieving trustworthy and accurate findings.

2.8 The Significance of Data Quality

Data quality is the basis for all AI projects. High-quality data is essential for training models to reliably discriminate between healthy and sick plants in the context of plant disease detection. The capacity of the model to generalize and make wise evaluations is improved by having clean, well-labeled, and diversified datasets. Biased or inaccurate data might provide inaccurate forecasts and unreliable results (Picon et al. 2020).

2.9 Data Sources

Image Databases: Images are used to record illness symptoms and signs. These photographs encompass a range of settings, lighting types, and plant species, ensuring the model's durability and flexibility (Mohanty et al. 2016).

Sensor Networks: Information from environmental sensors, such as relative humidity, temperature, and soil moisture, is used to improve disease prediction models. In turn, prediction accuracy is improved by linking these environmental elements to illness incidence (Mohanty et al. 2016).

Historical Information: The construction of disease prediction algorithms is aided by historical information on outbreaks of disease, weather trends, and agricultural practices. Disease risks can be predicted with the use of patterns and trends derived from previous data (Mohanty et al. 2016).

2.10 Techniques for Pre-processing Data

Data cleaning is necessary since raw data frequently contains noise, inaccuracies, and discrepancies. To maintain the integrity of the dataset, duplicates must be eliminated, errors must be fixed, and missing values must be filled in.

2.10.1 Data Augmentation

By performing changes like spinning, trimming, and scaling on pre-existing photos, augmentation techniques artificially enhance dataset size. Through this method, overfitting is decreased, and model generalization is improved (Das et al. 2020).

2.10.2 Standardization and Normalization

By making pixel values in photos the same scale, features are guaranteed to be comparable. Convergence of the model is hastened by standardized procedures, which focuses data near zero with a unit standard deviation.

2.10.3 Feature Extraction

Simplifying the complexity of data and enabling more efficient model learning are two benefits of extracting relevant features from pictures, such as patterns or color histograms.

2.10.4 Balancing Classes

Class balance is achieved by either oversampling minority classes or undersampling dominant classes.

2.11 Data Collection and Pre-processing Challenges

2.11.1 Labeling Complexity

Accurately classifying pictures as healthy or ill requires knowledge. The symptoms of several disorders may overlap, making accurate identification difficult (Das et al. 2020).

2.11.2 Imbalanced Data

Imbalanced classes result from the fact that sick samples frequently exceed healthy ones by a large margin. This imbalance may affect how well the model can identify illnesses (Das et al. 2020).

2.11.3 Data Access and Privacy

Disseminating agricultural data presents privacy issues. Due to proprietary limitations and data ownership, accessing a variety of datasets may be difficult (Das et al. 2020).

2.11.4 Environmental Variability

Changes in the backdrop, imaging, and lighting can add noise and have an influence on the model's generalizability.

The steps of data gathering and pre-processing are essential in developing AI models for diagnosis of plant diseases. The dependability and accuracy of AI forecasts are fueled by high-quality data. AI models may recognize disease trends and generate accurate forecasts by utilizing resources like imaging databases, networks of sensors, and historical records together with efficient pre-processing algorithms. The promise of AI-based plant disease diagnosis is still transformational, even though issues like labeling intricacy, data disparity, and confidentiality must be addressed. Agriculture stands to gain from better disease management, improved crop yields, and sustainable farming methods by ensuring precise data collection and rigorous pre-processing.

2.12 Building and Training AI Models

2.12.1 *Comparative Analysis of Choosing Appropriate AI Algorithms for Plant Disease Diagnosis*

The adoption of artificial intelligence (AI) systems for plant-borne illness diagnostics has caused a notable revolution in the agricultural sector. The accuracy, effectiveness, and comprehension of the illness detection models are directly influenced by the choice of the appropriate AI algorithm. Convolutional neural networks (CNNs) and Random Forests have emerged as popular options among the several AI algorithms available. In the context of developing artificial intelligence models for plant pathology diagnostics, this section examines the concerns, benefits, and limits of these algorithms (Simonyan and Zisserman 2014).

2.12.2 *Convolutional Neural Networks (CNN)*

CNNs are a subset of deep learning algorithms that have completely changed the field of image analysis. They are naturally suited for plant disease diagnosis based on leaf photos since they excel at tasks requiring visual data. They imitate the human visual system by extracting characteristics from pictures in a hierarchical manner. With the use of this feature extraction capacity, they can identify subtle patterns and textures in leaves that point to the presence of illness (He et al. 2016). They automatically detect spatial hierarchies in pictures by identifying local characteristics and how they are combined to form more complex structures. This is crucial to distinguish between healthy and unhealthy plant tissues. As a starting point, CNNs may use pre-trained models on big datasets like ImageNet. Even with few photos of plant diseases, this transfer learning speeds up model convergence. CNNs have the ability to identify illnesses at different dimensions of leaves and resolutions because of their scale-invariant characteristics. Once these are learned, CNN-based models can identify diseases in real time, allowing for quick actions to stop the spread of the disease.

2.12.3 *Random Forests*

Random Forests is an ensemble learning method that combines multiple decision trees to make predictions. It has been widely used in various domains, including agriculture, due to its interpretability and robustness. Itprovides insights into feature importance, indicating which visual attributes contribute most to disease detection. This information can aid experts in understanding disease symptoms. By aggregating predictions from multiple decision trees, it reduces the risk of overfitting and improves generalization. It can handle missing data effectively, a crucial advantage in cases where datasets are not uniform.These algorithms can capture non-linear relationships between image features and disease presence, enabling them to identify complex patterns. It offers interpretable results, making it easier for domain experts to comprehend the decision-making process, and allowing them to recognize intricate patterns (Zhou et al. 2016).

2.12.4 Factors Affecting Algorithm Selection

The following factors strongly impair the selection of algorithm (Singh and Shrivastava 2018).

2.12.4.1 Data Accessibility

Due to their powerful feature extraction abilities, CNNs can be useful when a significant amount of labeled data is accessible.

2.12.4.2 Interpretability

Random Forests offer insights into feature significance and model decision-making if interpretability is a goal.

2.12.4.3 Computing Power

CNN training requires a large amount of computing power, which not all users may be able to afford. Random Forests use less computing power.

2.12.4.4 Real-time Requirements

CNN's prediction accuracy might be a determining factor if real-time illness diagnosis is required (Singh and Shrivastava 2018).

The choice of an AI system for diagnosing plant diseases relies on the application's goals, resources, and unique needs. When it comes to image-based analysis, CNNs stand out because of their outstanding precision and real-time capabilities, whereas Random Forests offer resilience and interpretability. The eventual decision may include a trade-off between precision, comprehension, and computational needs. Advancements in hybrid models and algorithm design may help to further hone disease diagnostic techniques as the area of AI develops, enabling sustainable and effective agriculture. The following figure describes the fundamental structure of plant disease identification (Fig. 2.1).

Building precise plant disease detection algorithms requires a well-annotated dataset as a base. This entails taking pictures of plants, classifying them as healthy or sick, and dividing the dataset into subgroups for training, validation, and testing. There are two types of commonly used data: image based and sensor based. The training data prepares the model for use, the validation data helps adjust hyperparameters, and the testing data assesses the model's efficacy on samples that have not yet been seen.

2.13 Image-based Plant Disease Diagnosis

PlantVillage is one of the renowned disease detection methods based on images. This tool, created by Penn State University researchers, uses AI and crowdsourcing to identify plant illnesses. A mobile app that is part of the system enables farmers to post pictures of their crops. The photos are analyzed by AI algorithms to spot probable ailments and provide therapy suggestions. Farmers in many locations have benefited from PlantVillage's rapid disease identification and control assistance, which have boosted crop yields and decreased losses.

Jung et al. (2023) developed a plant disease detection model using deep learning methods using 8121 healthy images of different kinds of crops. A total of 31,061 images of infected leaves included images of crops with bacterial spot either with

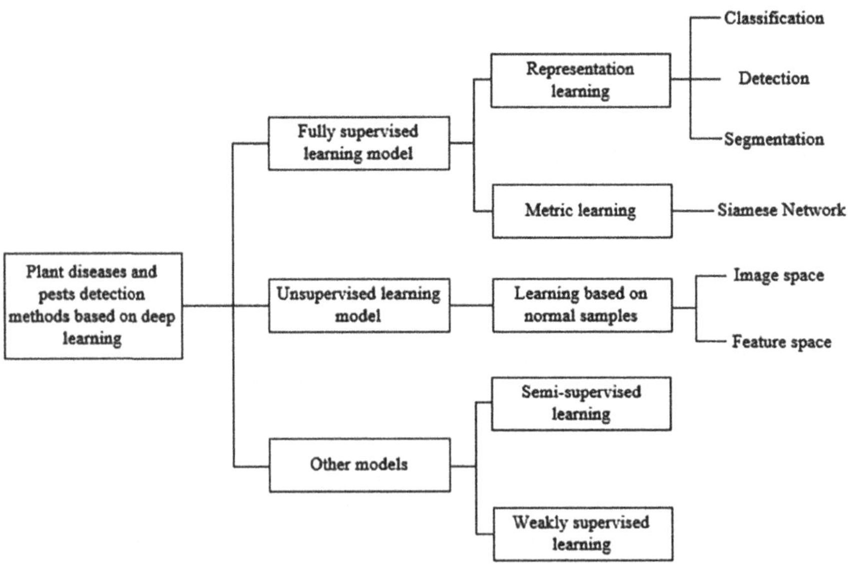

Fig. 2.1 Fundamental structure of plant disease identification

tomato mosaic virus or *Alternaria* sp., and potatoes infected by *Phytophthora infestans* were collected in this study. A total of 18,445 and 6481 photos of sick and healthy leaves, respectively, were kept after image filtering for further analysis to create a systematic AI detection model for plant disease. The image background was removed instead of annotating for training set in order to construct models based on the leaf's form or an image's lesion (Fig. 2.2).

2.13.1 Data Pre-processing and Training AI Models

Using stratified random sampling, the entire dataset was split 80:20, with training and test data, respectively. Using the training set, the model was built and validated. The model that had the best accuracy was chosen from CNN models that were trained earlier with high-precision scores in classification (Jung et al. 2023). It was important to employ more data in order to enhance the illness detection model's effectiveness. In order to gain extra picture data, data augmentation was done by modifying the training dataset that was already available. To create an ideal machine learning classification system independent of the angle, the original photos were resized to 224 pixels by rotating. The gathered data has a huge image size and the directions of the leaf midrib are all distinct. As data augmentation techniques, brightness or color changes were not selected since they likely interfere with the characteristics of disease or crop lesions. The sum was significant in step 1, since all data were used. In order to be utilized as analysis data for computing power, the

Fig. 2.2 Image selection (Source: Jung et al. 2023)

picture data was amplified 18 times and rotated 20 degrees. Each image was rotated 10 degrees for steps 2 and 3 in order to multiply the data 36 times.

2.13.2 *Systematic Algorithm for Image-based Disease Diagnosis*

The systematic algorithm for image-based disease diagnosis in plants has three key steps (Fig. 2.3).

In order to simulate human detection, a systematic detection model was built with successive submodels to identify crops, illness prevalence, and disease categorization. The CNN analysis method, one of the deep learning analysis approaches suited for image analysis, was used to build this model. Five pre-trained CNN models were used to fine-tune each submodel: AlexNet, ResNet50, GoogLeNet, VGG19, and EfficientNet. These performed well in the classification of plants and ranked first or second in the ImageNet Large Scale Visual Recognition Challenge. In step 1, five pre-trained CNN models were used to create a model to identify and categorize cultivars using pictures of a whole leaf from three crops in the Solanaceae family, independent of the presence or absence of disease. Step 2 included dividing the number of healthy and damaged leaves into two groups. Step 3 involved taking photos of infected leaves from step 2 to classify disease types using classification models for specific crop diseases. Only photos of a single disease type were available for the bell pepper, making it impossible to create a classification model that could discriminate between different intraspecies illnesses.

2.14 Model Evaluation

Calculations were made to determine accuracy, precision, recall, and F1-score in order to assess the classification model's performance. Values for true positive (TP), true negative (TN), false positive (FP), and false negative (FN) are produced after

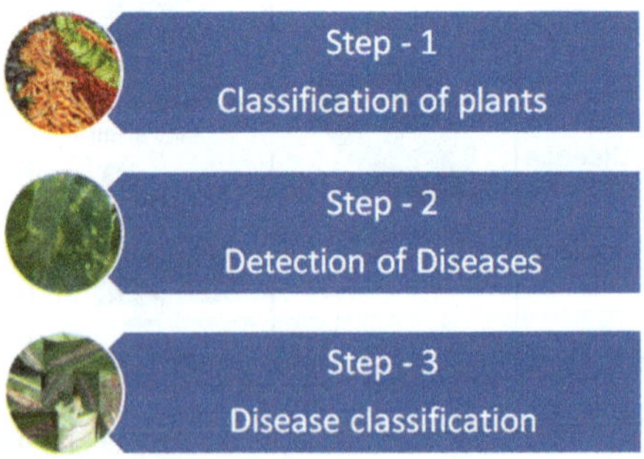

Fig. 2.3 Algorithm of image-based disease detection

generating a confusion matrix by comparing the model test result with the real condition. Li and Wang demonstrated customized DenseNet-77, an artificial neural network program for plant disease detection and classification that effectively identified and classified the infection among tomato, apple, grapes, rice, potato, etc. The model evaluation was determined by estimating accuracy, intersection of union, recall, and precision. The mean accurate precision and mean intersection of union scores for the CenterNet method were observed to be 0.99 and 0.93. The scores of the class-wise plant disease prediction are shown in Fig. 2.4.

The intersection of union (IOU) value was higher for the DenseNet-77 than YOLO or RetinaNet models. Parul et al. conducted a performance assessment between two types of CNN deep learning plant disease detection models using image-based classification with images from the PlantVillage database similar to Jung et al. (2023).

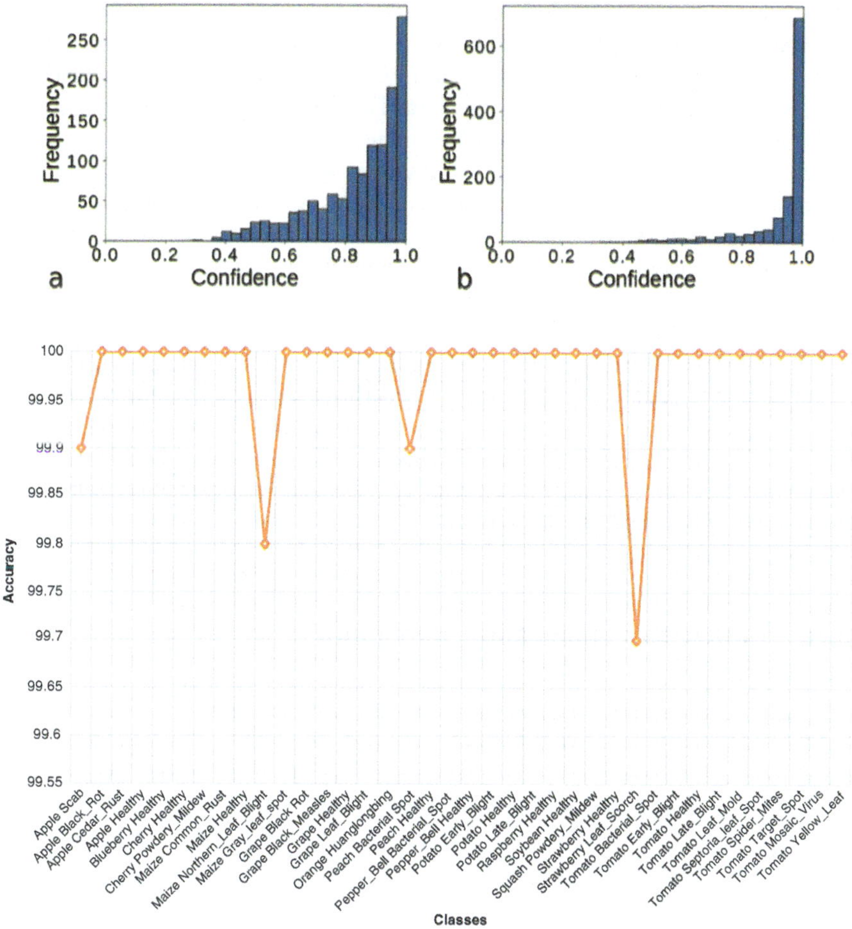

Fig. 2.4 Model evaluation of DenseNet-77 for detection of plant diseases

However, DenseNet-77 has 99% accuracy scores whereas the S-CNN model has 98.6% accuracy. The results given in Fig. 2.4 show that comparison of confidence scores among F-CNN and S-CNN neural network models reveals the higher confidence scores of S-CNN models over F-CNN. The mean IOU score of S-CNN models is 0.93, which is closer to the DenseNet-77 model. This infers that CNN models are better for image-based plant disease detection than other machine learning models.

2.15 Sensor Data-based Disease Diagnosis

The research study conducted by Molin and Kvarnheden (2010) presents a novel method for controlling apple scab disease using a wireless sensor network (WSN) and a decision support system.The goal of the project is to create a decision support system that forecasts apple scab primary infection periods using wireless sensor networks, allowing for prompt and precise disease control tactics. To track environmental factors important for the emergence of apple scab, the researchers set up a wireless sensor network in an apple orchard. This network gathered information on leaf moisture and temperature, two important variables affecting the spread of disease. The gathered information was subsequently included in a decision support system that offered forecasts for the times of initial infection. This study illustrates the effective use of wireless sensor networks and decision support systems in the management of apple scab disease. The technique enables farmers to make educated decisions and apply disease control measures at important time points by combining real-time data collecting and predictive modeling. This technologically advanced solution improves disease control tactics, optimizes resource consumption, and adds to the long-term viability of apple production.The study highlights the potential of technology, such as wireless sensor networks and decision support systems, to transform disease control techniques in agriculture. It shows how data-driven techniques may help farmers tackle illnesses more effectively, ensuring food security and agricultural sustainability in the long run.

2.16 Future Directions and Innovations

The field of plant disease diagnostics has a great deal of potential to be transformed by the quick development of artificial neural networks (AI) technology. Emerging trends are influencing the future environment of AI-driven illness diagnosis as innovation continues to advance. This section examines these developments, the incorporation of AI with robots and autonomous systems, and the significant influence AI is expected to have on sustainable agricultural methods and precision agriculture (Guidi and Salgado 2018).

2.16.1 *Explicit AI for Interpretability*

There is an increasing need for openness and interpretability as AI models become more complicated. Researchers are concentrating on creating methods that shed light on how AI models make decisions. Regions in pictures that are most important for the model's predictions are highlighted using explainable AI techniques like attention processes and saliency maps. In addition to increasing user confidence, this interpretable AI also makes it possible for domain specialists to comprehend the justifications for illness classifications. Therefore, explainable AI eliminates the disconnect between machine learning algorithms and human knowledge, increasing the technology's usability and influence (Ge and Li 2020).

2.16.2 *Fusion of Multi-modal Data*

Numerous elements, such as the environment, genetic data, and microbial interactions, have an impact on plant health. Integrating data from numerous sources to provide a full picture of plant health is one of the newest developments. AI models can offer a more comprehensive picture of disease dynamics by fusing visual data with environmental information from monitoring devices and genetic data. The accuracy and resilience of illness detection models are improved by the merging of multi-modal data, allowing for more accurate predictions and treatment advice.

2.16.3 *Limited Transfer Learning with Data*

Large-scale data labeling can be time- and resource-intensive when used to train AI models. The use of pre-trained models on larger datasets and their fine-tuning for plant disease diagnosis are emerging developments in transfer learning. Because less detailed tagged plant disease photos are required with this method, AI-driven diagnosis is more readily available to small-scale farmers and areas with scarce data resources. Transfer learning expedites the convergence of models and equips local communities to reap the rewards of AI technology (Tadesse et al. 2020).

2.16.4 *Active Learning Techniques*

Emerging trends call for the use of active learning techniques to enhance the learning process. By training the model's focus on the difficult situations that most contribute to its learning, active learning entails choosing the most instructive examples for labeling. By streamlining the training procedure, this iterative method decreases the overall annotation work needed to create reliable disease detection models (Sankaran et al. 2010). AI models are more effective and flexible to changing illness patterns thanks to active learning methodologies.

2.16.5 Distributed Learning for Decentralized Data

Federated learning seems to be a potential development in situations where confidentiality of information is an issue. This strategy keeps data local and enables the training of AI models across several devices or servers (Yang et al. 2019). Data privacy is maintained since each device adds to the algorithm's learning without transferring raw data. Federated learning is especially important for diagnosing plant diseases because it allows farmers to work together to create reliable models without jeopardizing private data (Torres-Sánchez et al. 2018).

The future of agriculture will be shaped by the growing trends in AI for diagnosing plant diseases, which will improve disease diagnostic precision, accessibility, and sustainability. Transfer learning democratizes AI adoption, explainable AI improves transparency, and multi-modal data fusion offers a full picture of plant health. Optimizing model performance and addressing data privacy issues are both goals of active learning and federated learning systems. The potential of AI-driven disease detection in agriculture is enormous as these developments converge, offering more robust crops, fewer losses, and a changed agricultural environment.

2.17 Integration of AI for Plant Disease Diagnosis Using Robotics, Drones, and Automated Farm Equipment

The landscape of plant disease detection is changing, as are agricultural practices, as a result of the combination of artificial intelligence (AI) and robots, drones, and autonomous farm equipment. Real-time, effective, and accurate solutions for disease diagnosis and management are provided by this confluence of technology. This section explains how AI-driven technologies might improve plant disease detection by fusing them with robots, drones, and autonomous machinery, emphasizing their advantages.

2.17.1 Robotics for Precision Plant Inspection

Robots powered by AI that have sensors and cameras can move around fields and scan plants for disease signs with unmatched precision. These autonomously moving robots cover enormous regions while collecting data and high-resolution photos for further analysis by AI systems. Real-time illness detection is enabled through the integration of AI and robotics.

2.17.2 Aerial Surveillance Using Drones

A bird's-eye view of agricultural fields may be obtained using drones that are fitted with specially designed cameras and sensors. The picture is processed by AI algorithms to find stress patterns, nutritional shortages, and illness signs. Drones quickly cover broad regions, enabling early detection and specialized solutions. Drones and AI working together improve disease surveillance effectiveness while reducing resource waste (Torres-Sánchez et al. 2018).

2.17.3 Autonomous Farm Machinery for Personalized Care

Sprayers and cultivators powered by AI can accurately target locations where disease outbreaks are occurring. These devices use treatments only when necessary, by evaluating information gathered by sensors and imaging systems, which minimizes the use of chemicals and their negative effects on the environment. This connection supports sustainable agricultural methods and improves resource efficiency (Torres-Sánchez et al. 2018).

2.17.4 Analyzing Real-time Data

Robotic systems, drones, and autonomous equipment continuously process data using AI algorithms. Instantaneous insights about disease prevalence, stressors, and plant health are provided by the analytics. Farmers are able to make quick decisions thanks to this real-time information, which successfully directs disease management tactics.

2.17.5 Improved Disease Surveillance

Continuous disease monitoring is made possible throughout the growing season by the integration of AI with robots, drones, and autonomous equipment. This preventative strategy helps with early diagnosis, enables prompt actions, and lessens the requirement for heavy pesticide applications.An important step toward more effective and long-lasting plant disease diagnostics is the integration of AI with robots, drones, and autonomous agricultural equipment. These innovations enable real-time monitoring of fields, fine-grained plant inspection, and targeted treatment. Agriculture is advancing toward a day where disease control is accurate, timely, and ecologically responsible by using the potential of AI-driven solutions.

2.18 AI Revolutionizing Sustainable Farming Methods and Precision Agriculture

Particularly in the domain of precision agriculture and sustainable agricultural methods, the application of artificial intelligence (AI) to agriculture has the potential to revolutionize the industry. AI-driven technologies have the potential to fundamentally alter how farmers manage their crops, make the most use of their resources, and protect the environment. This section explores how AI has the potential to transform precision agriculture and advance sustainability.With the help of AI, farmers can apply resources like water, fertilizer, and pesticides more effectively. Artificial intelligence (AI) algorithms develop specialized irrigation schedules and fertilizer application strategies by examining information collected by sensors, satellites, and drones. Through precise resource management, waste is reduced, resources are conserved, and agricultural production is increased.

Early warning systems for plant epidemics are provided by AI-driven disease detection algorithms, IoT devices, and drones. With the help of these technologies, quick action may be taken to prevent loss of crops and the requirement for heavy pesticide use. AI improves disease prediction accuracy, which supports resilient farming techniques. Huge datasets are analyzed by AI algorithms to produce insights that may be used to make decisions. Planting timetables, crop rotation plans, and the best times to harvest are all influenced by these observations. Data-driven decisions increase yields, reduce losses, and support environmentally friendly agricultural methods. AI-based image analysis systems are able to recognize and distinguish between weeds and crops. This makes it possible to apply herbicides in a targeted manner, using fewer chemicals and having a less negative environmental impact. Precision weed control encourages farming that is sustainable and protects crop health. AI-enabled sensors keep an eye on the moisture content of the soil, as well as aspects of the soil including its moisture content, pH, and nutrient levels. By analyzing this data, farmers may better target their fertilization programs to the unique requirements of their soils, avoiding overfertilization and leaching. This strategy minimizes environmental contamination while preserving the health of the soil. Based on the surrounding environment, AI systems can forecast insect outbreaks. With the use of this knowledge, farmers may manage pests by taking preventative steps and using natural predators rather than artificial pesticides, preserving the ecological balance. AI has enormous potential to transform sustainable agricultural methods and precision agriculture. With the use of data-driven insights provided by AI-driven technology, farmers can manage resources effectively and identify diseases and pests. In addition to greater output and fewer losses, this change offers a more sustainable method of farming that reduces the influence on the environment. AI's integration with agriculture provides a route to a future where global food production is more robust, effective, and ecologically responsible.

2.19 Conclusion

AI-based plant disease detection is more than just a technological achievement; it is a driving force behind agricultural revolution. It enables us to rethink how we approach disease concerns, enables farmers to make educated decisions, and cultivates a more sustainable and resilient agricultural future. We can enhance AI's influence, safeguard our food systems, and support the emergence of a wealthier agricultural world by embracing its promise and encouraging collaboration. The identification of diseases constitutes a quantum leap ahead. The trip through the investigation of AI's function in this critical sector brings us to a resounding conclusion: AI-based plant disease detection has the potential to transform how we perceive, manage, and safeguard our agricultural systems. AI's capacity to evaluate massive volumes of data—from photos to sensor readings—reveals hidden patterns and connections that the human eye may overlook. AI learns from previous data and improves its accuracy over time using machine learning techniques. This, in turn, improves disease detection and forecasting, resulting in more proactive agricultural management. One of the most important characteristics of AI-based plant disease detection is its capacity to detect diseases early.

References

Agrios GN (2005) Plant pathology, 5th edn. Academic Press

Das A, Pradhan B, Jha AK (2020) Integrating artificial intelligence with IoT in precision agriculture for sustainable crop production: a review. Comput Electron Agric 173:105370

Elad Y, Pertot I (2014) Climate change impacts on plant pathogens and plant diseases. J Crop Improv 28(1):99–139

Ge D, Li J (2020) A review on applications of deep learning in plant disease detection. Comput Electron Agric 177:105612

Guidi G, Salgado R (2018) Robotics and artificial intelligence in agriculture: current applications and future challenges. Agroecol Sustain Food Syst 42(7):702–722

He K, Zhang X, Ren S, Sun J (2016) Deep residual learning for image recognition. In Proceedings of the IEEE conference on computer vision and pattern recognition

Hughes G, Salathé M (2015) An open access repository of images on plant health to enable the development of mobile disease diagnostics. arXiv preprint arXiv:1512.03477

Jones DG (2009) Plant pathogens and principles of plant pathology. In: Plant pathology. Wiley-Blackwell, pp. 1–17

Jung M, Song JS, Shin AY (2023) Construction of deep learning-based disease detection model in plants. Sci Rep 13:7331. https://doi.org/10.1038/s41598-023-34549-2

Kamilaris A, Kartakoullis A (2021) Applications of machine learning in precision agriculture: a review. Precis Agric 22(3):397–425

Kamilaris A, Prenafeta-Boldú FX (2018a) Deep learning in agriculture: a survey. Comput Electron Agric 147:70–79

Kamilaris A, Prenafeta-Boldú FX (2018b) Deep learning in agriculture: a survey. Comput Electron Agric 147:70–90

Kamilaris A, Kartakoullis A, Prenafeta-Boldú FX (2021) A review on the practice of plant disease detection using convolutional neural networks. Inform Process Agric 8(1):11–28

LeCun Y, Bengio Y, Hinton G (2015) Deep learning. Nature 521(7553):436–444

Lucas JA (2008) Plant pathology and plant diseases. In: Encyclopedia of life sciences. John Wiley & Sons, pp. 1–8

Mohanty SP, Hughes DP, Salathé M (2016) Using deep learning for image-based plant disease detection. Front Plant Sci 7:1419

Molin J, Kvarnheden A (2010) Decision support system for forecasting primary infection periods of apple scab based on wireless sensor network. Comput Electron Agric 70(1):77–84

Picon A, Onelli E, Azzarello E, Giordano C, Masi E, Moscatiello R et al (2020) The role of data pre-processing in plant image analysis: a case study on image-based plant phenotyping. Front Plant Sci 11:1–17

Sankaran S, Mishra A, Ehsani R, Davis C (2010) A review of advanced techniques for detecting plant diseases. Comput Electron Agric 72(1):1–13

Savary S, Willocquet L, Pethybridge SJ, Esker P, McRoberts N, Nelson A (2019) The global burden of pathogens and pests on major food crops. Nat Ecol Evol 3(3):430–439

Schumann GL, D'Arcy CJ (2010) Essential plant pathology, 2nd edn. American Phytopathological Society

Simonyan K, Zisserman A (2014) Very deep convolutional networks for large-scale image recognition. arXiv preprint arXiv:1409.1556

Singh D, Shrivastava S (2018) An overview of convolutional neural networks: architectures, applications, and challenges. ArXiv Preprint arXiv:1710.09829

Tadesse T, Mwebaze E, Vossen G (2020) Deep learning techniques for plant disease detection and diagnosis. In: Deep learning and convolutional neural networks for medical image computing. Springer, pp. 315–339

Torres-Sánchez J, López-Granados F, De Castro AI (2018) Assessing the accuracy of mosaics from unmanned aerial vehicle (UAV) imagery for precision agriculture purposes in wheat. Precis Agric 19(5):770–786

Windstam ST, Schmale DG (2018) The complexity of diagnosing plant diseases. In: Plant disease diagnosis. Springer, pp. 1–18

Yang Q, Liu Y, Chen T, Tong Y (2019) Federated machine learning: concept and applications. ACM Trans Intell Syst Technol (TIST) 10(2):1–19

Zhou B, Khosla A, Lapedriza À, Oliva A, Torralba A (2016) Learning deep features for discriminative localization. In Proceedings of the IEEE conference on computer vision and pattern recognition

Chapter 3
Computer Vision-based Remote Care of Microbiological Data Analysis

Pritesh Kumar Jain and Sandeep Kumar Jain

Abstract The field of remote care of microbiological data analysis is rapidly evolving, utilizing computer vision to automate various tasks related to microorganism identification, classification, quantification, tracking, and presence detection in images. By employing this technology, healthcare providers can benefit from an improved and more efficient means to analyze microbiological data, leading to enhanced diagnosis, treatment, and prevention of infectious diseases.

This chapter delves into the challenges and prospects associated with utilizing computer vision in remote care of microbiological data analysis. Initially, an overview of the fundamental principles of computer vision is presented, along with its applicability to microbiological data analysis. Subsequently, the challenges associated with employing computer vision in this context are discussed, encompassing factors such as image variability, microorganism complexity, and the scarcity of extensive datasets. Furthermore, the chapter explores the potential of computer vision in enhancing the diagnosis, treatment, and prevention of infectious diseases.

To conclude, the chapter examines the future of computer vision-based remote care of microbiological data analysis. It contends that this technology holds immense potential in revolutionizing the diagnostic and treatment approaches utilized by healthcare providers in relation to infectious diseases.

Keywords Microbial · Computer vision · Healthcare · Diseases · Microorganism

3.1 Introduction

Microbiological data analysis plays a crucial role in comprehending the world of microorganisms, spanning various domains such as health, environment, and industry. It unveils the intricate microbial diversity, behaviors, and functions. Within the

P. K. Jain (✉) · S. K. Jain
Department of Computer Science and Engineering, Shri Vaishnav Vidyapeeth Vishwavidyalaya, Indore, Madhya Pradesh, India
e-mail: priteshkjain@svvv.edu.in; sandeepjain@svvv.edu.in

A. Khamparia et al. (eds.), *Microbial Data Intelligence and Computational Techniques for Sustainable Computing*, Microorganisms for Sustainability 47,
https://doi.org/10.1007/978-981-99-9621-6_3

healthcare sector, it aids in the diagnosis of diseases and tracking of epidemics, and contributes to the development of therapeutic interventions. Furthermore, environmental monitoring employs this analysis to assess the quality of air, water, and soil, thereby facilitating the detection of pollution and conservation efforts. The realm of food safety greatly relies on this analysis for the detection of hazards and ensuring quality control. Industries also make use of microbiological data analysis in areas such as bioproduction, agriculture, and bioremediation. Additionally, research and predictive modeling provide valuable insights into microbial genetics and behavior, ultimately contributing to product safety and playing a role in the field of biological engineering. As technology evolves, microbiological data analysis continues to drive scientific exploration.

3.2 The Role of Computer Vision

The remote management of microbiological data analysis, commonly known as digital microbiology or microbiome analysis, is undeniably a swiftly developing domain that utilizes computer vision and other advanced technologies to mechanize various facets of microbiological research (Zhang et al. 2022). This realm holds the potential to substantially augment the swiftness and precision of microbiological data analysis, rendering it more accessible and efficient. The following are a few pivotal facets of this realm:

3.2.1 Automatization of Microorganism Identification

Computer vision algorithms (Zhang et al. 2022) can be educated to recognize and discern diverse categories of microorganisms in images, comprising bacteria, viruses, fungi, and protozoa. This mechanization has the capacity to economize a significant amount of time in manual identification procedures.

3.2.2 Classification and Taxonomy

Automated systems have the capability to categorize microorganisms into distinct taxonomic groups based on their morphological characteristics or genetic indicators. This can facilitate the comprehension of the multitude of microorganisms that exist within a particular specimen.

3.2.3 *Quantification*

Computer vision can be employed to quantify the abundance of microorganisms within a given sample. This proves to be particularly advantageous in the realm of ecological research, as well as in the monitoring of microbial populations in diverse environments.

3.2.4 *Tracking and Behavior Analysis*

The tracking and analysis of the movement and behavior of microorganisms over a period of time is of utmost importance in the field of microbiology. Computer vision is capable of automating the tracking of individual cells or populations within a dynamic setting.

3.2.5 *Presence/Absence Detection*

Automated systems possess the ability to rapidly ascertain the presence or absence of specific microorganisms within a sample. This proves to be highly valuable in the domains of medical diagnostics, environmental surveillance, and food safety testing.

3.2.6 *High-throughput Screening*

One of the primary merits of employing computer vision in the analysis of microbiological data lies in its capacity to efficiently process a large quantity of samples within a short span of time, thereby enabling high-throughput screening and analysis.

3.2.7 *Integration with Other Technologies*

The integration of computer vision (Singh et al. 2020; Chen et al. 2023) with other technologies, such as genomics, metagenomics, and artificial intelligence, is a common practice in the field of digital microbiology. This approach aims to offer a comprehensive comprehension of microbial communities and their functionalities.

3.2.8 *Remote Monitoring*

The notion of "remote care" denotes the capacity of this technology to be employed in distant locations, such as field studies or areas with limited laboratory access. Consequently, it becomes feasible to monitor microbial ecosystems in real time.

3.2.9 *Applications*

Diverse domains benefit from the applications of this technology, encompassing healthcare (diagnosis and treatment monitoring), agriculture (soil and plant microbiomes), environmental science (water quality monitoring), and food safety (contaminant detection) (Dange et al. 2023).

Challenges encountered within this field encompass the assurance of accuracy and reliability of computer vision models, management of varied and intricate microbial communities, as well as the ethical and privacy concerns associated with data collection and analysis. However, with technological advancements and the availability of more data, the remote analysis of microbiological data holds tremendous potential in advancing our comprehension of microorganisms and their ecological functions.

3.3 Steps Required to Implement

There exists a plethora of tools and software that can be employed for the purpose of implementing computer vision in the context of remote care for the analysis of microbiological data.

Some of the most popular tools include (Rao et al. 2022):

OpenCV: OpenCV is a library of computer vision algorithms that is both free and open-source. It is accessible on a multitude of platforms, including Windows, macOS, and Linux.

TensorFlow: TensorFlow is a renowned open-source machine learning library. It has the potential to be utilized for the training and deployment of computer vision models.

PyTorch: PyTorch is another widely recognized open-source machine learning library. It bears similarities to TensorFlow, yet it has been engineered to possess enhanced flexibility and user-friendliness.

MATLAB: MATLAB is a commercial software package that is frequently employed for scientific computing. It encompasses an array of functions tailored to computer vision.

ImageJ: ImageJ is an image processing software that is both free and open-source. It serves the purpose of loading, viewing, and analyzing images.

The selection of programming language for the execution of computer vision in order to provide remote care for microbiological data analysis would be contingent on the particular tool or software being utilized. Nonetheless, Python has emerged as a prevalent language for computer vision and is compatible with the majority of aforementioned tools and software (Hu et al. 2021).

3.4 Microbiological Data

Microbiological data is data on microorganisms, such as bacteria, viruses, fungi, and parasites. This data can be collected from a variety of sources, such as clinical samples (e.g., blood, urine, stool), environmental samples (e.g., water, soil, air), and food samples.

Microbiological data can be collected using a variety of methods (Wang et al. 2020; Li et al. 2019), including:

Culture: Microorganisms are grown in a laboratory medium to allow them to multiply. This allows for the identification and quantification of microorganisms.

Serology: Antibodies to specific microorganisms are detected in the blood or other bodily fluids. This can be used to diagnose infections and to track the immune response to infection.

Molecular diagnostics: Nucleic acids (DNA and RNA) from microorganisms are detected and analyzed. This can be used to identify microorganisms, to determine their resistance to antibiotics, and to track the spread of disease.

Microbiological data is used in a variety of ways, including:

Diagnosis and treatment of infections: Microbiological data is used to diagnose infections and to determine the best course of treatment. For example, culture data can be used to identify the specific bacteria causing an infection, which can then be treated with the appropriate antibiotic.

Public health surveillance: Microbiological data is used to track the spread of disease and to identify outbreaks. For example, data on the types of bacteria causing pneumonia in hospitals can be used to identify and control hospital-acquired infections.

Food safety: Microbiological data is used to ensure the safety of food. For example, food companies test their products for harmful bacteria, such as Salmonella and *E. coli*.

Research: Microbiological data is used to study the biology of microorganisms and to develop new treatments and vaccines. For example, researchers are using microbiological data to develop new antibiotics that are effective against drug-resistant bacteria.

Microbiological data is essential for the diagnosis and treatment of infections, public health surveillance, food safety, and research.

Outlined below are several steps involved in the implementation of computer vision (Oh et al. 2018; Bohr and Memarzadeh 2020) for remote care of microbiological data analysis:

1. Collect a dataset of images of microorganisms: Assemble a dataset comprising a wide range of images of microorganisms so as to ensure that the computer vision model can effectively generalize to novel images.
2. Label the images in the dataset: Label the images within the dataset by identifying and categorizing the microorganisms present in each image.
3. Train a computer vision model on the labeled dataset: Train a computer vision model on the annotated dataset using various machine learning algorithms, including supervised learning, unsupervised learning, and reinforcement learning.
4. Deploy the computer vision model to a remote server: Deploy the computer vision model to a remote server, thereby enabling real-time analysis of microorganism images.

The implementation of computer vision for remote care of microbiological data analysis is a multifaceted endeavor. Nonetheless, the advantages offered by this technology are substantial and it is poised to witness greater adoption in the future.

3.5 Different Algorithms to Implement Computer Vision

Certainly, a variety of algorithms exist that can be employed to execute computer vision in the context of remote care for microbiological data analysis.

Several of the most widely utilized algorithms (Tripathi and Maktedar 2020; Tian et al. 2020) encompass:

Convolutional neural networks (CNNs):

CNNs are a type of deep learning algorithm that is well-suited for image analysis tasks. They are able to learn the spatial relationships between features in an image, which makes them effective at identifying and classifying microorganisms.

Support vector machines (SVMs):

SVMs are another type of machine learning algorithm that can be used for image analysis. They are able to learn the boundaries between different classes of images, which makes them effective at classifying microorganisms.

Random forests:

Random forests are a type of ensemble learning algorithm that combine the predictions of multiple decision trees. This makes them more robust to noise and outliers than single decision trees.

Gaussian mixture models (GMMs):

GMMs are a type of probabilistic model that can be used to segment images. They are able to identify clusters of pixels that belong to the same object, which makes them effective at identifying microorganisms.

K-means clustering:

K-means clustering is a simple unsupervised learning algorithm that can be used to segment images. It identifies k clusters of pixels in an image, and assigns each pixel to the cluster that it is closest to.

Table 3.1 Computer vision can vary depending on the specific algorithm and application

Parameter	Description
Image resolution	The number of pixels per unit of length in an image. Higher-resolution images provide more detail, which can be helpful for identifying and classifying microorganisms
Image size	The total number of pixels in an image. Larger images provide more data for the algorithm to learn from, which can improve the accuracy
Image contrast	The difference between the brightest and darkest pixels in an image. Images with high contrast will make it easier to identify and classify microorganisms
Image noise	The random variation of pixel values in an image. Images with low noise will be easier for the algorithm to process
Image background	The area of the image that does not contain microorganisms. Images with a simple background will make it easier for the algorithm to identify and classify microorganisms
Image pre-processing	The steps that are performed on an image before it can be analyzed. This may include steps such as resizing, cropping, and denoising
Microorganism type	The type of microorganism being analyzed. Different types of microorganisms may require different parameters
Microorganism concentration	The number of microorganisms per unit area in the image. Images with a high concentration of microorganisms may require different parameters than images with a low concentration
Microorganism morphology	The shape and size of the microorganisms. Different morphological features may require different parameters
Microorganism environment	The environment in which the microorganisms are found. Different environments may require different parameters

The selection of an algorithm will rely on the particular problem that is being addressed. For instance, CNNs are frequently employed for tasks such as identifying objects and dividing them into segments, whereas SVMs are frequently utilized for tasks such as categorization.

Various metrics, such as accuracy, precision, and recall, can be employed to compare the performance of different algorithms. Accuracy assesses the proportion of images that are correctly classified, precision evaluates the proportion of images that are classified as positive and are actually positive, and recall determines the proportion of positive images that are accurately classified (Table 3.1).

3.6 Challenges and Future Directions

The utilization of computer vision-based remote care for the analysis of microbiological data encounters numerous formidable limitations and obstacles. Microbiological data possesses inherent complexity, consisting of intricate structures and dynamic processes at the microscopic level. The accurate interpretation of this data through computer vision algorithms necessitates the advancement of sophisticated techniques for the processing of images.

Microorganisms exhibit a broad range of morphological diversity, encompassing various shapes, sizes, and characteristics. The training of computer vision models to reliably identify and differentiate between different microbial species and strains can be challenging due to this inherent variability.

Furthermore, the quality and resolution of microbial images can present problems. In remote or field settings, the acquisition of high-quality images of microorganisms may be hindered by factors such as suboptimal lighting conditions, variable magnification, and varying quality of imaging equipment.

In addition, sample preparation and staining techniques are crucial in the field of microbiology, but they can introduce further complexities to the analysis based on computer vision. Staining protocols have the potential to alter microbial morphology or introduce artifacts, thereby rendering accurate image analysis more challenging.

Moreover, the dynamic nature of microbiological processes, such as microbial growth and movement, can pose difficulties for real-time remote analysis. The continuous monitoring and tracking of microorganisms over time may necessitate advanced computer vision models capable of handling dynamic data streams.

3.7 Conclusion

In conclusion, while computer vision holds tremendous potential for the remote analysis of microbiological data, these challenges underscore the necessity for ongoing research and development to overcome the intricacies of this field and fully exploit the capabilities of computer vision in remote care and analysis of microbiological data.

References

Bohr A Memarzadeh K (2020) The rise of artificial intelligence in healthcare applications. In: Artificial intelligence in healthcare. Academic Press, pp 25–60

Dange BJ et al (2023) Grape vision: a CNN-based system for yield component analysis of grape clusters. Int J Intell Syst Appl Eng 11(9s):239–244

Hu et al (2021) Computer vision for medical imaging: a comprehensive guide. Springer

Li et al (2019) Computer vision for remote sensing: a comprehensive guide. Springer

Oh et al (2018) Telemedicine and telehealth systems: a comprehensive guide. Springer

Rao et al (2022) Microbiological data analysis: a comprehensive guide. Springer

Tian H et al (2020) Computer vision technology in agricultural automation—a review. Inform Process Agric 7(1):1–19

Tripathi MK, Maktedar DD (2020) A role of computer vision in fruits and vegetables among various horticulture products of agriculture fields: a survey. Inform Process Agric 7(2):183–203
Wang et al (2020) Deep learning for microbiological data analysis. Academic Press
Zhang, J., Li, C., Rahaman, M. M., Yao, Y., Ma, P., Zhang, J., ... & Grzegorzek, M. (2022). A comprehensive review of image analysis methods for microorganism counting: from classical image processing to deep learning approaches. Artificial Intelligence Review, 1-70.

Chapter 4
A Comparative Study of Various Machine Learning (ML) Approaches for Fake News Detection in Web-based Applications

Mahabub Hasan Mahalat, Sushree Bibhuprada B. Priyadarshini, Sandip Swain, Shobhit Sahoo, Atish Mohapatra, and Mangaldeep Das

Abstract The development of fake news detection and intervention methods is a result of the exponential spread of false news and its impact on justice, democracy, and public confidence. People's trust in the government, news media, news stories, social media, etc. has decreased as a result of widespread fake news. Hence, fake news has become a crucial problem in our society. In this chapter we have presented a comparative examination of detecting fake news. We have defined the negative behavior of bogus news and introduced detection methods. Many of such methods rely on identifying content and context aspects that suggest misrepresentation. We also looked at existing datasets that have been employed to categorize bogus news. Finally, we have suggested possible research avenues by applying various machine learning-based classification algorithms and analyzed bogus news in terms of various performance parameters like accuracy, precision, recall, and F1 score.

Keywords Accuracy · F1 score · Precision · Recall

4.1 Introduction

Social media has become a significant part of daily life today. Due to the exponential growth in users, it has become an indispensable requirement. In this context, Instagram is one of the most popular social media platforms with over 1 trillion active users, making it one of the most used social media platforms. The major benefit of online social media is the ability for people to communicate with one another effortlessly. This has opened up a fresh avenue for prospective attacks like identity theft and disinformation. According to a recent poll, there are more social media accounts established than there are active users. This has made it more

M. H. Mahalat · S. B. B. Priyadarshini (✉) · S. Swain · S. Sahoo · A. Mohapatra · M. Das
Computer Science & Information Technology, Institute of Technical Education and Research, Siksha 'O' Anusandhan Deemed to be University, Bhubaneswar, India
e-mail: mahabubhasanmahalat@soa.ac.in

A. Khamparia et al. (eds.), *Microbial Data Intelligence and Computational Techniques for Sustainable Computing*, Microorganisms for Sustainability 47,
https://doi.org/10.1007/978-981-99-9621-6_4

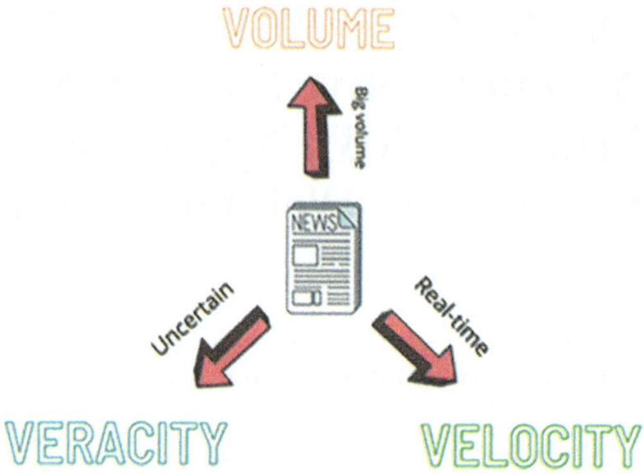

Fig. 4.1 Major features of data

challenging for social media companies to spot bogus accounts. Due to the prevalence of fraudulent information and marketing on social media, it is important to spot these fake accounts. The truth and intent of every remark are frequently intractable for computers alone, necessitating a collaborative effort between humans and technology (Radianti et al. 2016; Alkhodair et al. 2020).

Three major features associated with data are volume, veracity, and velocity, as illustrated in Fig. 4.1. Hence, identifying any new data is of paramount importance. Fake news has been circulating for several years and it will continue to do so; hence required procedures must be applied for eliminating such fake news. Articles, current events, posts, statements, assertions, and any other types of information linked to public interests and organizations are all examples of news.

Such news explanations raise social and political concerns. This concept is compatible with the majority of existing fake news studies and datasets. Although fake news is not a new phenomenon, questions such as "Why has it become a global topic of interest?" and "Why has it been increasingly attracting people's attention" are pertinent at this time. Furthermore, a large volume of false news is manufactured and distributed via the Internet, posing a potential threat to social communities and having a significant detrimental influence on Internet activities such as online shopping and social networking. During election campaigns, fake news has been blamed for increasing political polarization and partisan confrontation (Yi et al. 2003; Tapaswi et al. 2012).

Consider the COVID-19 pandemic that affected the entire world. Doctors and scientists were battling to find a cure, despite the fact that there were numerous therapies available on the Internet via social media, news articles, and so on. According to these articles, taking certain medications would cure COVID-19. People truly believed the "facts" and took the drugs without consulting a doctor, resulting in the spread of other disorders. We don't understand the context and jump to conclusions without understanding if what we have read is true or untrue. Figure 4.2 shows a scenario of news emerging from various sources (Radianti et al. 2016; Ranjan et al. 2003; MonaDiab et al. 2004; Rouse 2018; Dua and Du 2016; Huang n.d.; Researchgate.net 2018; Researchgate.net 2014; Sirikulviriya and Sinthupinyo 2011; Kevric et al. 2017; Parikh and Atrey 2018; Granik and Mesyura

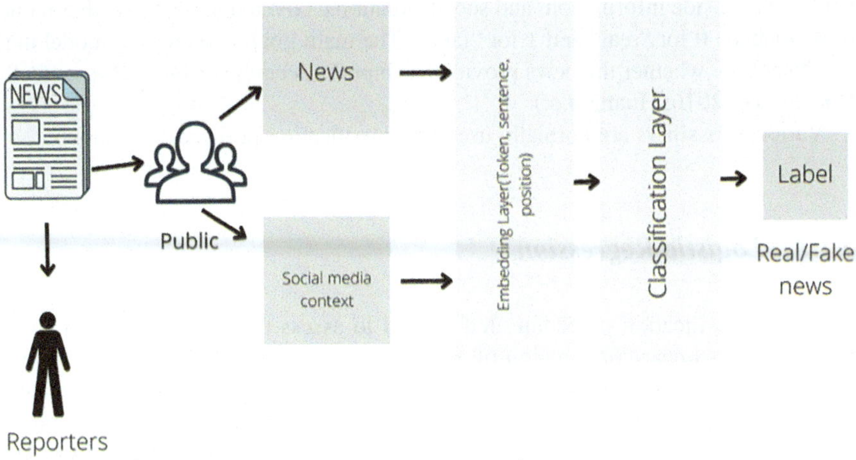

Fig. 4.2 News coming from various sources

2017; Gilda 2017; Jain and Kasbe 2018; Qin et al. 2018; Khanam and Ahsan 2017; Perez-Rosas et al. 2017; Aphiwongsophon et al. 2018; Kaur et al. 2019; Looijenga 2018; Anonymous 2018; Traore et al. 2017; Khanam and Ahsan 2018; Sharma et al. 2019; Shu et al. 2017; Khanam and Agarwal 2015; Zhang et al. 2020; Ludwig and Creation 2020; da Silva et al. 2019a, 2019b; Bovet and Makse 2019; Lazer et al. 2018; Raza and Ding 2022a, 2022b; Jain et al. 2019a, 2019b; Fan 2017; Shu et al. 2019; Rubin 2017; Kogan et al. 2019; Vosoughi et al. 2018).

This broad breadth of false news identification necessitates the use of highly developed computational quantification and visualization technologies. Previously, utilized algorithms such as naive Bayes, support vector machine, and random forests were ineffective in detecting bogus news. This study compares existing news identification approaches and enhances the effectiveness of previously used algorithms to correctly recognize whether the news is false or authentic (Yi et al. 2003; Tapaswi et al. 2012; Ranjan et al. 2003; MonaDiab et al. 2004; Rouse 2018; Antweiler and Frank 2005; Allcott and Gentzkow 2017; Ahern and Sosyura 2014; Mosseri 2016; Meyer 2017; Rashkin et al. 2017; Singh et al. 2017; Tacchini et al. 2017).

The most important reason for this study is that false news can be manufactured and distributed more quickly and cheaply than traditional news media such as newspapers and television. In today's society, social media is an integral part of daily life. Social media users are rapidly increasing on Instagram, one of the most popular social media platforms, which has over a trillion active users (Rashkin et al. 2017; Singh et al. 2017; Tacchini et al. 2017; Akoglu et al. 2013; Rubin et al. 2016). This has, in some ways, heightened awareness of potential attacks such as impersonation and misinformation. The suggested study would take the fake and real news datasets as input and determine whether the supplied input is true or false using various types of machine learning algorithms. The input may take the shape of news items, related side information, and social situations. Given the input, we obtain one of two labels: 0 for "real" and 1 for "fake." The main goal is to create a model that can determine whether the news provided as input is accurate or false (Rouse 2018; Dua and Du 2016; Huang n.d.).

Various classifiers are normally used for classification purposes, as follows.

4.1.1 Logistic Regression

This is a classification procedure that is used to assess the likelihood of an event occurring as a linear combination of a set of input features. Because it deals with probability approximation, the model is best suited for binary classification. Assume the logistic regression model calculates p for a linear combination of independent variables in order to establish the actual class label (Radianti et al. 2016; Alkhodair et al. 2020; Yi et al. 2003; Tapaswi et al. 2012; Ranjan et al. 2003; MonaDiab et al. 2004; Rouse 2018; Dua and Du 2016). The calculated expression can then be written as:

$$p(x) = \frac{1}{1 + e^{-(\beta_0 + \beta_1 x)}} \tag{4.1}$$

4.1.2 Decision Tree Classifier

This represents a supervised machine learning technique that uses a set of rules to make judgments in the same way that humans do. Every internal node of the decision tree assigns a condition or "test" to an attribute, and the connection is created based on the test conditions and the outcome. Finally, after computing all characteristics, the sheet node has a class label.

4.1.3 Random Forest Classifier

These are based on the idea of generating numerous decision tree algorithms, each of which generates a unique output. The Random Forest Classifier uses a variety of decision trees on different subsets of the input data to improve the dataset's predictive accuracy. The random forest employs the outcomes predicted by several decision trees. When applied to connected trees, the result will more or less resemble a single decision tree. Furthermore, uncorrelated decision trees can be generated using feature randomness and bootstrapping (Tapaswi et al. 2012; Ranjan et al. 2003; MonaDiab et al. 2004; Rouse 2018; Dua and Du 2016; Huang n.d.; Researchgate.net

2018; Researchgate.net 2014; Sirikulviriya and Sinthupinyo 2011; Kevric et al. 2017; Parikh and Atrey 2018; Granik and Mesyura 2017; Gilda 2017; Jain and Kasbe 2018; Qin et al. 2018; Khanam and Ahsan 2017; Perez-Rosas et al. 2017; Aphiwongsophon et al. 2018; Kaur et al. 2019; Looijenga 2018; Anonymous 2018; Traore et al. 2017; Khanam and Ahsan 2018; Sharma et al. 2019; Shu et al. 2017; Khanam and Agarwal 2015; Zhang et al. 2020; Ludwig and Creation 2020; da Silva et al. 2019a, 2019b; Bovet and Makse 2019; Lazer et al. 2018; Raza and Ding 2022a).

4.1.4 Linear Support Vector Classifier (SVC)

This method is based on arranging each data item as a point in a range of dimensions *n* (the number of accessible attributes), and the value of a certain property is the number of given coordinates. The SVM algorithm uses coordinates that correspond to the values of each feature in a set of n features to plot a data item in *n*-dimensional space. The hyperplane that was developed to split the two classes is used to classify the data. The goal of the SVM algorithm is to find the best line or decision boundary that can divide n-dimensional space into classes, allowing us to classify new data points fast in the future. This ideal decision is referred to as a hyperplane (Alkhodair et al. 2020; Yi et al. 2003; Tapaswi et al. 2012; Ranjan et al. 2003; MonaDiab et al. 2004; Rouse 2018; Dua and Du 2016).

4.1.5 Multinomial Naive Bayes (NB)

This algorithm, which is employed in a variety of machine learning applications, is based on the Bayes' theorem and assumes that it is free of predictors. Simply expressed, naive Bayes assumes that one function in a category has no bearing on another. For example, an apple is categorized as such if it is red in color, has swirls, and has a diameter of less than 3 inches. Whether these functions are dependent on one other or on distinct functions, and even if they are dependent on each other or on other functions, naive Bayes assumes that all of these functions have a different proof of the apples (Researchgate.net 2018; Researchgate.net 2014; Sirikulviriya and Sinthupinyo 2011; Kevric et al. 2017; Parikh and Atrey 2018; Granik and Mesyura 2017; Gilda 2017).

4.1.5.1 Naive Bayes Equation

$$(m|n) = (m|n)/(m)\ (n) \tag{4.2}$$

$$(m|n) = (m1|n) \times (m2|n) \times \ldots \times (m2|n) \times P(z) \tag{4.3}$$

where *P*(*m*|*n*) is the posterior probability, *P*(*m*|*n*) is the likelihood, *P*(*m*) is the class prior probability, and *P*(*m*) is the predictor prior probability.

The remainder of the chapter is organized as follows. The next section discusses the various threads of related work done in this context. Subsequently, Sect. 4.3 goes into more detail on this work. Section 4.4 details the performance evaluation of the various methods. Section 4.5 concludes.

4.2 Related Work

The study required to establish whether the provided content is real was done in (Parikh and Atrey 2018; Granik and Mesyura 2017; Gilda 2017; Jain and Kasbe 2018) by identifying various news sources and conducting the relevant analysis. The studies made use of models created around speech features and predictive algorithms that don't mesh with other current techniques. Similar to this, the works shown in (Granik and Mesyura 2017; Gilda 2017; Jain and Kasbe 2018) employ a classifier based on naive Bayes to recognize fake news. This method produced an accuracy of 74% when evaluated with different social media records and was employed as a software platform. The punctuation errors were disregarded, which reduced the paper's accuracy. Analogous to this, the studies described in (Jain and Kasbe 2018; Qin et al. 2018) computed a number of machine learning (ML) methods and investigated predictive accuracy.

The accuracy of several predictive models, including bounded decision trees, gradient improvement, and support vector machines, varied. Additionally, (Qin et al. 2018; Khanam and Ahsan 2017; Perez-Rosas et al. 2017; Aphiwongsophon et al. 2018; Kaur et al. 2019; Looijenga 2018) uses the naive Bayes classifier to describe and demonstrate how to incorporate bogus information identification into various social media sites. Through Facebook, Twitter, and other social networking sites, they were able to obtain news. Qin et al. (2018), Khanam and Ahsan (2017), and Perez-Rosas et al. (2017)s discuss locating erroneous information in the present moment. They draw on a novelty-based attribute and uses Kaggle for its data.

The mean accuracy of this pattern is 74.5%. Low-quality websites and advertising contribute to a lower resolution. We apply the methods in (Aphiwongsophon et al. 2018; Kaur et al. 2019) to locate Twitter spam senders. In this situation, models like the decision tree, clustering, and naive Bayes algorithms are used. Spam and fraudster identification accuracy averages 70 and 71.2%, respectively. Only a modest level of intermediate precision has been obtained by the models used to differentiate between spam and non-spam.

The methodology in (Looijenga 2018; Anonymous 2018) looked at a variety of ways to spot fake news. Only 76% of the language model is accurate. Improved accuracy can be obtained by using a predictive model (Looijenga 2018; Anonymous 2018). used machine learning approaches to detect fake news. Three frequently utilized methods are used in their research: support vector machines (SVM), neural networks, and naive Bayes. The discovery of fake news detection as a predictive analysis application was made in (Traore et al. 2017). To find fraudulent messages, the three phases of processing, feature extraction, and classification are used. The hybrid classification method used in this study was developed to reveal false information. The classification procedure combines k-nearest neighbors algorithm and random forests. The application of the proposed model is evaluated for precision and recall. The outcomes were enhanced by up to 8% when a mixed false message detection model was used (Khanam and Ahsan 2018; Sharma et al. 2019; Shu et al. 2017; Khanam and Agarwal 2015; Zhang et al. 2020; Ludwig and Creation 2020; da Silva et al. 2019a, 2019b; Bovet and Makse 2019; Lazer et al. 2018; Raza and Ding 2022a, 2022b; Jain et al. 2019a, 2019b; Fan 2017; Shu et al. 2019; Rubin 2017; Kogan et al. 2019; Vosoughi et al. 2018; Antweiler and Frank 2005; Allcott and Gentzkow 2017; Ahern and Sosyura 2014).

The work depicted in (Shu et al. 2017; Raza and Ding 2022a; Jain et al. 2019a) discussed fake news detection based on news content and social contexts, where a deep neural framework is used for fake news detection. The problem statement is identified into two unique challenges: (i) early detection of fake news and (ii) shortage of label. A unique transformer model is designed for the detection of fake news. The encoder and decoder blocks are used for the early detection of fake news. An effective supervision labeling scheme is used to resolve the label shortage issue. However, the paper doesn't touch upon enough algorithms, for example the classifier algorithms like logistic regression, support vector machine, etc., to carry out fake news detection.

Similar to this, some studies (Shu et al. 2017; Jain et al. 2019a) proposed a machine learning-based smart news system that uses the naive Bayes classifier, SVM, and natural language processing to identify bogus news. Based on the model used, the system in question can identify bogus news. Additionally, it offers some pertinent news on the subject, which is quite beneficial for any user. The prototype's effectiveness and precision can, however, be improved to a considerable extent. It currently has a 93.5% accuracy rate.

4.3 Work Done

Figure 4.3 depicts a high-level overview of our suggested design. We begin with news as input in the form of a Kaggle dataset. The news data are preprocessed. Then feature selection is carried out. Data is split into training and testing data. Later on we

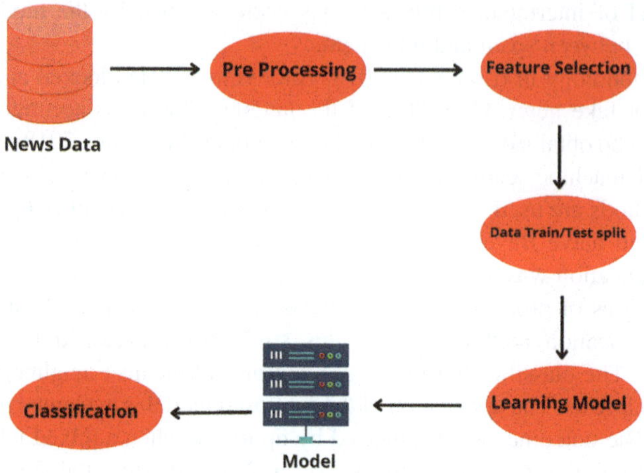

Fig. 4.3 Model for classification of data

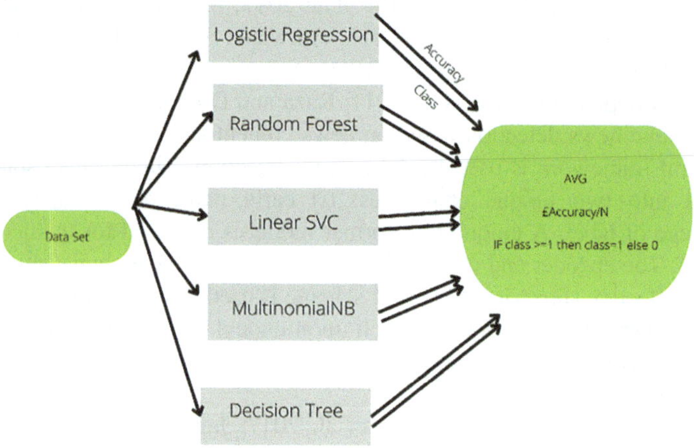

Fig. 4.4 Different methods applied on datasets

employ the learning model to perform classification. The news items, social media contexts, and side information are routed to the appropriate areas for data preprocessing. Then, from our preprocessed dataset, we extract various language features and choose the ones that are required. Then we divide our dataset into train and test datasets, with 80% of the dataset used for training and 20% used for testing. The train dataset is then utilized to train our model using different algorithms as shown in Fig. 4.4. The model is then evaluated after it has been trained. If there is a need to improve performance and accuracy, hyperparameters or model tuning can be used to help. The training is then evaluated on the test dataset, which uses multiple methods to determine whether the news is false or true.

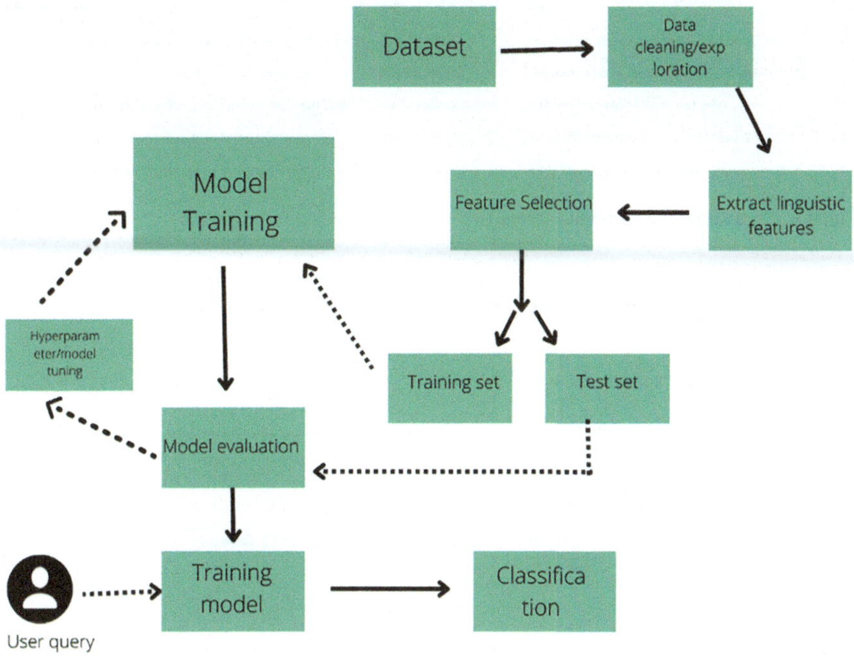

Fig. 4.5 A scenario of sequence of steps in proposed method

We obtained 20,800 distinct news data points from Kaggle for this study. When developing false news detection systems, the primary goal is to achieve the best feasible result by improving performance and accuracy. The numerous reasons why false news exists, its characteristics and nature, who spreads it, and how fake news spreads can be discovered through descriptive and in-depth examination. We used a comparative analysis strategy to create our model with several algorithms, yielding an accuracy of 96%. The project can be used to provide an ideal model that individuals can employ to determine whether or not the news they hear or see is fake. Figure 4.5 shows the sequence of steps in the proposed method.

The following are the steps involved in our proposed strategies:

Step 1.
We collected the news dataset from Kaggle which has 20,800 data points spread across five columns: - id, title, author, text, and label (as depicted in Fig. 4.6).

Step 2.
We imported and downloaded the necessary libraries and modules like *pandas*, *numpy*, *matplotlib*, *nltk*, and *sklearn*, required to construct our model.

Step 3.
We retrieved and assigned the collected news data from the comma-separated values (CSV) file to a variable as illustrated in Fig. 4.7.

id	title	author	text	label
0	House Dem Aide: We Didn't Even See Comey's Let...	Darrell Lucus	House Dem Aide: We Didn't Even See Comey's Let...	1
1	FLYNN: Hillary Clinton, Big Woman on Campus - ...	Daniel J. Flynn	Ever get the feeling your life circles the rou...	0
2	Why the Truth Might Get You Fired	Consortiumnews.com	Why the Truth Might Get You Fired October 29, ...	1
3	15 Civilians Killed In Single US Airstrike Hav...	Jessica Purkiss	Videos 15 Civilians Killed In Single US Airstr...	1
4	Iranian woman jailed for fictional unpublished...	Howard Portnoy	Print \nAn Iranian woman has been sentenced to...	1

Fig. 4.6 Spreading of data points across five columns collected from Kaggle

```
train_df = pd.read_csv('train.csv', header=0)
```

Fig. 4.7 Collected news data from CSV to variable

```
train_df = train_df.fillna(' ')
train_df['text'] = train_df['text'].str.strip()
train_df['raw_text_length'] = train_df['text'].apply(lambda x: len(x))
print(len(train_df[train_df['raw_text_length']==0]))
train_df.head()

116

train_df = train_df[train_df['raw_text_length'] > 0]
train_df.shape

(20684, 6)
```

Fig. 4.8 Performing Null imputation on the dataset

Step 4.

We performed null imputation on the dataset that replaces null values present in the dataset with empty space as shown in Fig. 4.8. Here 116 is the number of null rows. Since the number of null rows is lower compared to that of the dataset, we need to remove these rows.

Step 5.

Next we performed data preprocessing on the dataset, which included *removing special characters*, *expanding contractions*, *converting to lowercase*, *word-tokenization*, and *removing stopwords* as shown in Fig. 4.9a, Fig. 4.9b, and Fig. 4.9c respectively. The defined function preprocess_text removes special characters, expands the contractions (e.g., if the dataset contains text like "I'll" then it will be converted to "I will"), and converting to lowercase. Here word_tokenize on text_cols is performed, which splits a given sentence into a list of words (e.g., "Hello, how are you?" will be converted to ["Hello," ",", "how," "are," "you," "?"]). Here stopwords are removed from the text_cols (stopwords are the most commonly used words, e.g., "a," "an," "the," "in," etc.).

```
def preprocess_text(x):
    cleaned_text = re.sub(r'[^a-zA-Z\d\s\']+', '', x)
    word_list = []
    for each_word in cleaned_text.split(' '):
        try:
            word_list.append(contractions.fix(each_word).lower())
        except:
            print(x)
    return " ".join(word_list)
```

(a)

```
text_cols = ['text', 'title', 'author']

for col in text_cols:
    print("Processing column: {}".format(col))
    train_df[col] = train_df[col].apply(word_tokenize)
```

(b)

```
for col in text_cols:
    print("Processing column: {}".format(col))
    train_df[col] = train_df[col].apply(
        lambda x: [each_word for each_word in x if each_word not in stopwords])
```

(c)

Fig. 4.9 Data preprocessing: (**a**) removal of special characters, (**b**) word tokenization, (**c**) removal of stopwords

Step 6.

After data preprocessing, the dataset is then considered for model training as illustrated in Fig. 4.10a. Here label is taken as the target variable which will be detected by the model. Here Feature Extraction is done on the dataset as illustrated in Fig. 4.10b. In order to use text data for predictive modeling, the text must be parsed to remove certain words, which is called tokenization. The words in the dataset may be encoded as integers or floating-point values to be provided as input in model training algorithms. Count Vectorizer counts the number of unique words, limits vocabulary size, etc. The tf-idf transformer then assigns a unique value to each word as shown in Fig. 4.10c. This feature can be taken for model training, which will be easier for the model to detect perfectly.

4.4 Result Discussion

The hardware and software requirements for our research are listed in Table 4.1. We have split our dataset into training and testing dataset using train_test_split library where 80% of the dataset is taken for training and the remaining 20% is taken for testing. Here the function get_perf_metrics performs model training on the training dataset and testing on the test dataset on the basis of the algorithms passed as

```python
train_df['text_joined'] = train_df['text'].apply(lambda x: " ".join(x))
target = train_df['label'].values
```

(a)

```python
count_vectorizer = CountVectorizer(ngram_range=(1, 2))
tf_idf_transformer = TfidfTransformer(smooth_idf=False)

# fit train data to count vectorizer
count_vectorizer.fit(train_df['text_joined'].values)
count_vect_train = count_vectorizer.transform(train_df['text_joined'].values)

# fit ngrams count to tfidf transformers
tf_idf_transformer.fit(count_vect_train)
tf_idf_train = tf_idf_transformer.transform(count_vect_train)
```

(b)

```python
X_train, X_test, y_train, y_test = train_test_split(tf_idf_train, target,test_size = 0.2, stratify=target, random_state=2)
```

(c)

Fig. 4.10 Steps after data preprocessing: (**a**). model training, (**b**). feature extraction, (**c**) tf-idf transformer's use

Table 4.1 Hardware and software required for implementation

Hardware requirements		Software requirements
Processor	Intel(R) Core (TM) i5-8250 CPU @ 1.60 GHz 1.80 GHz	Python ReactJs Flask Visual Studio Code
RAM	8.00 GB (7.89 GB usable)	Jupyter notebook
System type	×64-based processor	64-bit operating system
Graphics card	NVIDIA GeForce 940MX	

parameters as shown in Fig. 4.11. In this context, training of the model is done using various algorithms like logistic regression, decision tree classifier, Linear SVC, multinomialNB, and Random Forest Classifier with hypertuning as illustrated in Fig. 4.12. We have assessed the performance in terms of various performance parameters such as accuracy, precision, recall, and F1 score.

From the applied algorithms, Linear SVC gives us a better accuracy of 96.1566% among all the approaches, as illustrated in Table 4.2, thereby ensuring its greater ability to predict the correctness as compared to other strategies. Table 4.2 also shows the attained precision, recall, F1 score, and training duration of various classification models. It can be seen that precision is attained at 99.7475% for multinomialNB, which is the highest value among all the chosen classifiers. Moreover, Linear SVC attained maximal recall value at 97.3774% and maximal F1 score

```
models_trained_list = []


def get_perf_metrics(model, i):
    # model name
    model_name = type(model).__name__
    # time keeping
    start_time = time.time()
    print("Training {} model...".format(model_name))
    # Fitting of model
    model.fit(X_train, y_train)
    print("Completed {} model training.".format(model_name))
    elapsed_time = time.time() - start_time
    paramsArr=["class_weight='balanced',solver='saga'","min_samples_split=4","C=2.0","alpha=2.0","max_depth=2.0"]
    # Time Elapsed
    print("Time elapsed: {:.2f} s.".format(elapsed_time))
    # Predictions
    y_pred = model.predict(X_test)
```

Fig. 4.11 Model training and testing

```
models_list = [LogisticRegression(class_weight='balanced',solver='saga'),
               DecisionTreeClassifier(min_samples_split=4),
               LinearSVC(C=2.0),
               MultinomialNB(alpha=2.0),
               RandomForestClassifier(max_depth=2.0)
               ]
```

Fig. 4.12 Various algorithms used for training of model

at 96.1861%. The training time is the least at 0.72 s for multinomialNB. The accuracy levels of various strategies are as depicted in Fig. 4.13. Fig. 4.14 shows a screenshot of an actual fake news item.

4.5 Conclusions and Future Direction

In this research, we have applied many classifiers and divided our model into two phases: characterization and detection. We have also created a web-based application for our suggested model that checks whether the news is fake or true as we enter the news body. We have used 80% of the dataset for training, and 20% was used for testing. Numerous classification models were employed to obtain accuracy, precision, recall, F1 score, and training time. It has been found that Linear SVC provides us with a greater accuracy of 96% whereas multinomialNB achieves precision at 99.74%, which is the maximal value of all the classifiers applied. Additionally, Linear SVC achieved a maximum F1 score of 96.18% and a maximum recall value of 97.37% among all the other classification strategies. Further, for multinomialNB, the training time is attained as the minimal at 0.72 s.

Although all the models perform well, our suggested model possesses a few downsides, such as the fact that it can only detect news in English. Multi-linguistic traits are not supported by our model and we have not trained it to understand more than one language.

Table 4.2 Results attained through various algorithms

Serial No.	Model used	Accuracy_Training_Set	Accuracy_Test_Set	Precision	Recall	F1 score	Training time (s)
1	Logistic regression	0.983441	0.944888	0.929611	0.962118	0.945585	19.38
2	Decision tree classifier	0.999275	0.898235	0.890744	0.906751	0.898676	434.05
3	Linear SVC	0.999940	**0.961566**	0.950237	**0.973774**	**0.961861**	5.68
4	MultinomialNB	0.899257	0.788011	**0.997475**	0.575522	0.729905	**0.72**
5	Random Forest Classifier	0.748776	0.733382	0.732264	0.731909	0.732086	9.31

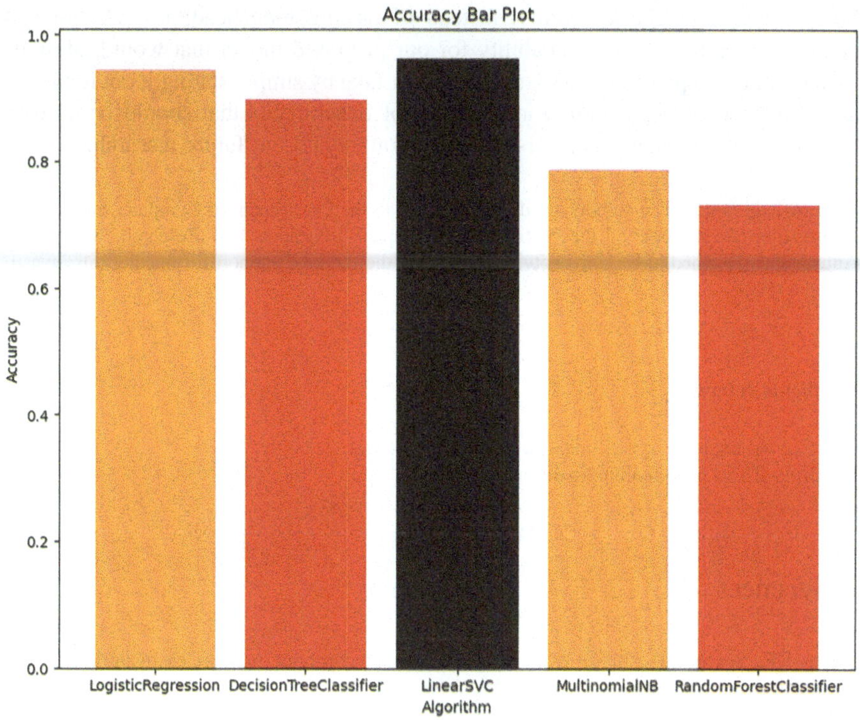

Fig. 4.13 Accuracy bar plot

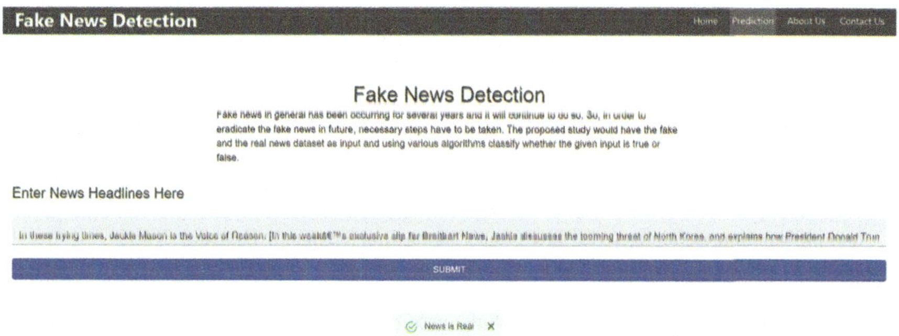

Fig. 4.14 Fake news detection

In addition, our model does not examine the context of news throughout the Internet. It only determines whether the news is true or not. It only notes the news based on the words, i.e., the news body's content. We hope to create a future system that can support more than one language. We would also like to apply this methodology to social media apps, which are the leading source of bogus news. Our plan is to enable the model to accurately detect the news, eliminating the need for a background cross-check. Finally, instead of writing long headlines, we hope to create a voice-based input capability for our proposed model that would allow us to determine if a piece of news is legitimate or fake by simply stating it out loud. We are also open to suggestions for the types of benchmarks that should be used to assess neural fake news detectors as a direction toward our future research.

Acknowledgments The authors are deeply grateful to the Department of Computer Science and Information Technology of the Institute of Technical Education and Research, Siksha 'O' Anusandhan Deemed to be University for providing the required support for making this investigation a success.

Conflict of Interest

There are no financial or non-financial conflicts of interest.
Funding There is no funding to declare.

References

Ahern KR, Sosyura D (2014) Who writes the news? corporate press releases during merger negotiations. J Finance 69(1):241–291
Akoglu L, Chandy R, Faloutsos C (2013) Opinion fraud detection in online reviews by network effects. In: ICWSM
Alkhodair SA, Ding SHH, Fung BCM, Liu J (2020) Detecting breaking news rumors of emerging topics in social media. Inf Process Manag 57:102018
Allcott H, Gentzkow M (2017) Social media and fake news in the 2016 election. J Econ Perspect 31(2):211–235
Anonymous (2018) Twente student conference on IT, Jun 6th, 2018, Enschede, The Netherlands. Netherlands. essay.utwente.nl
Antweiler W, Frank MZ (2005) Is all that talk just noise? The information content of internet stock message boards. J Finance 59(3):1259–1294
Aphiwongsophon S et al. (2018) Detecting fake news with machine learning method. 2018 15th international conference on electrical engineering/electronics, computer, telecommunications and information technology (ECTI-CON). Chiang Rai, Thailand, Thailand. IEEE
Bovet A, Makse HA (2019) Influence of fake news in twitter during the 2016 US presidential election. Nat Commun 10(1):1–14
Dua S, Du X (2016) Data mining and machine learning in cybersecurity. Auerbach Publications, New York
Fan C (2017) Classifying fake news. http://www.conniefan.com/2017/03/classifying-fake-news. Accessed 18 Feb 2018
Gilda S (2017) Evaluating machine learning algorithms for fake news detection. In: 15th student conference on research and development (SCOReD). IEEE, pp 110–115
Granik M, Mesyura V (2017) Fake news detection using naive Bayes classifier. In: First Ukraine conference on electrical and computer engineering (UKRCON). Ukraine. IEEE.
Huang T-Q (n.d.). https://www.researchgate.net/figure/Pseudo-code-of-information-gain-basedrecursive-feature-elimination-procedure-with-SVM_fig2_228366941 2018
Jain A, Kasbe A (2018) Fake news detection. In: 2018 IEEE international Students' conference on electrical, electronics and computer science (SCEECS). Bhopal, India. IEEE

Jain A, Shakya A, Khatter H, Gupta AK (2019a) A smart news system for fake news detection using machine learning. In: 2nd International conference on issues and challenges in intelligent computing techniques (ICICIT)
Jain A, Shakya A, Khatter H, Gupta AK (2019b) A smart system for fake news detection using machine learning. In: 2019 International conference on issues and challenges in intelligent computing techniques (ICICT), pp. 1–4, doi: https://doi.org/10.1109/ICICT46931.2019.8977659
Kaur P et al (2019) Hybrid text classification method for fake news detection. Int J Eng Adv Technol (IJEAT):2388–2392
Kevric J et al (2017) An effective combining classifier approach using tree algorithms for network intrusion detection. Neural Comput & Applic 28:1051–1058
Khanam Z, Agarwal S (2015) Map-reduce implementations: survey and performance comparison. Int J Comput Sci Inform Technol (IJCSIT) 7(4):119
Khanam Z, Ahsan MN (2017) Evaluating the effectiveness of test driven development: advantages and pitfalls. Int J Appl Eng Res 12:7705–7716
Khanam Z, Ahsan MN (2018) Implementation of the pHash algorithm for face recognition in secured remote online examination system. Int J Adv Sci Res Eng (IJASRE) 4(11):01
Kogan S, Moskowitz TJ, Niessner M (2019) Fake news: evidence from financial markets. https://ssrn.com/abstract=3237763
Lazer DMJ et al (2018) The science of fake news. Science 359(6380):1094–1096. https://doi.org/10.1126/science.aao2998. https://science.sciencemag.org/content/359/6380/1094.summary
Looijenga MS (2018) The detection of fake messages using machine learning
Ludwig K, Creation M (2020) Dissemination and uptake of fake-quotes in lay political discourse on Facebook and twitter. J Pragmat 157:101–118
Meyer R (2017) The rise of progressive 'Fake News'. Retrieved from The Atlantic: https://www.theatlantic.com/technology/archive/2017/02/viva-la-resistancecontent/515532/
MonaDiab et al. (2004) Automatic tagging of Arabic text: from raw text to base phrase chunks. Proceedings of HLT-NAACL 2004: short papers, Association for Computational Linguistics, Boston, MA, pp. 149–152
Mosseri A (2016) News feed FYI: Addressing hoaxes and fake news. Retrieved from Facebook newsroom
Parikh SB, Atrey PK (2018) Media-rich fake news detection: a survey. IEEE conference on multimedia information. Miami, FL: IEEE
Perez-Rosas V et al. (2017). https://www.researchgate.net/publication/319255985_Automatic_Detection_of_Fake_News
Qin Y et al (2018) Predicting future rumours. Chin J Electron 27(3):514–520
Radianti J et al. (2016) An overview of public concerns during the recovery period after a major earthquake: Nepal Twitter analysis. HICSS '16 Proceedings of the 2016 49th Hawaii International Conference on System Sciences (HICSS), Washington, DC, USA, IEEE, pp. 136–145
Ranjan et al. (2003) Part of speech tagging and local word grouping techniques for natural language parsing in Hindi. In Proceedings of the 1st international conference on natural language processing (ICON 2003). Semanticscholar
Rashkin H, Choi E, Jang J, Volkova S, Choi Y (2017) Truth of varying shades: analyzing language in fake news and political fact-checking. In: EMNLP
Raza S, Ding C (2022a) Fake news detection based on news content and social contexts: a transformer-based approach. Nat Publ Health Emerg Collection 13:335
Raza S, Ding C (2022b) Fake news detection based on news content and social contexts: a transformer-based approach. Int J Data Sci Analyt 13(4):335–362
Researchgate.net (2014). https://www.researchgate.net/figure/Pseudocode-for-KNNclassification_fig7_260397165
Researchgate.net (2018). https://www.researchgate.net/figure/Pseudocode-ofnaive-bayes-algorithm_fig2_325937073
Rouse M (2018). https://searchenterpriseai.techtarget.com/definition/machine-learning-ML

Rubin V (2017) Deception detection and rumor debunking for social media. Handbook of social media research methods

Rubin V, Conroy N, Chen Y, Cornwell S (2016) Fake news or truth? Using satirical cues to detect potentially misleading news. In: NAACLCADD

Sharma K, Qian F, Jiang H, Ruchansky N, Zhang M, Liu Y (2019) Combating fake news: a survey on identification and mitigation techniques. ACM Trans Intell Syst Technol (TIST) 10(3):1–42

Shu K et al (2017) Fake news detection on social media: a data mining perspective. ACM SIGKDD Explor Newslett 19(1):22–36

Shu K, Wang S, Liu H (2019) Beyond news contents: the role of social context for fake news detection. In: Proceedings of the twelfth ACM international conference on web search and data mining, pp. 312–320

da Silva FCD et al. (2019a) Can machines learn to detect fake news? A survey focused on social media. https://scholarspace.manoa.hawaii.edu/handle/10125/59713

da Silva FCD, Vieira, R., & Garcia, AC. (2019b). Can machines learn to detect fake news? A survey focused on social media. In: Proceedings of the 52nd Hawaii International Conference on System Sciences

Singh V, Dasgupta R, Sonagra D, Raman K, Ghosh I (2017) Automated fake news detection using linguistic analysis and machine learning. In: SBP-BRiMS

Sirikulviriya N, Sinthupinyo S (2011) Integration of rules from a random forest. International conference on information and electronics engineering, p 194, 198. Singapore, semanticscholar.org

Tacchini E, Ballarin G, Della Vedova M, Moret S, de Alfaro L (2017) Some like it hoax: automated fake news detection in social networks. CoRR, abs/1704.07506

Tapaswi et al. (2012) Treebank based deep grammar acquisition and part-of-speech tagging for Sanskrit m sentences. Software engineering (CONSEG), on software engineering (CONSEG). IEEE, pp 1–4

Traore et al. (2017) Detection of online fake news using N-gram analysis and machine learning techniques. International conference on intelligent, secure, and dependable systems in distributed and cloud environments. Springer International Publishing, pp. 127–138

Vosoughi S, Roy D, Aral S (2018) The spread of true and false news online. Science 359(6380): 1146–1151

Yi J et al. (2003) Sentiment analyzer: extracting sentiments about a given topic using natural language processing techniques. In Data mining, 2003. ICDM 2003. Third IEEE International Conference, pp 427–434. http://citeseerx.ist.psu.edu

Zhang J, Dong B, Yu Philip S (2020) Fakedetector: effective fake news detection with deep diffusive neural network. In: 2020 IEEE 36th international conference on data engineering (ICDE). IEEE

Chapter 5
Analytics and Decision-making Model Using Machine Learning for Internet of Things-based Greenhouse Precision Management in Agriculture

Ashay Rokade, Manwinder Singh, Anudeep Goraya, and Balraj Singh

Abstract As wired systems for smart farming are difficult to manage and install, wireless connectivity is currently taking their place. Smart farming with precision greenhouse technology is installed to improvise in managing the growth of agriculture and therefore observing different environments in precision agriculture. Numerous systems have been developed for control and remote monitoring of precision agriculture. But due to limited solutions, monitoring of greenhouse is not yet competent to deal with the agricultural growth on entirely control systems. For better farming growth control, smart farming with precision greenhouses must be applied, necessitating precision agriculture monitoring under various circumstances. Supervised machine learning techniques are used in intelligent agricultural systems to provide intelligent information farming systems with predictive data analysis of sensor parameters. Cloud layer, fog layer, edge layer, and sensor layer are four important parts of the proposed approach. The data needed for the sensor layer of the analytical model is collected using Internet of Things-based embedded system devices in two greenhouse facilities, with sensor parameters as inputs and corresponding actuators as outputs. Using classification and regression models, two distinct analytical models for intelligent and accurate farming were built. By modifying farming circumstances in accordance with plant requirements taken into account during experimentation, the primary goals of this analytics are to boost output and offer organic farming. A decision-making and analytics system was built at the fog layer using the support vector machine and artificial neural network, two

A. Rokade · M. Singh
School of Electronics and Electrical Engineering, Lovely Professional University, Phagwara, Punjab, India
e-mail: manwinder.25231@lpu.co.in

A. Goraya · B. Singh (✉)
School of Computer Science and Engineering, Lovely Professional University, Phagwara, Punjab, India
e-mail: balraj.13075@lpu.co.in

A. Khamparia et al. (eds.), *Microbial Data Intelligence and Computational Techniques for Sustainable Computing*, Microorganisms for Sustainability 47,
https://doi.org/10.1007/978-981-99-9621-6_5

supervised classification-based machine learning methods. MATLAB software's statistics and machine learning tools were used to analyze and interpret the experimental outcomes. Accuracy, sensitivity, specificity, and F-score are used to examine the confusion matrix-based metrics used in the performance evaluation of the suggested system. Based on the results of the experiments, the suggested method also proved to be the best at providing actuators with predictions and control.

Keywords Precision agriculture · Intelligent greenhouse · Internet of things · Machine learning · Smart farming · Fog computing

5.1 Introduction

Making plant agriculture a creative task, the land and the quality of the plants are now the crucial daily bounds for either money harvests or food crops. Poor farming knowledge and information about new techniques is a significant problem in modern agriculture. Our forefathers in the agricultural sector avoided using specialized technology for individual plant growth in favor of general natural phenomena. The introduction of technical machinery into the agricultural sector has made it possible to cultivate plants in settings far beyond the norm. This has led to the production of both higher yields and lower manure usage. The widespread use of fertilizers, defoliant, and water in plant crops is in line with their natural rationality (Chehri et al. 2020; Subahi and Bouazza 2020). In intensive nursery settings, growers often use agrochemicals in quantities that exceed the true yield demands, leading to ecological pollution and waste. When compared to predictions of total creation, the value of water and agronomics is low since crops are managed with a lot of induction rather than taking into account target estimations from advanced crop-checking technology. For instance, a sizable percentage of the working population relies on the farming sector for their living. In India, this number is particularly high, at 61% of the working population, respectively (Kour and Arora 2020; Chukkapalli et al. 2020).

Figure 5.1 shows the industry's significant export growth over the previous 12 months. In FY22, exports of marine products totaled $7.77 billion, exports of rice, basmati and non-basmati, totaled $6.98 billion USD, the entire value of buffalo meat exports was $3.30 billion, exports of sugar totaled $4.60 billion USD, tea exports were worth US$750.93 million, and coffee exports were worth $1020.80 billion USD.

Modern farming and horticultural production systems are undergoing significant technological progress, which has given rise to the terms "agriculture 4.0" and "Smart Agriculture" (SA) (El-Basioni and El-Kader 2020). Numerous innovative technologies, such as autonomous agricultural trucks, satellite infrastructure, and unmanned aerial vehicles (UAVs), will be linked to SA-based future scenarios. In particular, modern farmers will benefit greatly from adopting both technologies related to Precision Farming and the Internet of Things (IoT). In reality, a much more sensible and superior horticultural production framework is needed to

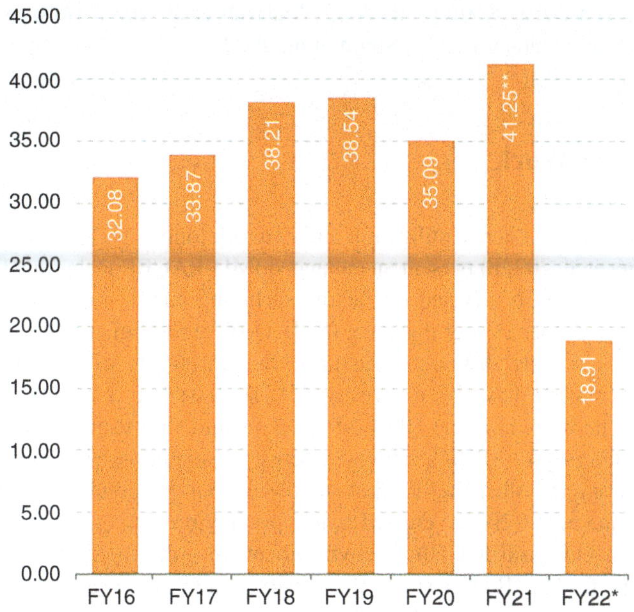

Fig. 5.1 Agriculture exports from India (US$ billion) (Source: India Brand Equity Foundation)

adequately meet a few impending problems, like rapid increase in population, atmospheric conditions, and common assets consumption. Thus, investigation and mechanical advancement can be the solutions to reduce these challenges. The advancements in agriculture (Alonso et al. 2020; Bandur et al. 2019; Carrasquilla-Batista and Chacon-Rodriguez 2019; Tan et al. 2020) in the twenty-first century helped to overcome the designing challenges with inventions (Prabha et al. 2018; Sadowski and Spachos 2018; Lakshmanna et al. 2022a; Rezk et al. 2021; Tageldin et al. 2020; Aliar et al. 2022; Sekaran et al. 2020; Singh et al. 2018; Roy et al. 2020; Hassan et al. 2021a). The advancement related to agriculture 5.0 is based on the concept that farms are using automated activities and emotionally supporting networks based on the freedom of individual choice, as outlined by the precision agriculture standards. As a result, it is likely that the ideas behind agriculture 5.0 will incorporate the usage of robotics and possibly even some types of artificial intelligence (Hassan et al. 2021b, 2021c, 2021d; Rokade and Singh 2021; Kadu and Singh 2021; Khalaf et al. 2021; Walia et al. 2021, 2022a, 2022b; Ali Al-Samawi and Singh 2022; Belsare and Singh 2022a). Ranches have typically relied on a large number of sporadic professionals to gather harvests and keep profits high. Since society has changed from an agricultural culture in which many people lived on homesteads to one in which many people live in urban settlements, ranches are facing the difficulty of a manpower shortage. AI-enhanced farming robots are one approach to the workforce crisis (Hassan et al. 2022a, 2022b, 2022c; Rokade et al. 2022a, 2022b;

Bolla and Singh 2022; Kadu et al. 2022; Marwah et al. 2022; Belsare and Singh 2022b; Singh and Gandam 2022; Singh et al. 2022).

5.2 Related Work

In order to maximize energy utilization and network longevity, K. Lakshmana et al. (Lakshmanna et al. 2022b) present an enhanced IMD-EACBR technique for wireless sensor networks (WSN). Based on factors including energy savings, detachment, node degree, and inter-cluster distance, the IAOAC algorithm chooses an appropriate aim that connects different structures. Several facets of the IMD-EACBR model's performance have been investigated. The last step is extensive testing of the proposed network utilizing all of NS-3.26's simulation features. Improvements in packet delivery ratio (PDR), latency, energy consumption, and number of dead nodes are among the other metrics that stand out from the simulation findings.

Lakshmanna et al. (2022a) created a new privacy-protecting multi-agent system (MAS) for an industrial IoT (IIoT) environment. In order to choose and build the right clusters for the IIoT system, they first created an expanded moth swarm algorithm-based clustering (EMSA-C) method. Additionally, a multi-agent system is implemented to provide encrypted exchanges between different clusters. The technique's potential is investigated via a comprehensive comparative study, with outcomes evaluated across a range of metrics. The investment required is substantial. The high cost of implementation is an evident issue for industrial IoT. Secure data storage and management connection failures are frequent because of the enormous volume of data generated by IoT devices. The simulation findings show that BDL-PPDT is superior to current approaches in terms of output. The provided BDL-PPDT method may only have a 98.15% success rate, but it nonetheless yields the best attainable result. The BDL-PPDT approach was demonstrated to be superior to the other existing methods across a number of criteria and is therefore advised based on the findings of the aforementioned data analysis.

Suma (2021) highlight the challenges and complications that may be encountered when integrating modern farming practices with older methods of production. The use of statistical and quantitative techniques can lead to revolutionary changes in our current agricultural system. The present and upcoming agricultural trends are provided through the systematic analysis.

In order to predict agricultural productivity and drought, Rezk et al. (2021) propose an IoT-based smart farming system and a powerful prediction technique called WPART that is based on ML techniques. The suggested approach is estimated using five different datasets. The findings showed that the suggested strategy outperformed the existing methods in categorizing and predicting drought and agricultural productivity. The outcomes revealed that the suggested strategy was the most effective at forecasting drought and measuring the yield of crops. The WPART approach outperforms state-of-the-art, gold standard algorithms with accuracy levels above 90%, across five datasets assessing drought categorization and crop yield.

Araby et al. (2019) collected various data samples from a variety of crop fields using a sensor network, transmitted it to an ML algorithm to produce an alarm message, and then showed it through graphical user interface.

Tageldin et al. (2020) have developed a number of ML strategies for predicting infection in plants. This study laid the groundwork for utilizing ML to forecast the presence of CLW in greenhouse crops. Weekly CLW data collection was conducted in a commercial hydroponic greenhouse for 2 years. Temperature and relative humidity readings were also taken continuously during the investigation. The XGBoost algorithm was found to be the most efficient one used throughout this research. This algorithm has attained accuracy in prediction of 84%. To guarantee a comprehensive dataset for future outcomes, the authors investigated the effect of several environmental factors on prediction precision.

In their comprehensive review of smart farming techniques and designs, Ahamed Aliar et al. (2022) cover all the bases. They also provide an in-depth analysis of various designs and viable recommendations for fixing the current state of smart farming.

Remote monitoring of rice paddies using deep learning and the IoT is proposed by Sethy et al. (2021). For rice leaf disease detection and nitrogen status assessment, the Vgg16 pre-trained network is being investigated. In this context, transfer learning and deep feature extraction are used to recognize photos. Support vector machines (SVM) have been introduced to identify pictures with the deep feature extraction method. Vgg16's transfer learning method achieves 79.86 and 84.88% accuracy, respectively, when used to recognize four distinct leaf diseases and to forecast nitrogen status. The Vgg16 deep features and the SVM findings have a 97.31 and 99.02% accuracy rate in recognizing four different leaf diseases and predicting nitrogen status, respectively. Additionally, an IoT and deep learning-based architecture is proposed for remote paddy field monitoring. The proposed prototype has advantages over the state of the art since it not only regulates temperature and humidity, but also monitors the extra two factors: the detection of nitrogen status and illnesses.

The Smart Agriculture approach proposed by Kaushik et al. (2019) includes monitoring the agricultural land and can greatly aid farmers in increasing output. The cloud-stored data, which contains details like temperature, moisture, and humidity that affect disease in an agricultural field, is subjected to a naive Bayes analysis.

An IoT-enabled agricultural monitoring prototype was proposed by Wong Hin Yong et al. (Siddiquee et al. 2022), which would use a variety of algorithms to monitor crops for a variety of purposes, including detection, quantification, maturity testing, and disease. The authors discuss IoT-enabled, intelligent farm monitoring solutions. Defects in vegetables have also been identified by the use of color threshold and color segmentation. All methods were designed and implemented using convolutional neural networks (CNNs), an ML technique. In order to determine which approach would be best for integrating into this agricultural monitoring system, MATLAB simulations have been used to compare traditional methods with

CNNs. This study found that CNNs outperformed other approaches and existing algorithms with an accuracy of 90% or higher, making it the preferred option.

Using the IoT to carry out a variety of outdoor tasks, Ferehan et al. (2022) created and launched a revolutionary wireless mobile robot. Agriculture, transportation, and water distribution can all benefit from this study's findings. The IoT and remote sensor system are used to create sophisticated agricultural frameworks in many regions of the world. Singh et al. (Singh and Verma 2022a, 2022b) discussed the various applications of machine learning (ML) and data processing using a big data platform. In this regard, one of the offshoots of intelligence is the practice of exactness. For a number of agricultural applications, researchers have created a wide range of checking and robotization frameworks. Data collection and sharing among ranch-based IoT devices will be simple with the help of WSN.

5.3 Development of IoT-based Smart Farming System

The proposed Smart Farming System for Data Analytics (SFSDA) Using ML Enabled Internet of Things will be divided into four layers for better understanding. The four basic architectural layers for a precisely regulated greenhouse management system are the cloud layer for data storage, fog layer for data processing, edge layer for data conversion and sampling, and sensor/device layer for data collection. In order to gather data on the greenhouse environment, including temperature, humidity, CO_2, soil moisture, and light intensity, a hardware prototype has been constructed. Figure 5.2 depicts the suggested paradigm for an intelligent farming system to be used in the greenhouse.

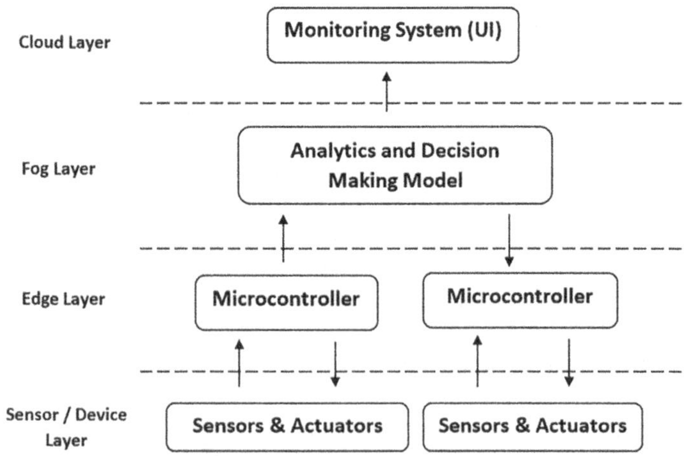

Fig. 5.2 Smart farming infrastructure for greenhouse management

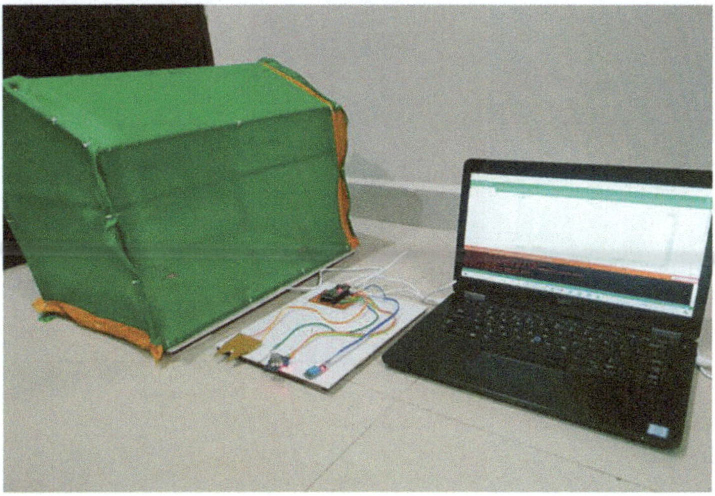

Fig. 5.3 Experimental test bench

The sensor layer is where all the field environment-related sensors and actuators are situated. The edge layer typically comprises a controller unit to which various sensors and actuators are connected in order to collect data for transmission to the fog layer. By leveraging edge-level data, the fog layer can build analytics and decision-making models and send actuator signals to the next layer. Finally, the upper layer provides a user interface dashboard which displays a graphical representation of sensor and actuator data. The proposed framework is unique in that it uses the IoT to aid farmers with greenhouse management. It is all done remotely so that farmers may monitor and adjust factors like soil moisture, CO_2, light, and temperature from afar.

The results of prototype tests utilizing the suggested experimental design, which took into consideration two harvests produced under various conditions, are shown in Fig. 5.3. The three main testing phases involved constructing core model-embedded systems for plant growth and feeding, automating actuators, and developing a sensor net for intelligent greenhouse monitoring. The suggested solution used an embedded system to precisely measure the soil moisture, temperature, CO_2, and plant light—all essential elements in a greenhouse's effective operation.

The climate conditions in the study region are tropical wet and dry, with hot, dry summers and mild to cool winters. Winter lasts from November to March, rainy season from July to October, and the summer season from March to June. The test field region, which is 0 feet (0 m) above sea level, experiences tropical wet and dry seasons. The district's average annual temperature of 30.63 °C (87.13 °F) is 4.66% higher than the national average for India. The test field area normally has 103.26 wet days (28.29% of the time) annually and receives about 120.15 millimeters (4.73 inches) of precipitation on average. The maximum temperature observed is 47 ° C and the lowest is 24.03 ° C.

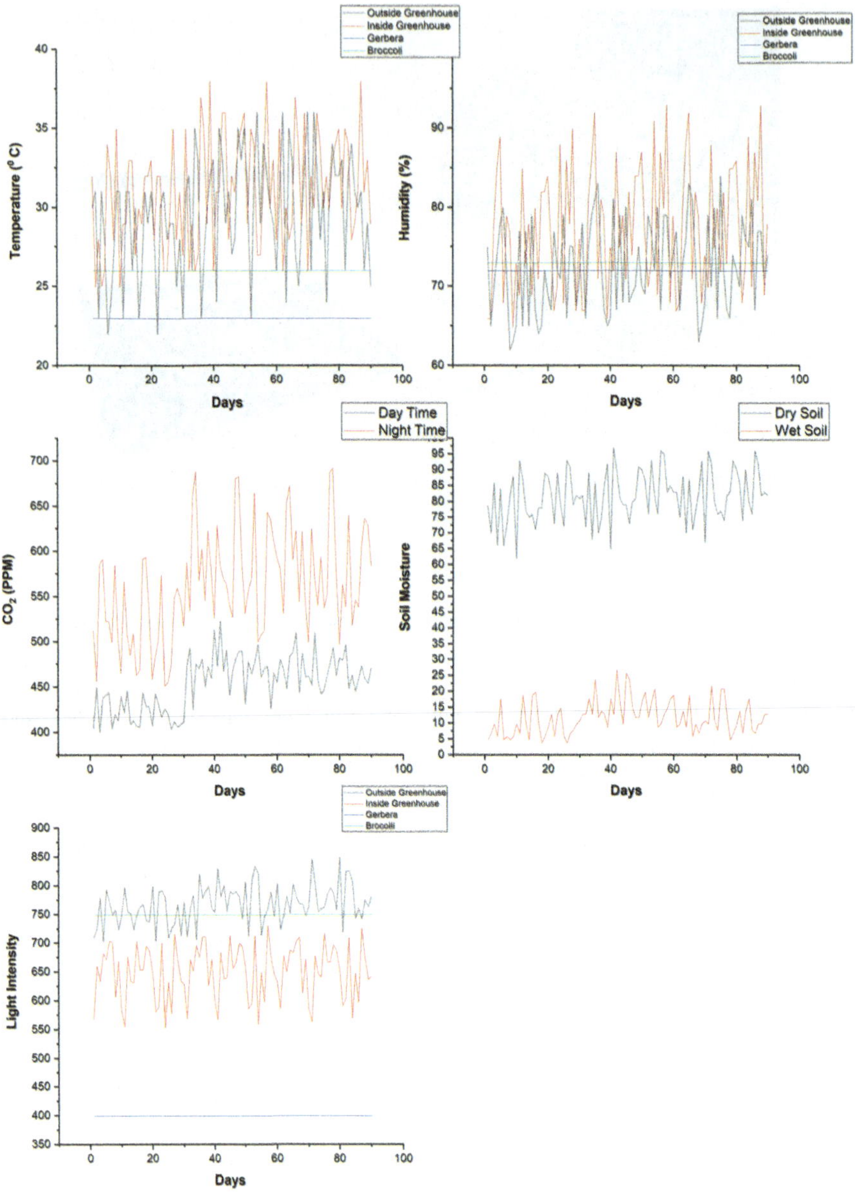

Fig. 5.4 Sensors reading in rainy season for greenhouse monitoring

Similarly, the data collection is processed for the rainy season for all the five parameters inside the greenhouse comparing reference data of gerbera and broccoli with actual data in Fig. 5.4.

As shown in the graph, the temperature of the greenhouse is quite low at night and high in the day time. It will require much more data monitoring and parameter controlling over the season as the temperature requirement for both gerbera and broccoli is low for both day and night time.

The graphs conclude that winter is the best season for cultivation of gerbera and broccoli. The research works for the cultivation of crops in any season using the greenhouse management system and by applying ML algorithms to control the atmosphere of the greenhouse for smooth development of the crop.

The proposed approach was established at the third layer of our proposed model, i.e., the fog layer for a data analytics system employing various ML methods. The SVM and multilayer perceptron machine learning algorithms were chosen because of their theoretical and implementation advantages which suit the system dataset perfectly to obtain the precise output. The classification and regression techniques were adopted and analyzed to acquire the expected results. The actual data and expected data are compared for both techniques. The system is developed to control the various devices like pump, fan, and light within the greenhouse, making the greenhouse a smart one.

The training and testing of data is done in this chapter using the MATLAB 2021 version. The accuracy, sensitivity, specificity, latency, F-score, and least root square mean (LSME) are calculated and observed.

5.4 Experimental Results and Discussion

A confusion matrix is used to gauge the system's effectiveness. While root-mean-square-error (RMSE) is computed for regression models using the confusion matrix parameters, accuracy, sensitivity, specificity, and F-score are computed for classification models. Sensitivity quantifies a model's ability to recognize positive class members accurately, while specificity gauges how well it can recognize negative class members. In some applications, one statistic could be more important than another. The accuracy of a model is assessed using the F score, a machine learning evaluation statistic. It takes into consideration a model's recall and precision scores. The accuracy statistic shows how often a model predicts correctly over the entire dataset. When your class is divided unevenly, the F grade is typically more helpful than accuracy. When false positive and false negative costs are roughly similar, accuracy performs best. In cases where the costs of false positives and false negatives differ greatly, it is advisable to incorporate both precision and recall.

The recommended approach makes use of an embedded system to conduct a reliable analysis of CO_2, soil moisture, temperature, and plant light in greenhouse operations. As the measurements are taken under various environmental conditions, they are all monitored in real time on a personal computer through serial transmission. As data is published from nodes to Adafruit's broker, it can be seen in the Adafruit IO Cloud dashboard for remote monitoring of all these sensors. The user may then subscribe to this information in order to obtain updates as they occur.

Table 5.1 Performance evaluation of classification model

	Accuracy (%)		Sensitivity (%)		Specificity (%)		F-score (%)	
Attributes	SVM	MLP	SVM	MLP	SVM	MLP	SVM	MLP
Pump	86.66	96.66	81.81	100	100	87.5	95.74	96.49
Light	96.66	100	100	100	90.90	100	95.95	100
Fan	90	96.66	94.11	94.11	84.61	100	89.88	98.76

Table 5.2 Performance evaluation of regression model

	PPV (%)		NPV (%)		FNR (%)		FPR (%)	
Attributes	SVM	MLP	SVM	MLP	SVM	MLP	SVM	MLP
Pump	100	95.65	66.66	100	18.18	0	0	12.5
Light	95	100	100	100	0	0	9.09	0
Fan	88.88	100	91.66	92.85	5.88	5.88	15.38	0

Table 5.3 The result of comparative analysis

Ref.	Accuracy (%)	Sensitivity (%)	Specificity (%)	F-score (%)	Latency (s)	RMSE
Lakshmanna et al. (2022a)	NI	93.59	94.63	96.30	NI	NI
Rezk et al. (2021)	NI	NI	NI	NI	NI	0.2431
Tageldin et al. (2020)	87.35	87.33	95.76	87.17	224.6	NI
Aliar et al. (2022)	90	NI	NI	NI	NI	NI
Sekaran et al. (2020)	NI	NI	NI	NI	NI	0.02726
Proposed work	**97.77**	**98**	**98.83**	**98.41**	**6.49**	**0.0615**

NI: Not investigated in research

Table 5.1 shows the categorization model based on a confusion matrix. The MLP method outperforms the SVM in every metric taken into account, according to the data in Table 5.1.

Similarly, the other performance evaluation parameters are evaluated from confusion matrix attributes, PPV, NPV, FNR, and FPR as shown in Table 5.2. It has been found that the suggested system can effectively classify with greater PPV and NPV values as well as with minimal FNR and FPR ratio values.

The final section of Table 5.3 presents a comparison to the most recent state-of-the-art techniques. The results show that the proposed classification and regression model for intelligent and precision smart farming in greenhouses produces better results when accuracy, sensitivity, and specificity of the classification model are compared with the root-mean-square-error (RMSE) of the regression model.

The results show that the proposed system works with high precision by implementing the proper resource utilization and comparing system parameters like accuracy, sensitivity, specificity, F-score, latency, and RMSE. The average values of the MLP classification and SVM regression algorithms were used to extract the findings from the suggested work in order to produce superior outcomes.

Table 5.4 Comparative analysis of various existing studies

Ref.	Technique(s)/models used	Gap/feature	System accuracy
Tan et al. (2020)	Rasbery Pi & Open Cv	Only focuses on automating pH control	92%
Prabha et al. (2018)	Bayesian network and ML	Less data to analyze in the datasets	66%
Sadowski and Spachos (2018)	ANN and fuzzy logic	It is an expensive system	89%
Proposed work	SVM and MLP	Low-cost system	97.77%

A comparison of different models that have already been developed is shown in Table 5.4.

In Table 5.4 it is shown that researchers have taken the initiative toward precision agriculture and the proposed model considered numerous factors, e.g., crop sensors and technology, to make an effective smart greenhouse management system which uses machine learning-enabled Internet of Things. This proposed system can be easily used and handled by farmers for crop wellness.

5.5 Conclusion

This study proposes a method through which an intelligent greenhouse's automation is upgraded by using IoT technologies and concepts. A user can now control and monitor data transfer between a device and a fog layer, and vice versa, using real-time sensor data. In order to improve agricultural output, the IoT concept is applied to the system by centralizing data storage and processing in a reliable cloud. The precision in data rectifies the proper resource utilization. Use of IoT lowers maintenance expenses.

The proposed system will correctly monitor and adjust greenhouse characteristics such soil moisture, carbon dioxide levels, temperature, humidity and light to assist farmers in boosting production. The model is validated by using data from actual greenhouses to determine the optimum soil moisture, carbon dioxide, temperature, humidity and light for producing broccoli and gerbera. The proposed monitoring system will be used for any crop which can be cultivated inside the greenhouse, which results in disease-free and large production of the crop.

Utilizing supervised ML methods based on classification and regression, the special greenhouse system is designed for intelligent, accurate control. The recommended approach can be applied in a smart agricultural environment that uses an IoT-based paradigm for decision-making. Classification and regression models are the two types of analytics used to create programs for smart and accurate farming. Both SVM and MLP can be utilized for these modeling purposes.

Using classification and regression-based supervised ML algorithms, the effectiveness of the smart farming system is ultimately assessed in terms of intelligent and precision farming. The outcomes showed that MLP performed better than SVM and other state-of-the-art classification methods. The MLP system's accuracy is 97.77%, sensitivity is 98%, specificity is 98.83%, and F-score is 98.41% with a lower error rate achieved by the system. The suggested technique also provided the most accurate predictions to actuators and the most precise control.

References

Ali Al-Samawi MA, Singh M (2022) Effect of 5G on IOT and daily life application. 2022 3rd International conference for emerging technology (INCET), pp 1–5, doi: https://doi.org/10.1109/INCET54531.2022.9823983

Aliar AAS, Yesudhasan J, Alagarsamy M, Anbalagan K, Sakkarai J, Suriyan K (2022) A comprehensive analysis on IoT-based intelligent farming solutions using machine learning algorithms. Bull Electric Eng Inform 11(3):1550–1557, ISSN: 2302-9285. https://doi.org/10.11591/eei.v11i3.3310

Alonso RS, Sittón-Candanedo I, Casado-Vara R, Prieto J, Corchado JM (2020) Deep reinforcement learning for the management of software-defined networks in smart farming. 2020 International conference on Omni-layer intelligent systems (COINS), Barcelona, Spain, pp 1–6. doi: https://doi.org/10.1109/COINS49042.2020.9191634

Araby A et al. (2019) Intelligent IoT monitoring system for agriculture with predictive analysis. In 2019 8th International conference on modern circuits and systems technologies (MOCAST), pp. 1–4, doi: https://doi.org/10.1109/MOCAST.2019.8741794

Bandur D, Jaksic B, Bandur M, Jovic S (2019) An analysis of energy efficiency in Wireless Sensor Networks (WSNs) applied in smart agriculture. Comput Electron Agric 156:500–507. https://doi.org/10.1016/j.compag.2018.12.016

Belsare KS, Singh M (2022a) Various frameworks for IoT-enabled intelligent waste management system using ML for smart cities. In Mobile computing and sustainable informatics, pp 797–817. Springer, Singapore, doi: https://doi.org/10.1007/978-981-19-2069-1_55

Belsare K, Singh M (2022b) An intelligent internet of things (IoT) based automatic dry and wet medical waste segregation and management system. In 2022 International Conference on Augmented Intelligence and Sustainable Systems (ICAISS), pp. 1113–1119. IEEE

Bolla S, Singh M (2022) Energy harvesting technique for massive MIMO wireless communication networks. J Phys Conf Ser 2327(1):012059. IOP Publishing. https://doi.org/10.1088/1742-6596/2327/1/012059

Carrasquilla-Batista A, Chacon-Rodriguez A (2019) Standalone fuzzy logic controller applied to greenhouse horticulture using internet of things, 2019 7th international engineering, sciences and technology conference (IESTEC), Panama, Panama, pp 574–579, doi: https://doi.org/10.1109/IESTEC46403.2019.00108

Chehri A, Chaibi H, Rachid S, Hakem N, Wahbi M (2020) ScienceDirect "A framework of optimizing the deployment of IoT for precision agriculture" Industry-NC-ND license (https://creativecommons.org/licenses/by-nc-nd/4.0/) Peer-review under responsibility of KES International

Chukkapalli S, Mittal S, Gupta M, Abdelsalam M, Joshi A, Sandhu R, Joshi K (2020) Ontologies and artificial intelligence systems for the cooperative smart farming ecosystem. IEEE Access 8: 164,045

El-Basioni BMM, El-Kader SMA (2020) Laying the foundations for an IoT reference architecture for agricultural application domain. IEEE Access 8:190,194–190,230. https://doi.org/10.1109/ACCESS.2020.3031634

Ferehan N, Haqiq A, Wazih M (2022) Intelligent farming system based on intelligent internet of things and predictive analytics. J Food Qual 2022:1–8. https://doi.org/10.1155/2022/7484088

Hassan M, Singh M, Hamid (2021a) Survey on NOMA and spectrum sharing techniques in 5G, 2021 IEEE International Conference on Smart Information Systems and Technologies (SIST), pp. 1–4. doi: https://doi.org/10.1109/SIST50301.2021.9465962

Hassan M, Singh M, Hamid K (2021b) Overview of cognitive radio networks. J Phys Conf Ser 1831

Hassan M, Singh M, Hamid K (2021c) Survey on advanced spectrum sharing using cognitive radio technique. In ICT systems and sustainability. Springer, Singapore, 639–647

Hassan M, Singh M, Hamid K (2021d) Impact of power and bandwidth on the capacity rate and number of users in SC-NOMA. Harbin Gongye Daxue Xuebao/J Harbin Inst Technol 53(9): 118–124. http://hebgydxxb.periodicales.com/index.php/JHIT/article/view/726. Accessed 15 Feb 2022

Hassan M, Singh M, Hamid K (2022a) Review of NOMA with spectrum sharing technique. In ICT with intelligent applications. Springer, Singapore, pp 135–143. doi: https://doi.org/10.1007/978-981-16-4177-0_16

Hassan M, Singh M, Hamid K, Saeed R, Abdelhaq M, Alsaqour R (2022b) Modeling of NOMA-MIMO-based power domain for 5G network under selective rayleigh fading channels. Energies 15(15):5668. https://doi.org/10.3390/en15155668

Hassan M, Singh M, Hamid K, Saeed R, Abdelhaq M, Alsaqour R (2022c) Design of Power Location Coefficient System for 6G downlink cooperative NOMA network. Energies 15(19): 6996. https://doi.org/10.3390/en15196996

Kadu, Singh M (2021) Comparative analysis of e-health care telemedicine system based on internet of medical things and artificial intelligence. 2021 2nd International conference on smart electronics and communication (ICOSEC), pp 1768–1775, doi: https://doi.org/10.1109/ICOSEC51865.2021.9591941

Kadu A, Singh M, Ogudo K (2022) A novel scheme for classification of epilepsy using machine learning and a fuzzy inference system based on wearable-sensor health parameters. Sustainability 14(22):15079. https://doi.org/10.3390/su142215079

Kaushik N, Narad S, Mohature A, Sakpal P (2019) Predictive analysis of IoT based digital agriculture system using machine learning. Int J Eng Sci Comput 9(3)

Khalaf OI, Ogudo KA, Singh M (2021) A fuzzy-based optimization technique for the energy and spectrum efficiencies trade-off in cognitive radio-enabled 5G network. Symmetry 13(1):47. https://doi.org/10.3390/sym13010047

Kour V, Arora S (2020) Recent developments of the internet of things in agriculture: a survey. IEEE Access 8:129,924–129,957. https://doi.org/10.1109/ACCESS.2020.3009298

Lakshmanna K, Kavitha R, Geetha BT, Nanda AK, Radhakrishnan A, Kohar R (2022a) Deep learning-based privacy-preserving data transmission scheme for clustered IIoT environment. Comput Intell Neurosci 8:2022

Lakshmanna K, Subramani N, Alotaibi Y, Alghamdi S, Khalafand OI, Nanda AK (2022b) Improved metaheuristic-driven energy-aware cluster-based routing scheme for IoT-assisted wireless sensor networks. Sustainability 14(13):7712

Singh M, Navdeep Kaur Jhajj, and Anudeep Goraya (2022) IoT-enabled wireless mobile ad-hoc networks: introduction, challenges, applications: review chapter. Internet of Things, pp 121–134

Marwah, Kour GP, Jain A, Malik PK, Singh M, Tanwar S, Safirescu CO, Mihaltan TC, Sharma R, Alkhayyat A (2022) An improved machine learning model with hybrid technique in VANET for robust communication. Mathematics 10(21):4030. https://doi.org/10.3390/math10214030

Prabha R, Sinitsambirivoutin E, Passelaigue F, Ramesh MV (2018) Design and development of an IoT based smart irrigation and fertilization system for chilli farming, 2018 International conference on wireless communications, signal processing and networking (WiSPNET), Chennai, pp 1–7, doi: https://doi.org/10.1109/WiSPNET.2018.8538568

Rezk NG, Hemdan EED, Attia AF et al (2021) An efficient IoT-based intelligent farming system using machine learning algorithms. Multimed Tools Appl 80:773–797. https://doi.org/10.1007/s11042-020-09740-6

Rokade, Singh M (2021) Analysis of precise green house management system using machine learning based internet of things (IoT) for smart farming. 2021 2nd International conference on smart electronics and communication (ICOSEC), pp. 21–28, doi: https://doi.org/10.1109/ICOSEC51865.2021.9591962

Rokade A, Singh M, Arora SK, Nizeyimana E (2022a) IOT-based medical informatics farming system with predictive data analytics using supervised machine learning algorithms. Comput Mathemat Methods Med:8434966, 15 pages. https://doi.org/10.1155/2022/8434966

Rokade A, Singh M, Malik PK, Singh R, Alsuwian T (2022b) Intelligent data analytics framework for precision farming using IoT and regressor machine learning algorithms. Appl Sci 12(19): 9992. https://doi.org/10.3390/app12199992

Roy SK, Singh M, Sharma KK, Bhargava C, Singh BP (2020) Mathematical modelling of simple passive RC filters using floating admittance technique. 2020 IEEE international conference for innovation in technology (INOCON), pp 1–6, doi: https://doi.org/10.1109/INOCON50539.2020.9298230

Sadowski S, Spachos P (2018) Solar-powered smart agricultural monitoring system using internet of things devices, 2018 IEEE 9th annual information technology, electronics and Mobile communication conference (IEMCON), Vancouver, BC, pp 18–23, doi: https://doi.org/10.1109/IEMCON.2018.8614981

Sekaran K, Meqdad MN, Kumar P, Rajan S, Kadry S (2020) Intelligent agriculture management system using internet of things. TELKOMNIKA (Telecommun Comput Electron Contr) 18: 1276–1285. https://doi.org/10.12928/TELKOMNIKA.v18i3.14029

Sethy P, Behera S, Pandey C, Narayanand S (2021) Smart paddy field monitoring system using deep learning and IoT. Concurr Eng Res Appl 29(1). https://doi.org/10.1177/1063293X21988944

Siddiquee KN-E-A, Islam M, Singh N, Gunjan V, Yong W, Huda M, Naik D (2022) Development of algorithms for an IoT-based intelligent agriculture monitoring system. Wireless Commun Mob Comput 2022:1–16. https://doi.org/10.1155/2022/7372053

Singh M, Gandam A (2022) Link restoration and relay node placement in partitioned wireless sensor network. Design Dev Effic Energy Syst 4:101–117. https://doi.org/10.1002/9781119761785.ch7

Singh B, Verma HK (2022a) Dawn of big data with Hadoop and machine learning. In: Agarwal P et al (eds) Machine learning and data science: fundamentals and applications. Wiley, pp 47–65

Singh B, Verma H (2022b) EMM: extended matching market based scheduling for big data platform Hadoop. Multimed Tools Appl 81:34,823–34,847. https://doi.org/10.1007/s11042-021-11283-3

Singh M, Kumar M, Malhotra J (2018) Energy efficient cognitive body area network (CBAN) using lookup table and energy harvesting. J Intellig Fuzzy Syst 35(2):1253–1265

Subahi A, Bouazza KE (2020) An intelligent IoT-based system design for controlling and monitoring greenhouse temperature. IEEE Access:1. https://doi.org/10.1109/ACCESS.2020.3007955

Suma V (2021) Internet of Things (IoT) based intelligent agriculture in India: an overview. J ISMAC 03(01):1–15, http://irojournals.com/iroismac/. https://doi.org/10.36548/jismac.2021.1.001

Tageldin A, Adly D, Mostafa H, Mohammed H (2020) Applying machine learning technology in the prediction of crop infestation with cotton leafworm in greenhouse. Int J Appl Earth Observ Geoinform 105:102608. https://doi.org/10.1101/2020.09.17.301168

Tan E-K, Chong Y-W, Niswar M, Ooi BY, Basuki A (2020) An IoT platform for urban farming, 2020 International seminar on intelligent technology and its applications (ISITIA), Surabaya, Indonesia, pp. 51–55, doi: https://doi.org/10.1109/ISITIA49792.2020.9163781

Walia GS, Singh P, Singh M (2021) Localizing Mobile nodes in WSNs using dragonfly algorithm. In 2021 International conference on computing sciences (ICCS), pp. 223–227. IEEE

Walia, G. S., Singh, P., Singh, M., Abouhawwash, M., Park, H. J. et al. (2022a). Three-dimensional optimum node localization in dynamic wireless sensor networks. CMC-Comput Mater Continua, 70(1), 305–321. doi:https://doi.org/10.32604/cmc.2022.019171

Walia, Singh G, Singh P, Singh M (2022b) Localizing mobile nodes in WSNs using neural network algorithm. Mater Today Proc. https://doi.org/10.1016/j.matpr.2022.06.153

Chapter 6
DistilBERT-based Text Classification for Automated Diagnosis of Mental Health Conditions

Diwakar and Deepa Raj

Abstract Mental health disorders present a substantial global health challenge, underscoring the critical importance of timely diagnosis and intervention for effective treatment. However, the proliferation of textual data within online mental health communities has underscored the need for automated methods that can classify and diagnose individuals based on their expressed thoughts and emotions. Existing diagnostic practices for these conditions primarily rely on labour-intensive and time-consuming clinical assessments and interviews conducted by healthcare professionals. Notably, a key challenge in this context is the absence of robust physiological indicators for mental disorders. This paper addresses this challenge by proposing a DistilBERT-based text classification approach for the automated diagnosis of mental health conditions. In addition, recognizing the emerging landscape of interdisciplinary exploration into the gut brain axis, we acknowledge the potential role of gut bacteria, the microbiome, and microorganisms in influencing mental health. Our research specifically focuses on three distinct mental health disorder conditions: anxiety, borderline personality disorder (BPD), and autism. To ensure the robustness of our approach, we curated a balanced dataset, comprising 500 samples for each of these three classes. This diligent effort yielded a noteworthy achievement, with our model attaining a remarkable accuracy rate of 96%.

Keywords BERT · Classification · Mental disorder · Deep learning · Transformers

6.1 Introduction

Mental health disorders, encompassing conditions such as anxiety, borderline personality disorder (BPD), and autism, affect millions of individuals worldwide, impacting their daily lives and overall well-being. Timely diagnosis and personalized treatment are pivotal for improving the quality of life for those affected by these

Diwakar (✉) · D. Raj
Babasaheb Bhimrao Ambedkar University (A Central University), Lucknow, India, Uttar Pradesh

A. Khamparia et al. (eds.), *Microbial Data Intelligence and Computational Techniques for Sustainable Computing*, Microorganisms for Sustainability 47,
https://doi.org/10.1007/978-981-99-9621-6_6

conditions. The growing popularity of online mental health communities and social media platforms has led to an unprecedented volume of user-generated text data. Individuals share their thoughts, emotions, and experiences related to mental health openly on these platforms. While this creates an opportunity for early detection and support, it also presents a formidable challenge—how to effectively analyze and classify this vast textual data to assist in the diagnosis of mental health disorders. Emerging studies have begun to uncover intriguing connections between the composition of gut bacteria, known as the gut microbiome, and mental health conditions such as anxiety and depression. Although this paper primarily delves into text classification, it is situated within the broader landscape of mental health research, where interdisciplinary exploration into the gut brain axis holds promise for new insights and therapeutic approaches.

In this paper, we present a novel approach based on DistilBERT, a lightweight variant of the BERT (Bidirectional Encoder Representations from Transformers) model.

DistilBERT has demonstrated exceptional capabilities in natural language understanding and can efficiently process large volumes of text data. Unlike the original BERT model, DistilBERT offers similar performance while significantly faster and requiring fewer computational resources. This makes it an ideal choice for real-time or resource-constrained applications, such as mental health disorder diagnosis. The proposed methodology aims to address the challenges associated with the diagnosis of mental health disorders using online text data. Used a dataset of three classes namely Anxiety, BPD, and Autism to train and evaluate our model. Through rigorous experimentation, we achieved an impressive accuracy of 96%, showcasing the effectiveness of our approach in accurately classifying mental health conditions from user-generated textual content. This high level of accuracy demonstrates the potential of DistilBERT-based text classification in assisting healthcare professionals and support organizations in identifying individuals who may benefit from early intervention and specialized care. The fundamental architecture of the BERT model is displayed in Fig. 6.1. In the following sections, we provide an in-depth description of our methodology, simulation setup, and detailed results in mental health diagnosis.

6.2 Related work

The classification of mental health disorders from textual data has been a subject of increasing interest in recent years. Researchers have recognized the potential of natural language processing (NLP) and machine learning techniques in aiding mental health diagnosis and support. In this section, we review the existing literature on mental health disorder classification and NLP-based approaches, providing insights into the advancements and challenges in this field.

Previous studies have explored various methodologies for classifying mental health disorders based on textual content. Peng et al. (2019) introduced a multi-

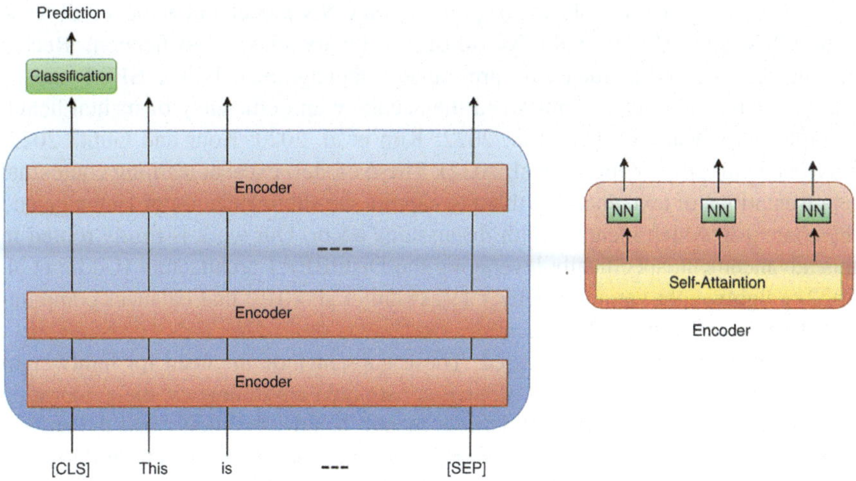

Fig. 6.1 In the fundamental architecture of BERT, which is commonly employed for various natural language processing tasks, including classification, the special tokens [CLS] and [SEP] play a critical role. These tokens have distinct purposes within the model. The [CLS] token is placed at the beginning of the input sequence, followed by the content to be classified, and the [SEP] token is added at the end. Notably, the [CLS] token takes on the responsibility of encapsulating the entire sequence, making it particularly vital for classification tasks in the BERT framework

kernel SVM approach, which achieved a better accuracy of 83.46% for depression recognition. Amanat et al. (2022) presented a powerful model combining LSTM with two hidden layers and a large bias, alongside an RNN with two dense layers, to predict depression from text. This model achieves an impressive accuracy of 99% with binary classification and has the potential to help safeguard individuals against mental disorders and suicidal tendencies. Researchers have employed traditional machine learning algorithms, such as Support Vector Machines (SVM) and Random Forests, as well as deep learning architectures, including Convolutional Neural Networks (CNNs) and recurrent models like Long Short-Term Memory (LSTM) networks. These approaches have shown promise in distinguishing between different mental health conditions but often require substantial feature engineering and may struggle with capturing the nuanced semantics of text.

Nova (2023) proposed a powerful model combining LSTM with two hidden layers and a large bias, alongside an RNN with two dense layers, to predict depression from text. This model achieved 77% with the LightGBM model for classification based on the text content. Kour and Gupta (2022) proposed a hybrid CNN-biLSTM model and achieved a 94.28% accuracy. Uddin et al. (2022) presented Long Short-Term Memory (LSTM)-based.

RNN to efficiently identify self-perceived symptoms of depression in text data, outperforming traditional word frequency-based methods. The approach offers the potential for improving mental healthcare technologies like intelligent chatbots. Murarka et al. (2020) utilized the RoBERTa model for mental illness and achieved

89% of accuracy. Kim et al. (2020) presented a CNN model and achieved 96.96% with CNN and 94.91% with XGBoost for binary class classification. Recent advancements in NLP, fueled by pre-trained language models like BERT and its variants, have significantly improved the accuracy and efficiency of mental health disorder classification (Diep et al. 2022; Kim et al. 2020; Kour and Gupta 2022; Vaswani et al. 2017; Zeberga et al. 2022). These models excel at learning contextual representations of text, enabling them to capture the subtle nuances of language and the context in which mental health discussions occur. Our work is firmly rooted in this advancement, specifically leveraging the DistilBERT architecture (Uddin et al. 2022) a lightweight variant of BERT (Vaswani et al. 2017) that balances computational efficiency with performance.

Several gaps remain unaddressed. These gaps include the need for lightweight models that can be deployed in resource-constrained environments, the exploration of diverse datasets representing multiple mental health conditions, and the assessment of model performance in real-world scenarios. Our study aims to bridge these gaps by presenting a comprehensive approach that combines a lightweight model with a diverse dataset, resulting in accurate and practical classification for improved mental health diagnosis and support.

6.3 Dataset and Prepossessing

6.3.1 Dataset

To conduct our research on automated mental health disorder classification, we utilized a comprehensive dataset sourced from the Kaggle repository (Ameer et al. 2022). This dataset comprises a diverse collection of posts from online mental health communities, offering a rich and varied source of text data for our analysis. The dataset consists of 500 samples from each class, representing a wide range of user-generated content (Adha 2022). The dataset includes posts related to various mental health conditions, with labels corresponding to three distinct classes: Anxiety, Borderline Personality Disorder (BPD), and Autism. Each post contains textual content, often accompanied by user-generated titles.

6.3.2 Prepossessing

To prepare our dataset for analysis, an extensive preprocessing pipeline is applied, comprising several crucial steps as shown in Table 6.1. Additionally, these preprocessing methods standardized and cleaned the text data, ensuring its suitability for subsequent analysis.

Table 6.1 Text preprocessing methods

Method	Description
URL Removal	Elimination of web links using regular expressions
Newline Removal	Removal of newline characters for consistent formatting
Lowercasing	Conversion of text to lowercase for case-insensitive processing
HTML Tag Removal	Removal of HTML tags using BeautifulSoup
Whitespace cleanup	Reduction of extra whitespaces to a single space
Bracket Removal	Removal of various bracket types
Quote Removal	Elimination of single and double quotes
Special Character Removal	Removal of non-semantic special characters
Chars with Digits Removal	Elimination of character–digit combinations
Digit Removal	Removal of standalone digits
Contractions Expansion	Expansion of contractions to full forms
Stopword Removal	Elimination of common English stopwords
Lemmatization	Reduction of words to their base forms (lemmas)

6.4 Methodology

The choice of DistilBERT (Uddin et al. 2022) as the Natural Language Processing (NLP) model for this study is grounded in its remarkable performance and efficiency. DistilBERT, a lightweight variant of the BERT (Bidirectional Encoder Representations from Transformers) model, offers several advantages for our task of mental health disorder classification. DistilBERT provides comparable performance to the original BERT model but with significantly fewer parameters (66 million), making it computationally efficient and suitable for real-time or resource-constrained applications, which are crucial in the context of mental health support. Furthermore, DistilBERT excels in capturing the contextual understanding of the text, allowing it to comprehend the nuanced semantics of user generated content related to mental health discussions. This contextual awareness is essential for accurate classification. Leveraging pretrained representations on massive text corpora, similar to BERT, DistilBERT is equipped with a broad understanding of language and concepts. Fine-tuning such a pretrained model on our mental health dataset enhances its classification capabilities.

To train and evaluate our DistilBERT-based model, we employed a standard data splitting technique, dividing our dataset into training and validation sets. In this process, 80% of the dataset is allocated to the training set, enabling our model to learn from a diverse range of textual data representing different mental health conditions. The remaining 20% of the dataset served as the validation set, ensuring an evaluation of the model's performance on unseen data and assessing its generalization capabilities. Our training procedure adhered to established best practices in deep learning and NLP. We utilized the AdamW optimizer for parameter updates during training, combining the advantages of the Adam optimizer with weight decay to ensure robust convergence. A learning rate of $2e^{-5}$ is selected based on empirical experimentation and fine-tuning, striking a balance between rapid convergence and

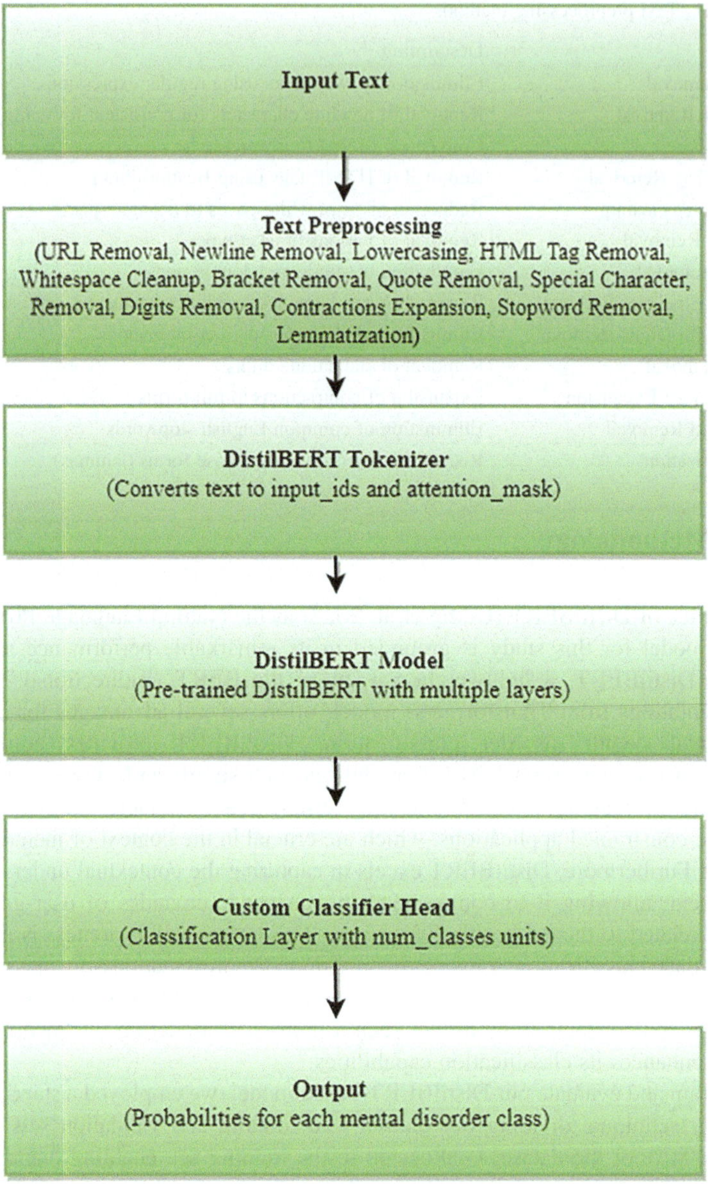

Fig. 6.2 Architecture for the classification of mental health disorders

stability. The model underwent training over 20 epochs, a choice made after careful monitoring of training loss and validation performance to prevent overfitting. The architecture of the model is displayed in Fig. 6.2.

To further enhance DistilBERT's performance in classifying mental health disorders, we conducted model fine-tuning. This process involved iterative adjustments of hyperparameters, such as the learning rate, batch size, and dropout rates. Ad-ditionally, we explored strategies to address the class imbalance, including oversampling and weighted loss functions, to ensure fair and accurate classification across all mental health conditions. Our choice of DistilBERT as the Natural Language Processing (NLP) model for this study is grounded in its remarkable performance and efficiency. DistilBERT, a lightweight variant of the BERT (Bidirectional Encoder Representations from Transformers) model, offers several advantages for our task of mental health disorder classification. DistilBERT provides comparable performance to the original BERT model but with significantly fewer parameters, making it computationally efficient and suitable for real-time or resource-constrained applications, which are crucial in the context of mental health support. DistilBERT achieves its efficiency by employing knowledge distillation. It is trained to mimic the behaviour of the larger BERT model by learning from its outputs. The core architecture of DistilBERT consists of a stack of transformer layers, each composed of multi-head self-attention mechanisms and feedforward neural networks. Mathematically, it can be represented as follows in Eq. (6.1).

$$\begin{aligned}&\text{DistilBERT}(\mathbf{X})\\&\quad=\text{LayerNorm}(\text{MultiHeadSelfAttention}(\mathbf{X})+\mathbf{X}+\text{FeedForward}(\mathbf{X})+\mathbf{X})\end{aligned} \tag{6.1}$$

This equation represents the forward pass through one transformer layer in DistilBERT. $\mathbf{X}$ represents the input embeddings, and the operations include multi-head self-attention and feedforward neural networks. Furthermore, DistilBERT excels in capturing the contextual understanding of the text, allowing it to compre hend the nuanced semantics of user-generated content related to mental health discussions. This contextual awareness is essential for accurate classification. The contextual understanding in DistilBERT is achieved through multi-head self-attention mechanisms, which mathematically compute weighted sums of input embeddings based on their relevance to each other. This is expressed as follows Eq. (6.2):

$$\text{SelfAttention}(\mathbf{X})=\text{softmax}\left(\frac{\mathbf{X}W_Q(\mathbf{X})^T}{\sqrt{d_k}}\right)\mathbf{X}W_V(\mathbf{X}) \tag{6.2}$$

Here, $\mathbf{X}$ is the input, and W_Q and W_V are learned weight matrices. The softmax function calculates attention scores, and the weighted sum provides contextualized embeddings.

Leveraging pretrained representations on massive text corpora, similar to BERT, DistilBERT is equipped with a broad understanding of language and concepts. Fine-tuning such a pretrained model on our mental health dataset enhances its classification capabilities. Pretraining involves training a transformer model to predict masked words in a large text corpus. The pretrained model is characterized by a vocabulary

V, input embeddings E, and transformer parameters Θ. Fine-tuning on our dataset involves minimizing a classification loss L by using stochastic gradient descent. This is expressed as Eq. (6.3):

$$\min_{\Theta} \frac{1}{N} \sum_{i=1}^{N} L(\text{DistilBERT}(\mathbf{X}_i; \Theta), y_i) \tag{6.3}$$

where N is the number of training examples. $\mathbf{X}_i$ is the ground truth label for the ith training example. L is a suitable classification loss function, such as cross-entropy.

In the training phase of DistilBERT, the primary objective is to minimize a classification loss function L over a labelled dataset. Given a dataset of N training examples $\{(\mathbf{X}_i, y_i)\}$, where $\mathbf{X}_i$represents the input text and y_i represents the ground truth label.

6.4.1 Batch Training

During training, it is common to use batch training, where the dataset is divided into smaller batches of size B. The optimization process updates the model's parameters using gradients computed over these batches. Mathematically, the update step for the model parameters Θ can be expressed as Eq. (6.4):

$$\Theta \leftarrow \Theta - \alpha \cdot \frac{1}{B} \sum_{i=1}^{B} \nabla_{\Theta} L(\text{DistilBERT}(\mathbf{X}_i; \Theta), y_i) \tag{6.4}$$

Θ represents the model parameters. α is the learning rate. B is the batch size. ∇_{Θ} denotes the gradient with respect to the model parameters

6.4.2 Hyperparameter Tuning

In the fine-tuning process, various hyperparameters are tuned to optimize the model's performance. These hyperparameters include:

Learning Rate (α)**:** In this work, the learning rate (α) is set to value of $2e^{-5}$ using empirical experimentation. This learning rate controls the step size in gradient descent and plays a crucial role in convergence speed and stability.

Batch Size (B)**:** We utilized a batch size B of 32 during training. The choice of batch size affects training efficiency and memory usage.

Dropout Rates: Dropout rates were applied as a regularization technique in our project, with a specific dropout rate of 0.2. These rates were tuned to control the amount of dropouts during training and mitigate overfitting.

6.4.3 Algorithm

The BERT-based mental health classification algorithm involves rigorous text pre-processing and model setup, followed by data splitting, training, and evaluation, culminating in classification results for mental health assessment as shown in Algorithm 6.1.

Algorithm 6.1 BERT-based Mental Health Classification Algorithm Step 1: Data Preprocessing

1. Remove URLs.
2. Eliminate newline characters.
3. Convert text to lowercase.
4. Remove HTML tags.
5. Clean up extra whitespaces.
6. Remove brackets, quotes, and special characters.
7. Expand contractions.
8. Remove stopwords.
9. Apply lemmatization.
10. Remove character digit combinations and digits.

Step 2: Model Initialization

1. Initialize a DistilBERT model for sequence classification.
2. Tokenize and encode text data.
3. Convert labels to numerical values.

Step 3: Data Splitting

1. Split data into training and validation sets.

Step 4: Model Training

1. Initialize AdamW optimizer (learning rate: 2 × 10^(−5)).
2. Create data loaders for training and validation.
3. Train the model for 20 epochs

Step 5: Model Evaluation

1. Evaluate the trained model on validation data.
2. Calculate accuracy, precision, recall, and F1 score.

Step 6: Output

1. Trained DistilBERT model for mental health post-classification.
2. Classification report on validation data.

6.4.4 Evaluation Standards

1. **Accuracy:** Accuracy is a commonly used evaluation metric for classification tasks. It measures the proportion of correctly classified instances among all instances. Mathematically, accuracy can be expressed as:

$$\text{Accuracy} = \frac{\text{NumberofCorrectPredictions}}{\text{TotalNumberofPredictions}} \tag{6.5}$$

2. **Precision**: measures the accuracy of positive predictions. It calculates the proportion of true positive predictions (correctly predicted positive instances) among all instances predicted as positive. Mathematically, precision can be expressed as

$$\text{Precision} = \frac{\text{TruePositives}}{\text{TruePositives} + \text{FalsePositives}} \tag{6.6}$$

3. **Recall:** Recall also known as sensitivity or true positive rate, quantifies the model's ability to identify positive instances correctly. It calculates the proportion of true positive pre- dictions among all actual positive instances. Mathematically, recall can be expressed as

$$\text{Recall} = \frac{\text{TruePositives}}{\text{TruePositives} + \text{FalseNegatives}} \tag{6.7}$$

4. **F1 Socre:** The F1 score is the harmonic mean of precision and recall. It provides a balanced measure that takes both false positives and false negatives into account. Mathematically, the F1 score can be expressed as:

$$\text{F1 score} = \frac{2 \cdot \text{Precision} \cdot \text{Recall}}{\text{Precision} + \text{Recall}} \tag{6.8}$$

5. **Confusion Matrix (CM):** A confusion matrix is a table that summarizes the model's classification performance. It provides counts of true positives, true negatives, false positives, and false negatives. From the confusion matrix, various metrics like precision and recall can be calculated.

6.5 Simulation Results and Discussion

In this section, we outline the key aspects of our simulation setup.

6.5.1 Model Training

The model is trained over 20 epochs with the AdamW optimizer, utilizing a learning rate of $2e^{-5}$, and during training, cross-entropy loss is computed. The training process yielded promising results, as evidenced by the classification report. The model achieved an overall accuracy of 96%. Notably, it exhibited excellent performance in classifying Anxiety, achieving a precision of 94% and recall of 99%, highlighting its proficiency in identifying this specific mental health subreddit. These findings underscore the model's effectiveness in categorizing posts related to various mental health conditions. The average loss over the epochs is presented in Table 6.2.

6.5.2 Text Data Distribution Analysis

In our analysis, utilized kernel density estimation (KDE) with seaborn to visualize the distribution of total words and characters in the dataset. These KDE plots, rendered in a winter-themed colour palette, shed light on the composition of text data across different mental health subreddits. The shaded KDE plots provide insights into potential differences in content length among the analyzed subreddits as shown in Fig. 6.3.

Table 6.2 Average loss over epochs

Epoch	Average Loss
1/20	0.9574
2/20	0.2829
3/20	0.1102
4/20	0.0553
5/20	0.0381
6/20	0.0272
7/20	0.0278
8/20	0.0093
9/20	0.0049
10/20	0.0037
11/20	0.0030
12/20	0.0025
13/20	0.0022
14/20	0.0019
15/20	0.0016
16/20	0.0014
17/20	0.0013
18/20	0.0011
19/20	0.0011
20/20	0.0010

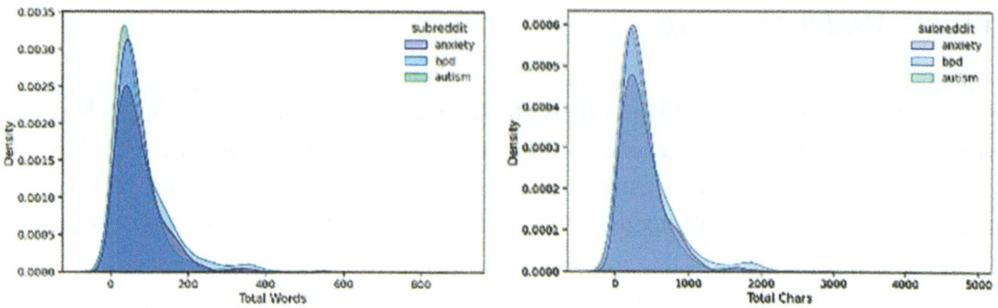

(a) KDE Plot of Total Words (b) KDE Plot of Total Characters

Fig. 6.3 Text data characteristics analysis. (**a**) KDE plot of total words. (**b**) KDE plot of total characters

Fig. 6.4 Confusion matrix

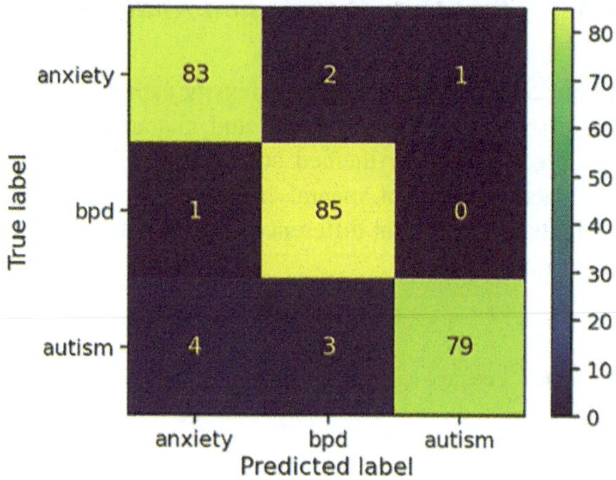

6.5.3 Confusion Matrix

In the paper, we performed resampling on the validation data to ensure an equal number of samples in each class, addressing class imbalance. The resulting confusion matrix plot illustrates the model's classification performance, with labels 'anxiety,' 'BPD,' and 'autism' prominently displayed for easy interpretation in Fig. 6.4.

6.5.4 Visualizing Word Clouds

The code generates word cloud visualizations for 'anxiety,' 'bpd,' and 'autism' subreddits, offering insights into their prevalent terms. Each word cloud showcases

Fig. 6.5 Visualizing word clouds

Table 6.3 Performance comparison of classification methods

Method	Precision	Recall	F1 Score	Accuracy
TF-IDF	0.85	0.93	0.89	0.91
TinyBERT	0.96	0.90	0.93	0.91
DistilBERT	**0.94**	**0.97**	**0.95**	**0.96**

frequently occurring words, aiding in understanding the distinctive textual content of these mental health communities. Figure 6.5 depicts the visualization of word clouds.

In this comprehensive analysis of mental health-related discussions on Reddit, our study integrated advanced natural language processing techniques and machine learning to categorize and understand user-generated content across three major mental health subreddits: 'anxiety,' 'BPD' (borderline personality disorder), and 'autism.' By utilizing a DistilBERT-based classification model, we achieved high accuracy in classifying posts into these categories. Additionally, we addressed class imbalance issues through resampling techniques, ensuring robust model performance. Furthermore, we visualized word clouds to provide intuitive insights into the most prevalent terms within each subreddit, shedding light on the unique language patterns and themes that characterize these online mental health communities. It's worth noting that we conducted comparative experiments with other methods such as TF IDF with SVM and TinyBERT, and the results, along with detailed comparisons, are presented in Table 6.3, demonstrating the effectiveness of the DistilBERT model in capturing nuanced textual features. This research not only enhances our understanding of how individuals discuss mental health on social platforms but also demonstrates the potential for machine learning in assisting mental health professionals and researchers in monitoring and providing support in online communities.

6.6 Conclusion

In this study, we harnessed advanced natural language processing techniques to analyze mental health discussions across 'anxiety,' 'BPD' (borderline personality disorder), and 'autism' subreddits. Leveraging the DistilBERT model, we achieved an impressive 96% accuracy in classifying posts, showcasing its efficacy in capturing nuanced text features. Addressing class imbalance concerns and visualizing word clouds provided valuable insights into these online mental health communities.

Looking ahead, expanding this analysis to include more subreddits, incorporating sentiment analysis, and developing real-time support systems are promising directions. Enhancing model interpretability and collaborating with mental health experts can further leverage this research for practical applications in mental health support.

References

Adha K (2022) Mental disorders identification (Reddit), *Kaggle*, Version 2, 4579285

Amanat A et al. (2022) Deep learning for depression detection from textual data. Electronics 11 (5):676

Ameer I et al. (2022) Mental illness classification on social media texts using deep learning and transfer learning. In: arXiv preprint arXiv:2207.01012

Ansari G, Garg M, Saxena C (2021) Data augmenta- tion for mental health classification on social media. In: arXiv preprint arXiv:2112.10064

Diep B, Stanojevic M, Novikova J (2022) "Multi-modal deep learning system for depression and anxiety detection". In: arXiv preprint arXiv:2212.14490

Kim J et al (2020) A deep learning model for detecting mental illness from user content on social media. Sci Rep 10(1):11,846

Kour H, Gupta MK (2022) An hybrid deep learning approach for depression prediction from user tweets using feature-rich CNN and bi-directional LSTM. Multimed Tools Appl 81(17): 23,649–23,685

Murarka A, Radhakrishnan B, Ravichandran S (2020) Detection and classification of mental illnesses on social media using RoBERTa. In: arXiv preprint arXiv:2011.11226

Nova K (2023) Machine learning approaches for automated mental disorder classification based on social media textual data. Contemp Issu Behav Social Sci 7(1):70–83

Peng Z, Qinghua H, Dang J (2019) Multi-kernel SVM based depression recognition using social media data. Int J Mach Learn Cybern 10:43–57

Sanh V et al. (2019) DistilBERT, a distilled version of BERT: smaller, faster, cheaper and lighter. In: arXiv preprint arXiv:1910.01108

Tran T, Kavuluru R (2017) Predicting mental conditions based on "history of present illness" in psychiatric notes with deep neural networks. J Biomed Inform 75:S138–S148

Uddin MZ et al (2022) Deep learning for prediction of depressive symptoms in a large textual dataset. Neural Comput Appl 34(1):721–744

Vaswani A et al. (2017) Attention is all you need. In Advances in neural information processing systems, vol 30

Zeberga K et al (2022) A novel text mining approach for mental health prediction using Bi-LSTM and BERT model. Comput Intell Neurosci 2022:7893775

Chapter 7
An Optimized Hybrid ARIMA-LSTM Model for Time Series Forecasting of Agricultural Production in India

Babita Pandey, Arvind Shukla, and Aditya Khamparia

Abstract The role of agriculture in a country's national income is very important. The zero-hunger sustainable goal of the United Nations cannot be achieved without substantial increase in agricultural production. The value of agricultural production shows the health of agriculture and its prediction has long been a challenge for academicians due to highly uncertain weather conditions as well as emergence of microbacteria due to variable soil and weather conditions. This chapter is focused on the prediction methodology for measuring accurate prediction of the value of agricultural production. The data of the gross value-added series for the years 1950–2021 for the agriculture and allied sectors has been taken for prediction. The traditional model autoregressive integrated moving average (ARIMA) and machine learning and the deep learning-based long short-term memory (LSTM) model have been compared with an optimized hybrid model comprising characteristics of both models. The efficiency of the models was checked by using root mean square error (RMSE). The result was compared with other studies on hybrid models. It has been reported that the hybrid model is more appropriate in comparison to both ARIMA and LSTM. The hybrid model provides the lowest RMSE among the three models. The ARIMA model gives lower RMSE in comparison to LSTM. The hybrid model provides 11% more accurate prediction in comparison to ARIMA. The prediction result of the hybrid model is highly significant and gives within the 95% confidence interval. The results are important and useful for the prediction of the value of agricultural production and policy formulation to meet the sustainable development goal of zero hunger.

B. Pandey (✉)
Babasaheb Bhimrao Ambedkar (Central) University, Lucknow, Uttar Pradesh, India

Department of Computer Science, Babasaheb Bhimrao Ambedkar (Central) University, Lucknow, Uttar Pradesh, India

A. Shukla
Babasaheb Bhimrao Ambedkar (Central) University, Lucknow, Uttar Pradesh, India

A. Khamparia
Department of Computer Science, Babasaheb Bhimrao Ambedkar University, Amethi, India

A. Khamparia et al. (eds.), *Microbial Data Intelligence and Computational Techniques for Sustainable Computing*, Microorganisms for Sustainability 47,
https://doi.org/10.1007/978-981-99-9621-6_7

Keywords Time series · Agriculture · Micro bacterial · Machine learning · ARIMA · LSTM

7.1 Introduction

Agriculture is the most important sector of the Indian economy. It is vital not only for feeding the people of the country but also for economic development. It helps in the development of the other sectors of the economy through its forward and backward linkages. Agricultural production is one of the important indicators for measuring the development of agriculture. The value of the agricultural production shows the health of agriculture. However, its prediction has long been a challenge for academicians due to highly uncertain weather conditions as well as emergence of microbacteria due to variable soil and weather conditions. The prediction of agricultural production will help policy makers to devise approaches to meet the growing demand for food due to increase in the population. Across the globe around 815 million people suffer from chronic hunger and one in three people is malnourished (UNDP 2023). It is one of the sectors that needs the greatest focus to achieve zero hunger. The production of agriculture depends upon the availability of irrigation, rainfall, yield, soil fertility, natural calamities, microbes, and climatic conditions. Considering many of the fixed factors, agricultural production requires gradual improvement in the land qualities and seed quality. Thus, agricultural production has certain limitations. However, Indian agricultural production has increased due to technological development from approximately 50 million tons to more than 300 million tons, but despite all these improvements, the Sustainable Development Report 2023 placed India in the category of the countries with average prevalence of undernourishment at 25–40%during 2020–2022. It has also been reported that this prevalence is attributed to the interplay of two opposing forces: income and inflation. Income is increased but its effect is eroded due to price inflation (UNDP 2023). Therefore, prediction of agricultural production and its values is imperative. Prediction of the value of agricultural products is quite cumbersome due to the volatility of income and prices. The nonlinear and complicated trends of the agricultural market make it difficult to accurately predict the value of agricultural products (Hauchet-Bourdon 2011). The share of gross value added (GVA) of agriculture and allied sectors in the total economy has decreased during the last 3 years from 20.3 in 2020–2021 to 19.0 in 2021–2022 and subsequently to 18.3 in 2022–2023 (PIB 2023). Considering the sustainable development goals, decreasing shares of agriculture and allied sectors in GVA poses significant challenges. Prediction, therefore, will help in policy formulations for achieving the sustainable development goals.

There are various methods of prediction. Autoregressive moving average (ARMA) and autoregressive integrated moving average (ARIMA) are the traditional methods used for prediction with univariate time series data. With the advent of new techniques like machine learning (ML) and deep learning (DL), preference is starting to be given to the techniques of neural networks that include support vector machines (SVM), Random Forest (RF), recurrent rural networks (RNN), and long short-term memory (LSTM). The main aim of this chapter is to discuss the techniques that can be used for the prediction of agriculture output in more accurate ways among the existing traditional techniques vis a vis available new ML and DL techniques. The ARIMA and LSTM techniques, considering the space limitation of the work, were tested for prediction. Different views on the prediction by these models are available. Prediction with ARIMA is efficient in a shorter period but as the time period increases the precision rate decreases. However, LSTM is quite efficient in making predictions even if there is less previous data (Hua 2020). The ARIMA model was devised primarily by Box and Jenkins (Box and Jenkins 1976). The ARIMA model has been transformed into various versions like seasonal ARIMA, called SARIMA, and multiplicative SARIMA, called MSARIMA (Kalpakis et al. 2000). Although the ARIMA model is widely used for prediction of univariate time series linear models, its optimal selection of parameters requires substantial computational time for suitable model selection, and this is considered a disadvantage (Nguyen et al. 2019). However, Sima et al. observed that LSTM is superior to ARIMA as it reduces the error rate by 85% (Siami-Namini et al. 2018). In contrast, Deng et al. found that the hybrid model synthesizes the characteristics of all individual models and therefore provides better predictions in comparison to the ARIMA and LSTM models (Deng et al. 2023). Naveen et al. evaluated the models for forecasting the price of washed coffee and found that the hybrid models can combine both linear and non-linear time series and hence improve the efficiency of forecasting (Naveena et al. 2017). Deng et al. also reported similar results that hybrid models provide better forecast in comparison to ARIMA or LSTM alone (Deng et al. 2023). They found that hybrid models provide the lowest root mean square error, mean absolute percentage error (MAPE), and mean absolute error (MAE), corroborating the findings of Khazee et al. (Khazee et al. 2019). Similarly, Deve et al. noted that the hybrid model provides better results. They formulated the hybrid model by integrating LSTM for non-linear components and ARIMA for the linear components of the data set (Dave et al. 2021). Kumar et al. also created a hybrid model that combines the non-Gaussian time series, i.e., beta seasonal ARIMA, with the LSTM and concluded on the basis of an experimental data set that this hybrid model provides better prediction accuracy (Sunil and Yadav 2023).

Prediction of the value of agricultural products is quite cumbersome due to volatility in income and prices. The non-linear and complicated trends of the agricultural market make it difficult to accurately predict the value of agricultural products (Hauchet-Bourdon 2011). Tae-woong and Seok proposed an Seasonal long short-term memory (SLSTM) model for the prediction of the sales of agricultural products (Yoo and Oh 2020).

Thus, this chapter is largely focused on the prediction methods of agricultural production using the techniques available for predicting based on time series data. The chapter is further divided into three sections, namely Materials and Methods (Sect. 7.2), Results and Discussion (Sect. 7.3), and Conclusion (Sect. 7.4). Section 7.2 elaborates the technique and methodology used for the prediction of the value of agricultural production. Section 7.3 describes the results of the analysis and issues linked with these results. Section 7.4 is a sketch of the findings of the research analysis.

7.2 Materials and Methods

7.2.1 Data

We used the gross value added by the agriculture fishing and mining at current prices on the base price year of 2011–2012. The data set from 1950 to 2021–2022 is available from National Income Statistics. The choice is based on achieving higher accuracy over a greater time period, which will help in the prediction of agricultural production.

7.2.2 Stationary Test

The time series data set must be stable for further analysis including prediction. Therefore, primarily the stationarity of the data set has been checked using the augmented Dickey–Fuller (ADF) test. The augmented Dickey–Fuller test is preferred over the Dickey–Fuller test to remove the possibility of autocorrection in the data test before predicting unit root hypothesis.

7.2.3 Autoregressive Integrated Moving Average (ARIMA)

The ARIMA model is used for prediction. The ARIMA model is a special version of the ARMA model incorporating non-stationary series that become stationary through differencing. It is manifested as ARIMA (p,q,d) wherein the p is the number of autoregressive terms, q is the number of differences required to make the series stationary, and d is the number of terms used for taking moving average. The stationarity of the data set has been checked using the augmented Dicky Fuller test, which is preferred over the Dickey–Fuller test to remove the possibility of autocorrection in the data test before predicting unit root hypothesis. The following ARIMA model is estimated:

$$y_t = \mu + y_{t-1} + \varepsilon_t$$

$$\Delta_{y_t} = Y_t - y_{t-1} = (1 - L)Y_t = \mu + \varepsilon_t$$

$$C(L)\left[(1 - L)^d Y_t\right] = \mu + D(L)\varepsilon_t$$

where $C(L)$ and $D(L)$ are polynomials in the lag operator and $(1 - L)^d Y_t = \Delta^d Y_t$ is the dth difference of Y_t.

$$\Delta^d Y_t = \mu + \gamma_1 \Delta^d Y_{t-1} + \gamma_2 \Delta^d Y_{t-2} + \ldots\ldots + \gamma_p \Delta^d Y_{t-p} + \varepsilon_t - \theta_1 \epsilon_{t-1} - \ldots\ldots\ldots\ldots\ldots - \theta_q \epsilon_{t-q}.$$

C(L) $Y_t = \varepsilon_t$. The Y_t denotes the value of the agricultural product in t time period. $\gamma_1, \gamma_2, \gamma_3 \ldots$ are the coefficients of the autoregressive (AR) model. $\theta_1, \theta_2, \theta_3 \ldots$ are the coefficients of the moving average. The *d*th difference produces the stationary process and is called the order of integration. This implies that the series will be called integrated of order one, i.e., $I(1)$, if the first difference of non-stationary series produces stationary series (Greene 2003). The ADF test is applied to check the stationarity with the null hypothesis that $\gamma = 0$. The rejection of the null hypothesis suggests that the series does not have a unit root and hence may be treated as stationary. The only limitation of the ARIMA model is its ability to capture only the linear relationship among the different time period values of the variable.

7.2.4 Long Short-term Memory (LSTM)

LSTM is a type of recurrent neural network (RNN) used in deep learning and machine learning. It was firstly developed by Hochreiter and Schmidhuber (Hochreiter and Schmidhuber 1997). LSTM is an efficient and gradient-based algorithm that provides constant error terms without explosion or vanishing of error flow. Its network topology carries three layers: input, hidden, and output layers. The hidden layer connects the remaining two layers through memory cells and gate units. There are three gates: forget gate, input gate, and output gate. Gate units work as the valve and controls the flow of neuronal information. All three gates are controlled by the sigmoid unit. The computation at each gate is shown in Table 7.1. LSTM has an ability to hold the properties of non-linearity of data while performing predictions. Therefore, the LSTM has an advantage over ARIMA.

Table 7.1 Computation at each gate in LSTM

Computation at gate	Architecture of LSTM
Forget gate: $f_t = \sigma(X_i * U_t + H_{t-1} * W_f)$	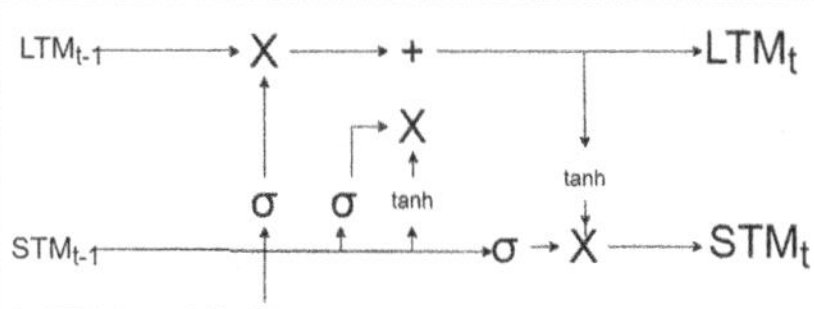
$C_{t-1} * f_t = 0. \ldots$ If $f_t = 0$	
$C_{t-1} * f_t = 0. \ldots$ If $f_t = 1$	
Input gate: $i_t = \sigma(X_t * U_i + H_{t-1} * W_i)$	
$N_t = \tanh(X_t * U_c + H_{t-1} * W_c)$	
$C_t = f_t * C_{t-1} + i_t * N_t$	LSTM architecture
Output gate: $O_t = \sigma(X_t * U_o + H_{t-1} * W_o)$	
$H_t = \tanh(C_t)$; output = Softmax (H_t)	

X Input, *U* Input weight matrix, *H* Hidden state, *W* Weight matrix associated with the hidden state, *t* Time tamp, *i* Input gate, *f* Forget gate, *c* Cell state, *o* Output gate, C_t Long-term memory

7.2.5 Optimized Hybrid ARIMA-LSTM

A two-phase hybrid model is developed. In the first phase ARIMA and LSTM are deployed independently for prediction. Optimal selection of parameters of ARIMA, i.e., (p, d, q), and LSTM (number of hidden layers, number of neurons in each layer) requires substantial computational time for suitable model selection and hence degrades the performance of the model. In this work, optimized ARIMA parameters are identified using a random search technique. In the second phase, the output of each model is compared with actual output and the smallest one is retained as final prediction at time stamp t (Eqs. (7.1) and (7.2)).

$$\mathrm{HP_i} = \min.\{\text{for each } t \; (\text{abs } |\mathrm{P\_AR_t} - \mathrm{Act_t}|, \text{Abs } |\mathrm{P\_LS_t} - \mathrm{Act_t}|)\} \tag{7.1}$$

$$\mathrm{HP} = \left(\mathrm{HP_1} \ldots\ldots\ldots\ldots\ldots\ldots\ldots\ldots\ldots\ldots\ldots \mathrm{HP_{length\ of\ sequence}}\right) \tag{7.2}$$

where the hybrid model prediction: HP; prediction by ARIMA: P_AR; Actual prediction: Act; prediction by LSTM: P_LS; and t: time stamp. The result showed that the optimum root mean square error (RMSE) of 39,035 is obtained by the hybrid model. The hybrid model is shown in Fig. 7.1. The left branch of Fig. 7.1 shows the ARIMA model computation flow and the right branch shows the LSTM architecture.

7.3 Results and Discussion

The experimentation was done in Python using spider editor. Python libraries like keras and tensor were used for prediction through LSTM models. This study used the statsmodels Python module for estimation of many different statistical models including ARIMA for the analysis. The agricultural production shown in Fig. 7.2 depicts the value of agricultural production over the period. The value of production increased steeply after 1990 and 2000. The effect of liberalization may be the reason behind such growth, including the impact of the Green Revolution. The figure shows that the series is non-stationary. The method of differencing has been adopted for making it stationary. The results of the ADF test are shown in Table 7.2. It shows that the data are non-stationary up to the second-order differencing. The p value for the ADF test statistics is higher than 0.05, depicting the failure to reject the null hypothesis that the data is non-stationary (as shown in Figs. 7.3 and 7.4). The third differentiating provides the $P < 0.5$ for the estimates of the augmented Dickey–Fuller test implying that the data set is stationary with third-order integration (as shown in Fig. 7.5). The negative test statistics are also indicative of stationary time series. The value of the lags used is the same as used for the second-order differencing. Thus, the autoregressive integrated moving average (ARIMA) model is suitable to use for the prediction. The optimized parameter (order) of ARIMA is (0, 3, 2). Table 7.3 shows the result of the ARIMA model. We found that the

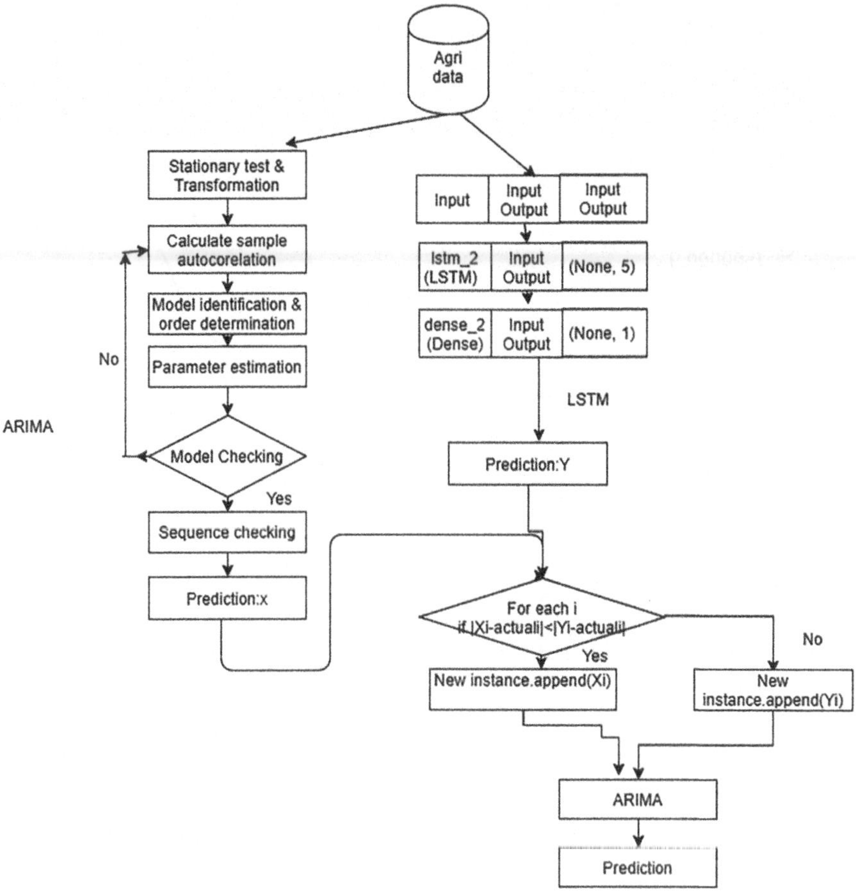

Fig. 7.1 Hybrid model

ARIMA statistics are 99% significant. This shows the approximately 9.9% growth at the lag of 2 years observed over the period, which implies that the value of the product doubled over almost 22 years. The residual errors seem fine with near zero mean and uniform variance as shown in Fig. 7.6.

We applied an LSTM model for the same data set to check the efficiency and accuracy of the prediction. The LSTM architecture comprised one input-one LSTM with five units and one dense layer output. A batch size of 72 is used for each batch of predictions and true outputs, and the mean squared error is calculated. The optimizer Adam is used. Further, it is observed that the residuals obtained through ARIMA and LSTM show a non-linear relationship. Therefore, the hybrid model was applied on the data set and observed that the prediction accuracy of the hybrid model is better than the ARIMA and LSTM. The result showed that root mean square error (RMSE) for ARIMA is 59,627. Around 3.7% MAPE implies the model is about 96.3% accurate in predicting the next 15 observations. The RMSE for the hybrid

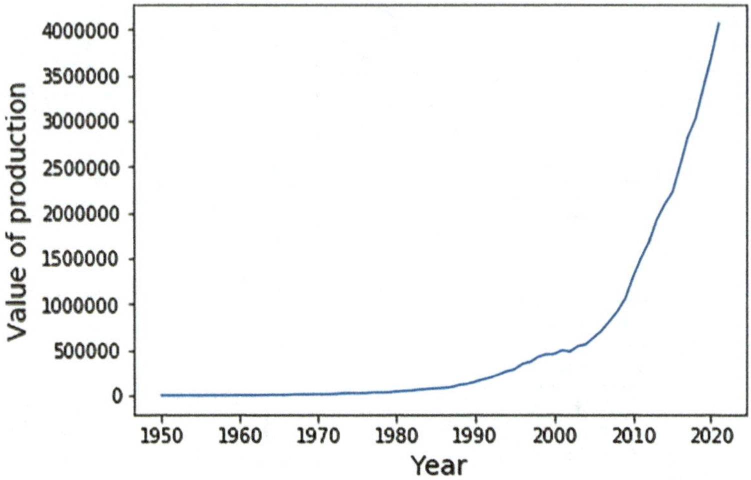

Fig. 7.2 Agri data

Table 7.2 Augmented Dickey–Fuller test results

First differentiation	Second differentiation	Third differentiation
ADF test statistic 5.878493 *P*-Value 1.000000 # Lags used 12 # Observations used 58 Critical Value (1%) –3.548494 Critical value (5%) –2.912837 Critical value (10%) –2.594129 Is the time series stationary? False	ADF test statistic 0.861765 *P*-Value 0.992571 Lags used 11.00 Observations used 58 Critical value (1%) –3.548494 Critical value (5%) –2.912837 Critical value (10%) –2.594129 Is the time series stationary? False	ADF test statistic –5.071515 *P*-Value 0.000016 # Lags used 11.00 # Observations used 57 Critical value (1%) –3.550670 Critical value (5%) –2.913766 Critical value (10%) –2.594624 Is the time series stationary? True

model is 52,734. The lowest RMSE of the hybrid model shows that the model has an advantage in predicting the value of agricultural production over the ARIMA and LSTM. Figures 7.7, 7.8, 7.9 show the comparative prediction of each model. The prediction line of the LSTM model shows the higher deviation in comparison to the prediction line of the ARIMA from the actual. The same is clearly visible in the individual graphs shown in Fig. 7.7 and in the overlapping graph depicted in Fig. 7.8. In economic analysis, the data science forecasting tools are assessed on the basis of their ability to produce good forecasts, i.e., the forecast that will be closer to the actual values.

For better visibility and comparison, an overlapping graph showing the prediction line of all three models, ARIMA, LSTM, and hybrid, is presented in Fig. 7.8. The accuracy in the prediction by the hybrid model over the two models is clearly visible

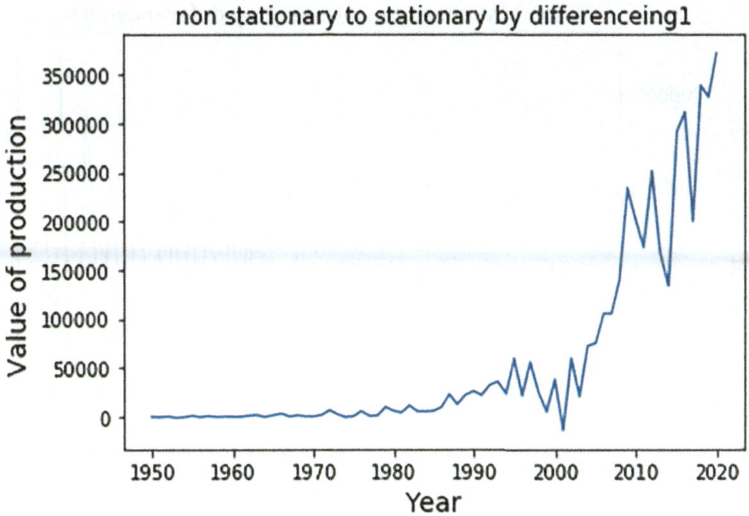

Fig. 7.3 After first differentiation

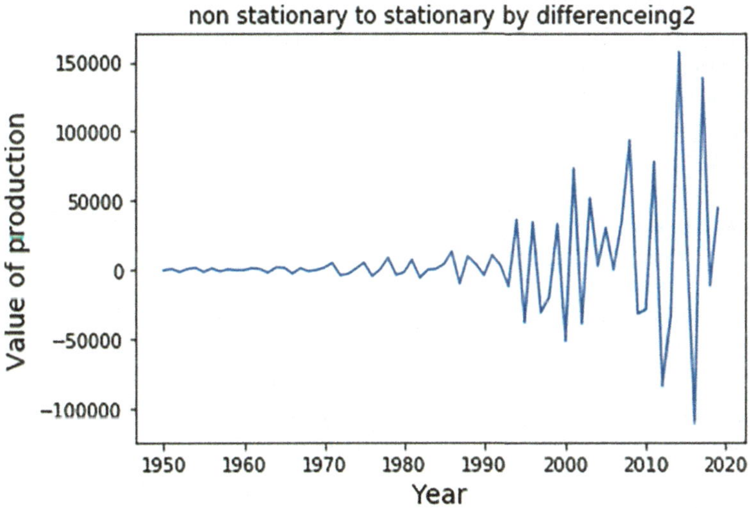

Fig. 7.4 After second differentiation

in the figure. RMSE square suggests that the hybrid model provides 11% more accurate results in comparison to ARIMA models. Several studies reported similar results for the prediction. Yung et al. reported 20% accuracy in the prediction under the hybrid model in comparison to the LSTM model for time series prediction (Yung et al. 2022). Purohit et al. have also observed that the price of agricultural products can be better predicted through the hybrid models (Purohit et al. 2021). Figure 7.9 shows that the prediction accuracy is up to a 95% confidence interval.

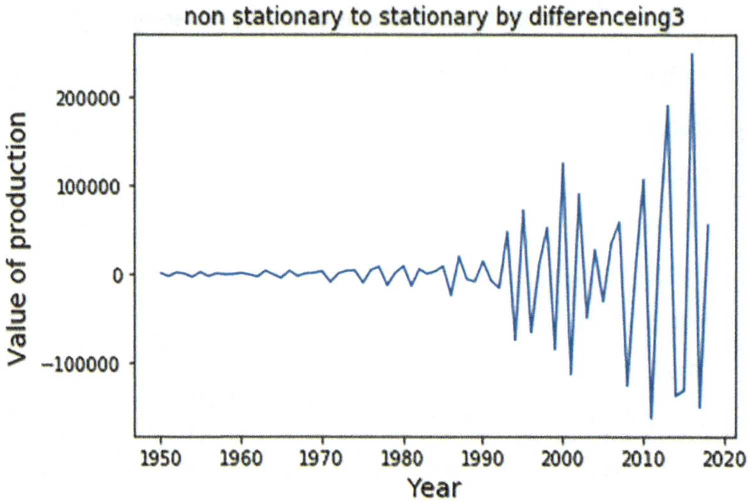

Fig. 7.5 After third differentiation

Table 7.3 ARIMA results

Dep. variable: y			No. observations: 47			
Model: ARIMA (0, 3, 2)			Log likelihood: −438.684			
Date: Wed, 06 Sep 2023			AIC: 883.367			
Time: 14:07:30			BIC: 888.720			
Sample: 0			HQIC: 885.352			
−47						
Covariance Type: opg						
	Coef	**Std err**	*z*	*P* > |z|	**[0.025**	**0.975]**
ma.L1	−1.8354	0.111	−16.567	0.000	−2.053	−1.618
ma.L2	0.9970	0.115	8.632	0.000	0.771	1.223
sigma2	2.404e+07	8.79e−09	2.73e+15	0.000	2.4e+07	2.4e+07
Ljung-Box (L1) (Q): 2.81			Jarque–Bera (JB):38.27			
Prob (Q): 0.09			Prob(JB): 0.00			
Heteroskedasticity (H): 72.47			Skew: 0.79			
Prob(H) (two-sided): 0.00			Kurtosis: 7.29			

7.4 Conclusion

We observed that the hybrid model comprising the linear and non-linear characteristics of the data is more appropriate for predicting the value of the agricultural production. The value of agricultural production is an outcome of the price of the product and volume of the production. The production level itself is dependent on the innumerable factors and the price itself is endogenous to the level of production and the demand. Therefore, the existence of non-linearity is intuitive to the analysis.

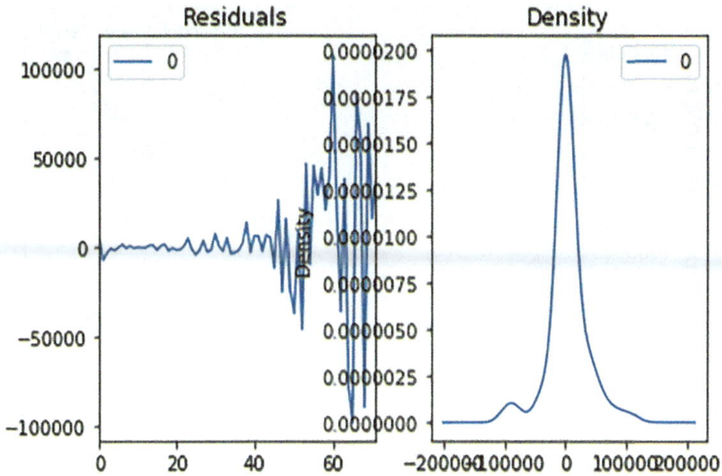

Fig. 7.6 Residuals density

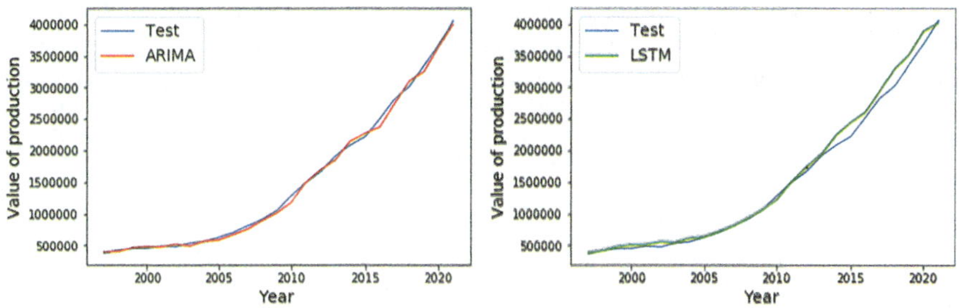

Fig. 7.7 Left graph shows comparative view of prediction of ARIMA and actual; right graph shows the comparative view of prediction of LSTM and actual

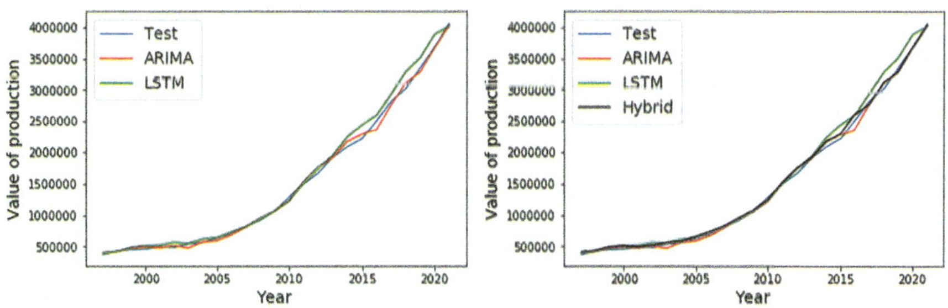

Fig. 7.8 Left graph shows comparative view of prediction of ARIMA, LSTM, and actual; right graph shows the comparative view of prediction of ARIMA, LSTM, hybrid, and actual observation

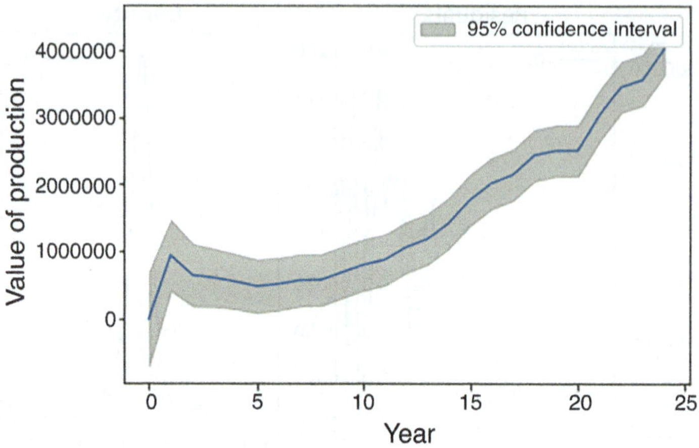

Fig. 7.9 Prediction accuracy

The study reports that the hybrid model based on the algorithm of the LSTM using the residuals of the ARIMA models is more suitable for the prediction of the value of the agricultural production. Considering the limitations of this chapter, the predictions analysis requires more complicate models that may bring together all the factors affecting the demand and supply of the agricultural product for more accurate predictions.

References

UNDP (2023) The sustainable development goals report special edition. UNDP

Hauchet-Bourdon M (2011) Agricultural commodity price volatility: an overview, OECD food, agriculture and fisheries papers, vol 52. OECD Publishing, Paris, France

PIB (2023, 21 March) "release id 1909213"

Hua Y (2020) Bitcoin price prediction using ARIMA and LSTM. In: E3S web of conferences 218, 01050, ISEESE 2020

Box B, Jenkins G (1976) Time series analysis: forecasting and control, Holden Day Series in timeseries analysis and digital signal Processing

Kalpakis K, Gada D, Puttagunta V (2000) Distance measure for effective clustering of ARIMA time series. In: IEEE Int Conf Data Min

Nguyen H, Naeem M, Wichitaksorn N, Pears R (2019) A smart system for short-term price prediction using time series models. Comput Electr Eng 76:339–352

Siami-Namini S, Tavakoli N, Namin AS (2018) A comparison of ARIMA and LSTM in forecasting time series. In: 17th IEEE international conference on machine learning and applications

Deng Y, Fan H, Wu S (2023) A hybrid ARIMA-LSTM model optimized by BP in the forecast. J Ambient Intell Humaniz Comput 14:5517–5527

Naveena K, Singh S, Rathod S, Singh A (2017) Hybrid time series modelling for forecasting the price of washed coffee (Arabica plantation coffee) in India. Int J Agric Sci 9(10):4004–4007

Khazee P, Bagherzadeh Z, Niazkhani PH (2019) Predicting the function of transplanyed kidney in long term care process: application of a hybrid model. J Biomed Inform 91

Dave E, Leonardo A, Jeanice M, Hanafiah N (2021) Forecasting Indonesia exports using a hybrid model ARIMA-LSTM. Proc Comput Sci 179:480–487

Sunil BK, Yadav N (2023) A novel hybrid model combining BSARMA and LSTM for time series forcasting. Appl Soft Comput 134

Yoo T-W, Oh I-S (2020) Time series forecasting of agricultural products' sales volumes based on seasonal long short-term memory. Appl Sci 10(22):8169

Greene WH (2003) Econometric analysis. Pearson

Hochreiter S, Schmidhuber J (1997) Long short-term memory. Neural Comput 9(8):1735–1780

Yung Y, Fan C, Xiong H (2022) A novel general-purpose hybrid model for time series forecasting. Appl Intell 52:2212–2223

Purohit SK, Panigarhi S, Sethy PK, Behera SK (2021) Time series forecasting of price of agricultural products using hybrid methods, applied artificial intelligence. Appl Artif Intell 35(15):1388–1406

Chapter 8
An Exploratory Analysis of Machine Intelligence-enabled Plant Diseases Assessment

Ashis Pattanaik, Agniva Bhattacharya, and Sushruta Mishra

Abstract Agriculture plays an important role in India due to population growth and increasing demand for food. Therefore, there is a need to increase crop yields. One of the significant effects of low crop yields is diseases caused by bacteria, fungi and viruses. This can be prevented and controlled by applying plant disease detection approaches. Machine learning techniques are used to identify plant diseases, mainly because they apply the information itself and provide better techniques for detecting plant diseases. Machine learning-based methods apply primarily to data dominance outcomes for specific tasks, and thus can be used to identify diseases. This approach provided a comprehensive overview of various techniques used for plant disease detection using artificial intelligence-based machine learning and deep learning techniques. Similarly, deep learning has become very important in the field of computer vision as it provides better performance results in plant disease detection. Advances in deep learning have been applied to many fields and have brought significant advances in machine learning and computer vision. This comparative study examines machine learning and deep learning techniques, and their performance and use in various research papers aims to show the effectiveness of deep learning and machine learning models. To prevent significant crop loss, leaf disease can be detected using captured images using deep learning techniques. Here we use Kaggle's PlantVillage dataset for study and analysis.

Keywords Agriculture · India · Population growth · Crop yields · Diseases · Bacteria · Fungi · Viruses · Plant disease detection · Machine learning

---

A. Pattanaik (✉) · A. Bhattacharya · S. Mishra
Kalinga Institute of Industrial Technology, Deemed to be University, Bhubaneswar, Odisha, India
e-mail: 2105779@kiit.ac.in; 21052221@kiit.ac.in; sushruta.mishrafcs@kiit.ac.in

A. Khamparia et al. (eds.), *Microbial Data Intelligence and Computational Techniques for Sustainable Computing*, Microorganisms for Sustainability 47,
https://doi.org/10.1007/978-981-99-9621-6_8

8.1 Introduction

The agricultural process commences with seed planting and concludes with crop harvest, encompassing challenges such as disease outbreaks, storage management, pesticide control, weed identification and management, and soil and water resource inadequacies. These categories have seen the integration of artificial intelligence (AI) and machine learning. AI advancements build on past learning, automating tasks through machine learning techniques like backpropagation, artificial neural networks, and convolutional neural networks, thereby advancing agricultural technology. The convergence of the Internet of Things (IoT), artificial intelligence (AI), and unmanned aerial vehicles (UAVs) is seamlessly harnessed to bolster agricultural domains in detecting and accurately reporting plant leaf diseases. In today's contemporary society, farming and agriculture have waned in appeal, largely due to the persistent challenges confronted by farmers on a daily basis. Consequently, the younger demographic is increasingly gravitating toward urban centers, seeking a more secure life and evading the impediments associated with agricultural pursuits. The effective safeguarding against plant diseases is intricately entwined with the imperative need for sustainable adaptations in both climate and farming practices (Loey et al. 2020). Research findings indicate that the dynamic of climate change has the potential to influence pathogenic stages and their progression rates. Moreover, alterations in host resistance mechanisms could instigate shifts in the physiological dynamics of host–pathogen interactions (Shruthi et al. 2019). In the present era, the reality that diseases now exhibit greater global mobility than ever before adds complexity to the scenario. The emergence of novel diseases in regions where they were previously absent underscores the inherent challenge of combating them in areas lacking local expertise to effectively address them (Bera et al. 2019). Plant diseases pose a dual peril: endangering worldwide food security and particularly devastating smallholder farmers who rely on thriving crops. Developing countries rely heavily on these farmers, with over 80% of their agricultural output stemming from their efforts (UNEP 2013). Irrespective of the methodology employed, precise disease identification upon initial emergence remains pivotal for effective disease control. Previously, agricultural extensions and local plant clinics aided disease identification. Now, digital resources supplement this process by offering online disease diagnosis information, capitalizing on global Internet accessibility. Moreover, mobile-based tools have swiftly gained traction, capitalizing on the unprecedented global adoption of mobile technology (ITU 2015). Smartphones, with their computational prowess, high-definition screens, and advanced built-in features like HD cameras, present innovative opportunities for disease identification. The projected global smartphone count is anticipated to range from 5 to 6 billion by 2020. In 2015, mobile broadband coverage was available to 69% of the world's population, showcasing a remarkable 12-fold surge in mobile broadband penetration from 2007 to 2015, reaching 47% (ITU 2015). The convergence of prevalent smartphone usage, advanced HD cameras, and powerful mobile processors creates an environment where automated disease diagnosis through image recognition can achieve an unprecedented reach, if technically viable. We showcase this feasibility by employing deep learning with 54,306 images of 14 crop species and 26 diseases (or health) from the PlantVillage project (Hughes and Salathé 2015). In various domains, deep neural networks have proven effective for end-to-end learning, exemplifying their versatility. They establish input-to-output relationships, like mapping diseased plant images to crop–disease pairs. Nodes in these networks act as mathematical functions, transforming inputs from incoming edges into numerical outputs through outgoing edges. Deep networks sequentially connect input and output layers through stacked nodes. The task is to craft a network where structure, functions (nodes), and edge weights accurately align input with output. Training deep neural networks involves refining parameters to enhance mapping through demanding computation, with recent advancements greatly enhancing efficiency through conceptual and engineering breakthroughs (LeCun et al. 2015; Schmidhuber 2015).

The main objectives of the work are as follows:

- AI-Powered Disease Detection: Develop a robust AI-based diagnostic system that accurately identifies plant diseases using machine learning techniques.
- Automated and Efficient Assessment: Streamline the diagnosis process through automation, facilitating rapid and effective plant health evaluation.
- Enhanced Crop Productivity: Improve crop yield by enabling early disease detection and intervention, contributing to healthier and more productive crops.
- Sustainable Agriculture: Promote environmentally friendly practices by reducing pesticide use through targeted disease identification.
- Holistic Integration: Create an adaptable, user-friendly interface that supports diverse crops, real-time monitoring, and practical application in agriculture.

8.2 Literature Review

In our evolving environment, the timely and precise recognition of diseases, encompassing early intervention, holds heightened importance. Diverse methods detect plant pathologies. While certain diseases lack immediate symptoms or are identified too late, meticulous scrutiny becomes necessary. Yet, many ailments exhibit visible signs, making professional evaluation the predominant detection approach. Skillful observation by plant pathologists is essential for identifying distinct symptoms and achieving precise plant disease diagnosis (Shirahatti et al. 2018). Variations in symptoms exhibited by diseased plants can lead to erroneous diagnoses, particularly challenging for non-experts compared to professional pathologists. Novice gardeners and seasoned specialists alike can gain substantial advantages from an automated system designed to identify plant conditions. The system utilizes visual cues and appearance, serving as a confirmation for disease diagnosis accuracy (Ganatra and Patel 2018). Advancements in computer vision present avenues for enhancing and fortifying the precision of plant protection, while also expanding the scope of precise agricultural computer vision applications. For the purpose of identifying and categorizing plant diseases, widely adopted techniques in digital image processing, such as color detection and thresholding, were employed (Ganatra and Patel 2018). Novel methodologies in deep learning are actively employed for identifying plant diseases, with convolutional neural networks (CNN) emerging as a prominent choice. Deep learning represents a contemporary machine learning paradigm, boasting exceptional outcomes in diverse realms like computer vision, pharmacy, and bioinformatics. This approach capitalizes on its ability to process raw data directly, bypassing the need for manual feature engineering (Ganatra and Patel 2018). Deep learning has garnered substantial success in both academic and industrial contexts due to two primary factors (Athanikar and Badar 2016). Firstly, the continual generation of vast datasets provides ample material for constructing intricate models. Secondly, the computational prowess of graphics processing units (GPUs) empowers the training and utilization of deep models through efficient parallel processing (Kumar and Raghavendra 2019). The paper by Ashourloo et al. delves into the detection of leaf rust disease using hyperspectral measurements. It explores various machine learning regression techniques for this purpose. Published in the Institute of Electrical and Electronics Engineers (IEEE) *Journal of Selected Topics in Applied Earth Observations and Remote Sensing*, the study aims to enhance disease detection through hyperspectral analysis (Ashourloo et al. 2016). The paper authored by Han, Haleem, and Taylor introduces an innovative computer vision-driven method for the automated detection and evaluation of crop diseases. Presented at the Science and Information Conference in July 2015, the approach utilizes computer vision techniques to identify and assess the severity of various crop diseases, thereby contributing to agricultural disease management (Han et al. 2015). The paper by Barbedo presents an innovative algorithm for semi-automatic segmentation of plant leaf disease symptoms via digital image processing. Published in *Tropical Plant Pathology* in 2016, the algorithm offers a novel approach to accurately segment and analyze disease symptoms on plant leaves using digital images (Barbedo 2016). The paper by Ramesh and Vardhan explores data mining methods applied to agricultural yield data in the *International Journal of Advanced Research in Computer Communication and Engineering* (2013). The study investigates techniques and applications of data mining in analyzing agricultural yield data, aiming to uncover insights and patterns within this context (Ramesh and Vardhan 2013). The research paper by Xia, Li, and Li introduces an intelligent diagnostic system for identifying wheat diseases using an Android phone. Published in the *Journal of Information & Computer Science* in December 2015, the paper presents a system designed to provide accurate diagnosis of wheat diseases through mobile technology (Xia et al. 2015). The paper by Kaur and Kang focuses on improving plant disease detection by enhancing the classifier support vector machine (SVM). It was presented at the IEEE third International Conference on MOOCs, Innovation, and Technology in Education (MITE) in 2015. The study aims to enhance the performance of SVM for more accurate plant disease detection (Kaur and Kang 2015). The research paper by E.C. Too and colleagues explores plant disease identification using deep learning models. It involves a comparative analysis of fine-tuning techniques for these models. Published in *Computers and Electronics in Agriculture* in 2019, the study delves into improving the accuracy of plant disease recognition through the refinement of deep learning approaches (Too et al. 2019). In the realm of plant species classification, Dyrmann et al. 2016 employed deep learning architectures. Their study introduced a technique capable of distinguishing between weeds and plant species using colored images. They harnessed convolutional neural networks (CNNs), evaluating their model on 10,413 images encompassing 22 weed and crop species. The CNN model demonstrated an accuracy of 86.2%, albeit encountering challenges in classifying specific plant species due to limited training samples for those categories. Plant disease classification has also seen the utilization of machine learning methods. In the work by Athanikar and Badar (2016), a neural network was employed to differentiate between healthy and diseased potato leaf images. The outcomes demonstrated the successful capability of the backpropagation neural network (BPNN) in identifying disease spots and accurately categorizing the specific disease type, achieving an impressive accuracy rate of 92%. The research paper by Sankur and Sezgin (2004) presents a comprehensive survey of image thresholding techniques and evaluates their quantitative performance. The study explores various methods for thresholding images and analyzes their effectiveness. The authors discuss a range of thresholding techniques and their applications, considering their advantages and limitations. The paper also delves into the quantitative evaluation of these methods, providing insights into their performance metrics. Overall, the paper serves as a valuable resource for understanding and assessing image thresholding techniques, contributing to the field of electronic imaging. Furthermore, in a study conducted by Wang et al. (2012), an empirical investigation was undertaken with the aim of developing an approach for achieving image recognition of plant diseases. This research involved the utilization of four distinct neural networks to differentiate between wheat stripe rust and wheat leaf rust, as well as between grape downy mildew and grape powdery mildew. These neural networks relied on extracted color, shape, and texture features from the disease images. The outcomes of the study demonstrated the successful utilization of image processing-based neural networks for the efficient identification and diagnosis of plant diseases. Furthermore, in their work from 2012, Samanta and colleagues introduce an image processing technique for identifying scab disease in potatoes. They gather images from diverse potato fields and subject them to enhancement processes. Through image segmentation, they isolate the specific areas of interest, which correspond to the regions affected by the disease. Ultimately, they employ a histogram-based approach to analyze these targeted regions and determine the disease stage (Samanta et al. 2018). The research paper by Han et al. (2015) presents a new computer vision-based method for automatically detecting and evaluating the severity of crop diseases. The approach was showcased at the Science and Information Conference. The authors introduce a technique that utilizes computer vision to identify and quantify the extent of disease in crops. This method likely involves innovative strategies for leveraging visual information to assess the severity of diseases impacting crops. The study contributes to the field of agriculture and technology by proposing a potentially advanced solution for addressing crop disease issues through automated visual analysis (Han et al. 2015). The research paper by Dey et al. (2016) focuses on the detection of leaf rot disease in betel vine (Piper betle) using image processing techniques. Presented at the International Conference on Computational Modeling and Security, the paper introduces a methodology to identify leaf rot disease in betel vine leaves. The authors utilize image processing methods to analyze visual characteristics associated with the disease. The study's findings demonstrate the effectiveness of their approach in detecting leaf rot disease accurately, showcasing its potential for disease diagnosis in agricultural contexts (Dey et al. 2016). The research paper titled "eAGROBOT-A robot for early crop disease detection using image processing" by Pilli et al. (2014) introduces the concept of an automated agricultural robot called eAGROBOT. The robot is designed for the early detection of crop diseases through the utilization of image processing techniques. The paper presents this innovative approach, highlighting the integration of electronics, communication systems, and image processing to develop a system that can identify and diagnose crop diseases at an early stage. The research emphasizes the potential of such technology in improving crop health monitoring and contributing to more efficient and sustainable agriculture practices. The paper was presented at the International Conference on Electronics and Communication Systems (ICECS) in 2014 (Pilli et al. 2014). The research paper titled "Using Deep Learning for Image-Based Plant Disease Detection" by S. P. Mohanty, D. P. Hughes, and M. Salathé was published in 2016 in the journal *Frontiers in Plant Science*. The paper focuses on leveraging deep learning techniques for the detection of plant diseases using images. The authors explore the potential of deep learning models, particularly convolutional neural networks (CNNs), to accurately identify plant diseases through image analysis. The study contributes to the field of agricultural technology by demonstrating the effectiveness of using advanced machine learning techniques for automating the diagnosis of plant diseases based on visual cues. The paper's findings underscore the significance of deep learning in improving disease detection accuracy, thereby aiding in timely and efficient disease management in agriculture (Mohanty et al. 2016). The paper by Soni (2018) presents an enhanced faster regional convolutional neural network for monitoring data centers. Published in the *International Journal of Advanced Research*, it focuses on improving efficiency and accuracy in data center monitoring using neural network techniques (Soni 2018). The paper by Ramcharan et al. (2017) presents deep learning techniques for detecting cassava diseases from images. It explores image-based methods for cassava disease identification, contributing to improved disease management in agriculture. Published in *Frontiers in Plant Science*, the research focuses on leveraging deep learning for accurate disease detection in cassava plants (Ramcharan et al. 2017). The paper by Fujita et al. (2016) presents a foundational study on a reliable and feasible plant diagnostic system. It was discussed at the 15th IEEE International Conference on Machine Learning and Applications. The research delves into creating a robust plant diagnosis approach, offering insights and practical implications (Fujita et al. 2016). The research paper by Huang (2007) focuses on the application of artificial neural networks to detect diseases in phalaenopsis seedlings. The study utilizes color and texture features to identify and diagnose plant diseases. By employing an artificial neural network, the research aims to enhance the accuracy and efficiency of disease detection in agricultural settings. The paper explores the potential of using computational methods to improve disease diagnosis in phalaenopsis seedlings, offering insights into the integration of technology and agriculture for improved plant health management (Huang 2007). The research paper by Harvey et al. (2014) explores the significant vulnerability of smallholder farmers in Madagascar to agricultural risks and climate change. The study emphasizes the extreme challenges these farmers face due to their dependence on agriculture for their livelihood. It investigates how changing climatic conditions exacerbate existing risks, impacting crop yields and food security. The research underscores the urgent need for targeted interventions and adaptive strategies to address the vulnerability of these farmers, considering the complex interplay of environmental, economic, and social factors. The findings highlight the importance of sustainable agricultural practices and support systems to enhance the resilience of Madagascar's smallholder farming communities in the face of ongoing climate challenges (Harvey et al. 2014). The research paper titled "ImageNet Large Scale Visual Recognition Challenge" by Russakovsky et al. (2015) presents a comprehensive overview of the ImageNet Challenge, a significant competition in computer vision. The paper discusses the challenge's objectives, dataset, evaluation metrics, and results. It highlights the rapid advancement in deep learning techniques and their impact on image classification and object recognition tasks. The authors emphasize the importance of large-scale datasets and high-capacity neural networks in achieving remarkable performance improvements. The paper's findings underscore the transformative influence of the ImageNet Challenge in driving innovation and progress within the field of computer vision (Russakovsky et al. 2015). The research paper by Sankaran et al. (2011) explores the application of visible-near infrared spectroscopy to detect huanglongbing (HLB) disease in citrus orchards. The study focuses on using spectral analysis to identify disease symptoms. The authors utilize computational and electronic agriculture techniques, demonstrating the potential of this approach for disease detection in citrus trees. The paper discusses the methodology, findings, and implications of using spectroscopy as a tool for disease identification in the context of citrus cultivation (Sankaran et al. 2011). The research paper titled "Going Deeper with Convolutions" by Szegedy et al. (2015) focuses on enhancing convolutional neural networks (CNNs) for computer vision tasks. The paper introduces a novel architecture called GoogLeNet, which employs deep and parallel convolutional layers to achieve better performance with fewer parameters. The key innovation is the utilization of "Inception" modules, which combine different kernel sizes within the same layer to capture diverse features at various scales. This approach enables more efficient and deeper networks while mitigating overfitting. The research paper's findings demonstrate that the proposed GoogLeNet architecture outperforms previous CNNs on benchmarks like the ImageNet dataset, showcasing the advantages of increased depth and the Inception modules for improving image recognition tasks (Szegedy et al. 2015). The research paper titled "A Comparative Study on Application of Computer Vision and Fluorescence Imaging Spectroscopy for Detection of Huanglongbing Citrus Disease in the USA and

Brazil" by Wetterich et al. (2012) explores the utilization of computer vision and fluorescence imaging spectroscopy for detecting the citrus disease known as huanglongbing (HLB) in both the USA and Brazil. The study compares the effectiveness of these two techniques in identifying HLB in citrus plants. The research investigates the potential of these technologies for disease detection and provides insights into their application in two different geographic regions. The paper was published in the *Journal of Spectroscopy* in 2013 (841738) and contributes to the field of spectroscopy and computer vision for disease diagnosis in agriculture, with implications for disease management and prevention in citrus cultivation. The study's findings offer valuable information for researchers, practitioners, and policymakers aiming to combat HLB and enhance citrus crop health in diverse contexts (Wetterich et al. 2012). The research paper by Zeiler and Fergus (2014) titled "Visualizing and Understanding Convolutional Networks" explores the visualization and comprehension of convolutional networks in the field of computer vision. Published in the European Conference on Computer Vision 2014 proceedings, the paper delves into methods for interpreting the behavior of deep convolutional neural networks (CNNs). The authors discuss techniques to visualize the learned features and activations within the CNN layers, shedding light on how these networks process and recognize visual patterns. The paper contributes to enhancing the interpretability and transparency of CNNs, offering insights into their inner workings and aiding researchers in understanding the mechanisms behind their impressive performance in image recognition tasks (Zeiler and Fergus 2014). In the research paper authored by Singh et al. (2015), the authors explore the application of machine learning techniques for efficient stress phenotyping in plants with high throughput. They address the challenge of accurately and rapidly assessing plant stress responses using advanced computational methods. The study emphasizes the potential of machine learning to analyze and interpret large-scale datasets, aiding in identifying stress patterns and facilitating agricultural advancements. The paper highlights the role of machine learning in enhancing plant stress research and its implications for future agricultural practices. Published in *Trends in Plant Science*, the work underscores the significance of technological innovation in plant biology (Singh et al. 2015). In the research paper titled "Metaheuristic optimization-based resource allocation technique for cybertwin-driven 6G on IoE environment," authored by D. Kumar Jain, S. Kumar Sah Tyagi, S. Neelakandan, M. Prakash, and L. Natrayan, published in the IEEE's *Transactions on Industrial Informatics* (2021), a novel approach to resource allocation for cybertwin-driven 6G networks in the Internet of Everything (IoE) environment is presented. The authors propose a metaheuristic optimization method to efficiently allocate resources, enhancing the performance of cybertwin applications in the dynamic IoE setting. The study's findings contribute to the advancement of resource management strategies in the context of emerging 6G technologies (Kumar Jain et al. 2021). In the research paper authored by B. Lurstwut and C. Pornpanomchai titled "Image analysis based on color, shape, and texture for rice seed (Oryza sativa L.) germination evaluation," published in the journal *Agriculture and Natural Resources* in 2017, the authors delve into an innovative approach to assessing the germination of rice seeds (Oryza sativa L.) through comprehensive image analysis techniques. The study addresses the critical need for accurate and efficient methods to evaluate seed germination, a pivotal process in agricultural productivity. The authors focus on the integration of color, shape, and texture analysis to provide a holistic assessment of seed germination. This novel approach aims to enhance the accuracy of germination evaluation and enable better decision-making in agricultural practices. Throughout the study, the authors meticulously detail the methodology employed. Image analysis algorithms are applied to capture and analyze various visual aspects of the germinating rice seeds. Color variations, morphological changes in shape, and textural alterations are quantitatively assessed. The authors gather extensive data and subsequently utilize statistical techniques to draw meaningful conclusions from the acquired information. The results of this research exhibit promising outcomes. The combined analysis of color, shape, and texture proves to be a potent tool in evaluating rice seed germination. The comprehensive approach not only enhances accuracy but also offers a more comprehensive understanding of the germination process. The paper discusses the implications of these findings for agricultural practices, underscoring the potential to optimize seed selection and crop management strategies. In conclusion, Lurstwut and Pornpanomchai's research paper showcases an innovative image analysis approach for assessing rice seed germination. By integrating color, shape, and texture analysis, the authors provide a comprehensive and accurate means of evaluating this crucial agricultural process. The findings contribute to advancing agricultural practices and hold the promise of improving crop yield and quality (Lurstwut and Pornpanomchai 2017). In the research paper titled "Smart Paddy Crop Disease Identification and Management using Deep Convolution Neural Network and SVM Classifier," authored by R. Rajmohan, M. Pajany, R. Rajesh, D. R. Raman, and U. Prabu and published in the *International Journal of Pure and Applied Mathematics* in 2018, the authors address the crucial issue of paddy crop disease identification and management through the integration of advanced machine learning techniques. The study focuses on utilizing a combination of deep convolutional neural networks (CNN) and support vector machine (SVM) classifier to effectively detect and manage diseases in paddy crops. Paddy crops are susceptible to various diseases that can significantly impact crop yield and quality. Early detection and management of these diseases are vital to ensure agricultural productivity and food security. The authors propose a novel approach that leverages the strengths of both deep learning and traditional machine learning. Deep CNNs are employed to process images of paddy leaves and extract relevant features that can aid in disease identification. CNNs are known for their capability to automatically learn hierarchical features from images, making them suitable for image recognition tasks. Following the feature extraction stage, the study integrates an SVM classifier into the framework. SVMs are known for their ability to classify data points into different classes based on their features. In this context, SVMs are applied to classify the extracted features and subsequently identify the specific disease affecting the paddy crops. The proposed methodology offers several advantages. By combining CNNs and SVMs, the approach harnesses the power of deep learning's feature extraction with SVM's classification capabilities. This enhances the accuracy and robustness of disease identification. The integration of advanced machine learning techniques offers a smart solution for timely disease management, helping farmers make informed decisions regarding pest control and crop protection measures. The research contributes to the agricultural domain by presenting an innovative strategy for paddy crop disease identification and management. The experimental results, outlined in the paper, demonstrate the effectiveness of the proposed approach in accurately identifying crop diseases based on leaf images. This research paper underscores the potential of machine learning technologies in revolutionizing agriculture by offering precise and timely disease detection mechanisms, thereby contributing to sustainable food production and economic stability in farming communities (Rajmohan et al. 2018). In the research paper titled "Rice Plant Disease Classification Using Transfer Learning of Deep Convolutional Neural Network," authored by V. K. Shrivastava, M. K. Pradhan, S. Minz, and M. P. Thakur, and published in the *International Archives of the Photogrammetry, Remote Sensing, and Spatial Information Sciences*, Volume 42 in 2019, the authors investigate the efficacy of transfer learning in the domain of rice plant disease classification. The study addresses the crucial issue of rice plant disease detection, which has significant implications for crop yield and food security. Traditional methods of disease identification often involve manual inspection and visual assessment, leading to delays and inaccuracies. To overcome these limitations, the researchers propose the application of deep convolutional neural networks (CNNs) with transfer learning. Transfer learning, a technique wherein a pre-trained neural network is fine-tuned for a specific task, is leveraged by the authors to enhance the accuracy of disease classification. The researchers choose a pre-trained CNN model and tailor it to recognize different types of rice plant diseases by retraining the model with a specific dataset containing images of healthy and diseased rice plants. The experimental methodology involves collecting a comprehensive dataset of rice plant images, including healthy plants and plants affected by various diseases. This dataset is utilized for training, validation, and testing of the CNN model. The authors employ various performance metrics to assess the model's accuracy in identifying different diseases accurately. The results of the study demonstrate the effectiveness of the proposed approach. The trained CNN model exhibits notable accuracy in classifying various rice plant diseases, outperforming traditional methods of disease detection. The application of transfer learning aids in overcoming the challenges of limited data and computationally intensive training, allowing for better generalization of disease patterns. The implications of this research are significant for the agricultural sector, as it offers a more efficient and accurate means of diagnosing rice plant diseases. Timely and accurate disease detection can help farmers take targeted measures to prevent the spread of diseases and optimize crop management strategies. In conclusion, V. K. Shrivastava, M. K. Pradhan, S. Minz, and M. P. Thakur's research paper explores the application of transfer learning using deep convolutional neural networks for rice plant disease classification. The study underscores the potential of this approach to revolutionize disease detection in the agricultural domain, contributing to improved crop health management and food security (Shrivastava et al. 2019). In the research paper authored by R. Islam and M. Rafiqul, titled "An image processing technique to calculate the percentage of disease-affected pixels of paddy leaf," published in the *International Journal of Computer Applications* in 2015, a novel approach for quantifying the extent of disease impact on paddy leaves through image processing is presented. The study addresses the critical concern of disease detection in agricultural crops, specifically focusing on paddy leaves. Timely and accurate identification of diseases is essential to ensure effective crop management and yield optimization. The authors propose an innovative image processing technique that offers a quantitative assessment of the percentage of disease-affected pixels on paddy leaves. The methodology centers around leveraging digital images of paddy leaves, captured using appropriate imaging equipment. These images are then subjected to a series of image processing steps to distinguish between healthy and disease-affected areas. The process involves segmenting the leaf images into distinct regions and isolating the portions exhibiting signs of disease. Through a combination of pixel intensity analysis and pattern recognition algorithms, the authors accurately quantify the ratio of disease-affected pixels to the total number of pixels in the leaf image. The experimental validation of the proposed technique is conducted to assess its effectiveness. A comprehensive dataset of paddy leaf images with varying degrees of disease manifestation is employed. By comparing the results obtained from the proposed method with manually annotated ground truth data, the authors demonstrate the technique's proficiency in accurately estimating the percentage of disease-affected pixels. The outcomes of this research hold significant implications for agricultural practices and crop management strategies. The automated and quantitative nature of the proposed image processing technique offers a streamlined approach to disease assessment, eliminating the subjective biases associated with manual inspection. This, in turn, enables timely interventions, facilitating improved disease control and ultimately enhancing crop yield. In conclusion, the paper authored by R. Islam and M. Rafiqul outlines an innovative image processing technique for evaluating the proportion of disease-affected pixels on paddy leaves. By harnessing the power of digital image analysis, this method presents a valuable contribution to the field of agricultural research and offers a promising avenue for the advancement of precision farming practices (Islam and Rafiqul 2015). In the research paper authored by P. K. Sethy, B. Negi, N. K. Barpanda, S. K. Behera, and A. K. Rath and published in 2018 in the journal *Cognitive Science and Artificial Intelligence* by Springer, the focus is on evaluating the severity of disease in rice crops utilizing machine learning and computational intelligence techniques. The study falls under the domain of applied sciences and technology, specifically categorized as "Springer Briefs." The primary objective of the research is to develop an effective method for measuring the severity of diseases that affect rice crops. Agricultural diseases have a significant impact on crop yield and food security, and accurate assessment of disease severity is crucial for implementing appropriate management strategies. The researchers employ advanced machine learning and computational intelligence methodologies to address this challenge. By leveraging the power of these techniques, they aim to create a model capable of accurately assessing the severity of diseases in rice plants. The integration of these modern technologies holds promise in providing a reliable and efficient solution to this pertinent agricultural concern. The paper outlines the methodology used in the study, which involves collecting data related to various parameters associated with the health of rice plants. These parameters are then processed using machine learning algorithms and computational intelligence techniques. The resulting model is designed to recognize patterns and correlations between the collected data and disease severity levels. The outcomes of the research showcase the successful implementation of the proposed approach. The developed model demonstrates a high degree of accuracy in assessing disease severity in rice crops. This achievement implies a significant step forward in early disease detection and management strategies for agricultural systems. In conclusion, P. K. Sethy, B. Negi, N. K. Barpanda, S. K. Behera, and A. K. Rath's2018 research paper contributes to the field of applied sciences and technology. Their study presents a comprehensive approach that combines machine learning and computational intelligence to accurately measure the severity of diseases in rice crops. The success of their methodology underscores the potential of these technologies in revolutionizing disease management within agriculture, thereby contributing to increased crop yield and food security (Sethy et al. 2018).

8.3 Working Methodology and Datasets

This study focuses on the domain of agriculture and plant health. It aims to develop an AI-based system for diagnosing plant diseases using technologies like TensorFlow and OpenCV. The research employs a diverse and well-annotated dataset of plant images. This dataset comprises images of healthy plants as well as plants afflicted by various diseases. The dataset covers a range of plant species and disease types to ensure robustness and accuracy in the AI model's diagnosis. Kaggle's datasets of PlantVillage were used for this study. The steps of the workflow model are highlighted in Fig. 8.1 and are discussed here.

- Image Preprocessing: Raw dataset images are processed using OpenCV, including resizing and normalization, to prepare input data for the AI model.
- CNN Architecture: A convolutional neural network (CNN) is developed using TensorFlow. CNNs excel in image recognition, capturing spatial features.
- Transfer Learning: Pre-trained CNN models (VGG, ResNet, Inception) are fine-tuned on the plant disease dataset, leveraging existing knowledge and saving training time.
- Data Augmentation: Training benefits from techniques like rotation, flipping, and zooming, enhancing model robustness by artificially expanding the dataset.
- Model Training and Validation: The CNN is trained on training data, validated on a separate set, and hyperparameters are tuned for optimal performance.

The research demonstrates AI's potential in transforming plant disease diagnosis for agricultural professionals through speed, accuracy, and accessibility

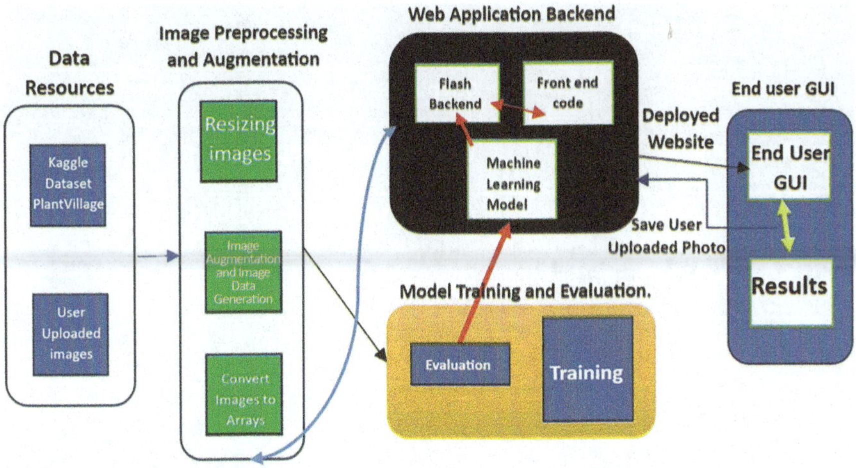

Fig. 8.1 Architectural model for plant disease detection using Tensorflow and OpenCV

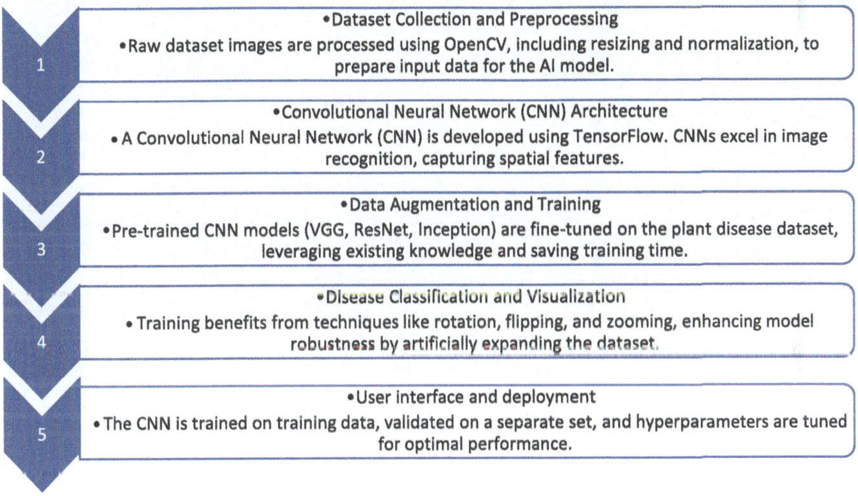

Fig. 8.2 Process flow diagram of the framework

improvements, facilitated by TensorFlow and OpenCV integration. The process flow diagram of the model is shown in Fig. 8.2.

The overall summary of the parameters taken into consideration is shown in Fig. 8.3. Different layers including max_pooling and dropout layers along with their output metrics are displayed. Configuration for activation function and batch normalization is also highlighted.

Table 8.1 illustrates the overall evaluation outcome of implementation. Different epoch combinations are considered and their corresponding values for accuracy,

```
Model: "sequential"
_________________________________________________________________
Layer (type)                Output Shape              Param #
=================================================================
conv2d (Conv2D)             (None, 100, 100, 32)      896
_________________________________________________________________
activation (Activation)     (None, 100, 100, 32)      0
_________________________________________________________________
batch_normalization (BatchNo (None, 100, 100, 32)     128
_________________________________________________________________
max_pooling2d (MaxPooling2D) (None, 33, 33, 32)       0
_________________________________________________________________
dropout (Dropout)           (None, 33, 33, 32)        0
_________________________________________________________________
conv2d_1 (Conv2D)           (None, 33, 33, 64)        18496
_________________________________________________________________
activation_1 (Activation)   (None, 33, 33, 64)        0
_________________________________________________________________
batch_normalization_1 (Batch (None, 33, 33, 64)       256
_________________________________________________________________
conv2d_2 (Conv2D)           (None, 33, 33, 64)        36928
_________________________________________________________________
activation_2 (Activation)   (None, 33, 33, 64)        0
_________________________________________________________________
batch_normalization_2 (Batch (None, 33, 33, 64)       256
...
Total params: 8,725,711
Trainable params: 8,722,703
Non-trainable params: 3,008
```

Fig. 8.3 Parameter metrics of the model

validation loss, and other metrics are validated. With 35 epochs, the model records the optimum accuracy rate as shown in Table 8.1.

8.4 Advantages and Constraints

Some Vital Benefits Offered by the Model in Plant Disease Diagnosis:

- Effective Feature Extraction: CNNs are adept at automatically extracting relevant features from plant images, capturing intricate patterns and textures indicative of diseases.

Table 8.1 Evaluation metrics with their values used in the model

Epoch number	step—loss	accuracy	val_loss	**val_accuracy**
1/35	0.0507	0.8543	0.2492	0.5650
2/35	0.0428	0.8785	0.2190	0.6306
3/35	0.0375	0.8896	0.1460	0.6936
4/35	0.0358	0.8944	0.2195	0.6580
5/35	0.0348	0.9020	0.3374	0.5609
6/35	0.0392	0.8882	0.2017	0.6129
7/35	0.0348	0.9024	0.1319	0.7415
8/35	0.0357	0.8986	0.3076	0.5404
9/35	0.0442	0.8744	0.1168	0.7524
10/35	0.0401	0.8872	0.6525	0.3694
11/35	0.0356	0.9024	0.5158	0.4104
12/35	0.0350	0.8955	0.1278	0.7579
34/35	0.0255	0.9290	0.4432	0.4651
35/35	0.0238	0.9311	0.0950	0.8098

- High Accuracy: CNNs' deep architecture enables them to learn complex relationships, resulting in accurate disease identification and reduced misdiagnoses.
- Generalization: Trained CNNs can generalize well to new, unseen images, making them adaptable to various plant species and disease types.
- Efficiency: Once trained, CNN models can process images quickly, allowing for rapid diagnosis and timely intervention.
- Automation: CNN-based systems automate diagnosis, reducing human effort and enabling non-experts to identify diseases.

A Few Constraints Visible in the Model for Plant Disease Diagnosis:

- Large Datasets Needed: CNNs require substantial labeled data for effective training, which might be challenging to acquire and annotate.
- Overfitting: Without proper regularization techniques, CNNs can overfit to the training data, resulting in poor performance on new images.
- Computationally Intensive: Training deep CNNs demands significant computational resources, making implementation and maintenance costly.
- Domain-Specific Knowledge: CNNs may lack the ability to provide insights into the underlying biology of diseases, requiring additional expert interpretation.
- Limited Interpretability: The internal workings of CNNs can be challenging to interpret, hindering the understanding of how decisions are made.
- Model Complexity: Complex CNN architectures might require experts for proper design, training, and fine-tuning.
- Overall, while CNNs offer remarkable potential for accurate and efficient plant disease diagnosis, careful consideration of their limitations and proper training strategies is essential for successful implementation.

8.5 Challenging Issues of the Framework

- Dataset Challenges: Obtaining a diverse and well-annotated dataset encompassing various plant species and diseases is complex. Ensuring precise annotations and representative samples affects model generalization.
- Generalization Complexity: Achieving model adaptability to varying conditions like lighting and setups is critical. Navigating new disease variations requires meticulous training and validation strategies.
- Imbalanced Classes: Uneven class distribution can skew model performance. Detecting rare diseases is hindered by limited data in certain categories.
- Deployment and User Interface: Designing an intuitive interface for farmers while maintaining real-time performance poses deployment challenges.
- Ethical Considerations: AI's entrance into agriculture raises privacy, ownership, and expertise concerns. Responsible AI usage and addressing societal implications are vital.

8.6 Conclusion

In conclusion, the integration of convolutional neural networks (CNNs) and artificial intelligence (AI) has ushered in a transformative era in plant disease diagnosis. This approach demonstrates remarkable accuracy, automating the detection of diseases by effectively learning intricate visual patterns from plant images. While challenges like data availability and model complexity exist, the potential benefits in terms of timely intervention, resource optimization, and enhanced agricultural practices are undeniable. As AI and CNN technologies continue to advance, their application in plant disease diagnosis holds the promise of revolutionizing the field, contributing to global food security and sustainable agriculture.

References

Ashourloo D, Aghighi H, Matkan AA, Mobasheri MR, Rad AM (2016) An investigation into machine learning regression techniques for the leaf rust disease detection using hyperspectral measurement. IEEE J Sel Top Appl Earth Obs Remote Sens 9(9):4344–4351

Athanikar G, Badar MP (2016) Potato leaf diseases detection and classification system. Int J Comput Sci Mob Comput 5(2):76–88

Barbedo JGA (2016) A novel algorithm for semi-automatic segmentation of plant leaf disease symptoms using digital image processing. Trop Plant Pathol 41(4):210–224

Bera T, Das A, Sil J, Das AK (2019) A survey on rice plant disease identification using image processing and data mining techniques emerging technologies in data mining and information security. Springer, pp 365–376

Dey AK, Sharma M, Meshram MR (2016) Image processing based leaf Rot disease, detection of betel vine (piper BETLEL.), international conference on computational modeling and security. Proc Comput Sci 85:748–754

Dyrmann M et al (2016) Plant species classification using deep convolutional neural network. Biosyst Eng 151:72–80

Fujita E, Kawasaki Y, Uga H, Kagiwada S, Iyatomi H (2016) Basic investigation on a robust and practical plant diagnostic system. In Proceedings of 2016 15th IEEE international conference on machine learning and applications (ICMLA), pp 989–992

Ganatra N, Patel A (2018) A survey on diseases detection and classification of agriculture products using image processing and machine learning. Int J Comput Appl 180:1–13

Han L, Haleem MS, Taylor M (2015) A novel computer vision-based approach to automatic detection and severity assessment of crop diseases. In Science and Information Conference (SAI), pp 638–644

Harvey CA, Rakotobe ZL, Rao NS, Dave R, Razafimahatratra H, Rabarijohn RH et al (2014) Extreme vulnerability of smallholder farmers to agricultural risks and climate change in madagascar. Philos Trans R Soc Lond B Biol Sci 369:20130089. https://doi.org/10.1098/rstb.2013.008

Huang KY (2007) Application of artificial neural network for detecting phalaenopsis seedling diseases using color and texture features. Comput Electron Agric 57:3–11. https://doi.org/10.1016/j.compag.2007.01.015

Hughes DP, Salathé M (2015) An open access repository of images on plant health to enable the development of mobile disease diagnostics. arXiv:1511.08060

Islam R, Rafiqul M (2015) An image processing technique to calculate percentage of disease affected pixels of paddy leaf. Int J Comput Appl 123:28–34

ITU (2015) ICT facts and figures—the world in 2015. International Telecommunication Union, Geneva

Kaur R, Kang SS (2015) An enhancement in classifier support vector machine to improve plant disease detection. In MOOCs, Innovation and Technology in Education (MITE), IEEE 3rd international conference on IEEE, pp. 135–140

Kumar Jain D, Tyagi SKS, Neelakandan S, Prakash M, Natrayan L (2021) Metaheuristic optimization-based resource allocation technique for cybertwin-driven 6G on IoE environment. IEEE Trans Indust Inform 18(7):4884–4892

Kumar SS, Raghavendra B (2019) Diseases detection of various plant leaf using image processing techniques: a review 2019 5th International Conference on Advanced Computing & Communication Systems (ICACCS), pp. 313–316

LeCun Y, Bengio Y, Hinton G (2015) Deep learning. Nature 521:436–444. https://doi.org/10.1038/nature14539

Loey M, ElSawy A, Afify M (2020) Deep learning in plant diseases detection for agricultural crops: a survey. Int J Serv Sci Manag Eng Technol 11:41–58

Lurstwut B, Pornpanomchai C (2017) Image analysis based on color, shape and texture for rice seed (Oryza sativa L.) germination evaluation. Agric Nat Resour 51:383–389

Mohanty SP, Hughes DP, Salathé M (2016) Using deep learning for image-based plant disease detection. Front Plant Sci 7:1419

Pilli SK, Nallathambi B, George SJ, Diwanji V (2014) eAGROBOT—a robot for early crop disease detection using image processing. In Electronics and Communication Systems (ICECS), International Conference on IEEE, pp 1–6

Rajmohan R, Pajany M, Rajesh R, Raman DR, Prabu U (2018) Smart paddy crop disease identification and management using deep convolution neural network and SVM classifier. Int J Pure Appl Mathem 118:255–264

Ramcharan A, Baranowski K, McCloskey P, Ahmed B, Legg J, Hughes DP (2017) Deep learning for image-based cassava disease detection. Front Plant Sci 8:1852

Ramesh D, Vardhan BV (2013) Data mining techniques and applications to agricultural yield data. Int J Adv Res Comput Commun Eng 2(9):3477–3480

Russakovsky O, Deng J, Su H, Krause J, Satheesh S, Ma S et al (2015) ImageNet large scale visual recognition challenge. Int J Comput Vis 115:211–252. https://doi.org/10.1007/s11263-015-0816-y
Samanta S, Pal DK, Palsamanta B (2018) Flood susceptibility analysis through remote sensing, GIS and frequency ratio model. Appl Water Sci 8:Article number: 66
Sankaran S, Mishra A, Maja JM, Ehsani R (2011) Visible-near infrared spectroscopy for detection of huanglongbing in citrus orchards. Comput Electron Agric 77:127–134. https://doi.org/10.1016/j.compag.2011.03.004
Sankur B, Sezgin M (2004) Survey over image thresholding techniques and quantitative performance evaluation. J. Electron Imag 13:146–165
Schmidhuber J (2015) Deep learning in neural networks: an overview. Neural Netw 61:85–117. https://doi.org/10.1016/j.neunet.2014.09.003
Sethy PK, Negi B, Barpanda NK, Behera SK, Rath AK (2018) Measurement of disease severity of rice crop using machine learning and computational intelligence. In: Gurumoorthy S, Rao BNK, Gao X-Z (eds) Cognitive science and artificial intelligence, Springer briefs in applied sciences and technology. Springer, Berlin, Germany, pp 1–11
Shirahatti J, Patil R, Akulwar P (2018) A survey paper on plant disease identification using machine learning approach 2018 3rd International Conference on Communication and Electronics Systems (ICCES), pp 1171–1174
Shrivastava VK, Pradhan MK, Minz S, Thakur MP (2019) Rice plant disease classification using transfer learning of deep convolution neural network. Int Arch Photogrammetry Remote Sensing Spatial Inform Sci 42
Shruthi U, Nagaveni V, Raghavendra B (2019) A review on machine learning classification techniques for plant disease detection 2019 5th International Conference on Advanced Computing & Communication Systems, (ICACCS), pp 281–284
Singh A, Ganapathysubramanian B, Singh AK, Sarkar S (2015) Machine learning for high-throughput stress phenotyping in plants. Trends Plant Sci 21:110–124. https://doi.org/10.1016/j.tplants.2015.10.015
Soni AN (2018) Data center monitoring using an improved faster regional convolutional neural network. Int J Adv Res Electric Electron Instrum Eng 7(4):1849–1853. https://doi.org/10.15662/IJAREEIE.2018.0704058
Szegedy C, Liu W, Jia Y, Sermanet P, Reed S, Anguelov D, et al. (2015) Going deeper with convolutions. In Proceedings of the IEEE conference on computer vision and pattern recognition.
Too EC et al (2019) A comparative study of fine-tuning deep learning models for plant disease identification. Comput Electron Agric 161:272–279
UNEP (2013) Smallholders, food security, and the environment. International Fund for Agricultural Development (IFAD), Rome
Wang H, Li G, Ma Z, Li X (2012) Image recognition of plant diseases based on backpropagation networks. Conference: image and signal processing (CISP), 2012 5th international congress doi: https://doi.org/10.1109/CISP.2012.6469998
Wetterich CB, Kumar R, Sankaran S, Junior JB, Ehsani R, Marcassa LG (2012) A comparative study on application of computer vision and fluorescence imaging spectroscopy for detection of huanglongbing citrus disease in the usa and brazil. J Spectrosc 2013:841738. https://doi.org/10.1155/2013/841738
Xia, Y.Q., Li, Y., Li, C.: Intelligent diagnose system of wheat diseases based on android phone. J Inform Comput Sci 12, 6845–6852, Dec (2015)
Zeiler MD, Fergus R (2014) Visualizing and understanding convolutional networks. In: Fleet D, Pajdla T, Schiele B, Tuytelaars T (eds) Computer Vision–ECCV 2014. Springer, pp 818–833

Chapter 9
Synergizing Smart Farming and Human Bioinformatics Through IoT and Sensor Devices

Sandeep Kumar Jain and Pritesh Kumar Jain

Abstract The fusion of human bioinformatics with smart farming has developed as a cutting-edge interdisciplinary field with enormous promise to transform both healthcare and agriculture. This chapter delves into the intersection of these fields, examining how sensor technology and the Internet of Things (IoT) enable data-driven decisions in both personalized healthcare and agricultural operations. While exploring the significance of IoT in gathering, sending, and evaluating data from human bioinformatics systems, we analyze how smart farming approaches might maximize crop yields and resource use. This convergence not only improves agricultural sustainability but also gives people more control over how they monitor and take care of their health.

Keywords Sensor · Personalized · Healthcare · Agriculture · Bioinformatics · Smart

9.1 Introduction

The nexus of smart farming and human bioinformatics presents a new strategy to tackle these intricate problems at a time when the globe is dealing with the challenges of a growing population, depleting natural resources, and rising health concerns. Utilizing IoT and sensor technology, smart farming develops data-driven precision agriculture systems that maximize resource efficiency and boost productivity. Similar technology is also used by human bioinformatics to track and examine health-related data for individualized medical insights. The symbiotic relationship between these two fields is examined in this chapter along with its potential to alter both agriculture and healthcare.

S. K. Jain (✉) · P. K. Jain
Department of Computer Science and Engineering, Shri Vaishnav Vidyapeeth Vishwavidyalaya, Indore, Madhya Pradesh, India
e-mail: sandeepjain@svvv.edu.in; priteshkjain@svvv.edu.in

A. Khamparia et al. (eds.), *Microbial Data Intelligence and Computational Techniques for Sustainable Computing*, Microorganisms for Sustainability 47,
https://doi.org/10.1007/978-981-99-9621-6_9

9.2 Smart Farming and IoT: Enhancing Agricultural Productivity

Precision agriculture (Routray et al. 2019; Zhang et al. 2021), commonly referred to as smart farming, is a contemporary method of farming that makes use of technology and data-driven strategies to maximize crop output and resource management. To collect real-time information regarding soil conditions, weather patterns, and crop growth, it involves the integration of numerous technologies, including sensors, drones, GPS, and data analytics. Making informed decisions about planting, irrigation, fertilizer, and pest management is then possible thanks to this data. Smart farming aims to maximize crop yields while improving agricultural efficiency, cutting waste, and ensuring sustainable practices (Ramesh et al. 2023).

The term "Internet of Things" (IoT) describes a system of networked physical objects, including machines, cars, and buildings, that are equipped with sensors, software, and network connectivity. Without the need for direct human involvement, these devices may gather and exchange data online. The IoT enables seamless connection between systems and devices, allowing them to cooperate and share information on their own (Al-Fuqaha et al. 2015). IoT devices are essential for gathering and transferring data for analysis and decision-making in the context of smart farming and human bioinformatics.

Electronic components known as sensor devices are used to monitor and detect many physical characteristics, including temperature, humidity, light, and pressure. As they offer real-time data that guides decision-making, these gadgets are essential to both smart farming and human bioinformatics. Agricultural fields, hospitals, houses, and wearable technology are just a few of the areas where sensors can be installed to collect pertinent data points that support data-driven insights.

By giving farmers previously unattainable insights and control over their operations, IoT and sensor device integration is changing the agriculture industry. These innovations support data-driven decision-making, resource optimization, and increased crop yields. Here, we look at how the Internet of Things (IoT) and sensor technology are changing agriculture through initiatives like resource optimization, precision farming, environmental monitoring, and data analysis.

9.2.1 IoT in Agriculture

Real-time monitoring of the environment, the health of the soil, and crop development is made possible by the integration of IoT devices into farming systems. Sensors gather information on variables like temperature, humidity, soil moisture, and nutrient concentrations. Weather stations that give farmers precise local weather information so they can decide how to sow, irrigate their fields, and manage pests are an example of this real-time data collecting. Through the creation of a network of connected data points by these IoT devices, conventional farming becomes a precision-based activity (Khan et al. 2020; Singh et al. 2019).

9.2.2 *Precision Agriculture*

Farmers may make informed decisions about irrigation, fertilization, and pest control by utilizing the data gathered by IoT devices. This focused strategy increases yields while reducing resource waste (Rezk et al. 2021). The transformational potential of IoT in agriculture is best illustrated by precision agricultural technologies, such as automated irrigation systems that modify water distribution based on soil moisture levels. As a result, productivity is increased while water use and environmental effect are decreased.

9.2.3 *Predictive Analytics*

The gathered data is processed using sophisticated machine learning and data analytics algorithms to discover patterns in crop health, predict disease outbreaks, and optimize planting schedules (Almalki et al. 2021). Farmers can forecast disease outbreaks or insect infestations and take preventive actions using historical data analysis and machine learning models. These forecasts are essential for avoiding crop losses and maintaining environmentally friendly farming methods.

9.2.4 *Resource Optimization*

IoT-driven smart farming reduces the negative effects of agriculture on the environment by maximizing water use, minimizing chemical use, and limiting soil erosion. Farmers may precisely apply fertilizers by closely observing the state of the soil and the nutrient content, reducing overuse that might be harmful to the environment. Similar to this, the combination of precision irrigation with IoT ensures that water is used effectively, cutting down on waste and protecting water resources.

9.3 Benefits of IoT and Sensor Technologies in Agriculture

Enhanced Efficiency: IoT and sensor devices remove guessing by supplying real-time data, enabling farmers to react quickly to changing circumstances (Syed et al. 2019).

Resource Conservation: By using pesticides, fertilizers, and water only when essential, precision agriculture lowers waste and benefits the environment while saving money.

Increased Yield and Quality: By keeping an eye on and optimizing environmental conditions, crops are healthier, yields are better, and product quality is higher.

Data-Driven Decisions: Data analytics allow for well-informed choices on resource allocation, planting, and harvesting, reducing risks and optimizing profits.

Sustainability: IoT-driven agriculture helps promote sustainable farming methods by maximizing resource use and minimizing environmental effect.

Risk Reduction: Timely action can be taken in the event that diseases, pests, or unfavorable weather conditions are discovered early.

9.4 Human Bioinformatics and IoT: Revolutionizing Healthcare

The gathering, archiving, analysis, and interpretation of biological data pertaining to human health and well-being are all part of human bioinformatics (Yoosefzadeh Najafabadi et al. 2023). It includes a variety of data types, such as genomic data, medical records, and wearable device data. These data sets are processed using bioinformatics tools and techniques in order to spot trends and derive valuable insights. Understanding unique health profiles, identifying illness risks, and creating individualized medical therapies are the ultimate goals.

9.4.1 IoT in Healthcare

Remote patient monitoring has been made possible by wearable technology and medical sensors, which allow for the continuous collection of health-related data such as vital signs and activity levels. A steady stream of health-related data is provided by wearable medical devices including glucose monitors, smartwatches, and fitness trackers. Individuals and healthcare professionals can monitor health trends and take appropriate action using this data that is communicated over IoT.

9.4.2 Personalized Health Insights

People can learn more about their health and well-being by analyzing the data collected by wearable technology. This proactive strategy makes it easier to spot anomalies and chronic illnesses early on. People learn more about their health by analyzing information on their heart rate variability, sleep patterns, and physical activity. Timely medical interventions and lifestyle changes can result from early discovery of abnormalities.

9.4.3 *Disease Management*

Healthcare professionals can monitor patients with chronic illnesses and modify treatment strategies in real time based on obtained data with the help of IoT-enabled bioinformatics tools. The use of remote monitoring is advantageous for patients with long-term illnesses like diabetes or hypertension. Real-time data is sent to healthcare professionals, allowing them to better manage diseases and adapt treatment programs to the needs of specific patients.

9.4.4 *Healthcare Accessibility*

Remote patient monitoring fills in gaps in healthcare access by bringing medical treatment to underserved groups across geographic boundaries. Healthcare solutions powered by the Internet of Things (IoT) bring medical care to isolated or rural locations where access to medical institutions may be limited. Wearable tech-enabled telemedicine provides consultations and diagnostics without requiring in-person visits.

9.5 Synergy Between Smart Farming and Human Bioinformatics

The chance to unlock synergies that go beyond their respective fields is fascinating given the convergence of smart farming and human bioinformatics (Anand et al. 2023) (Barbhuiya and Ahmad 2021). This technological synergy delivers significant advantages for agriculture, human health, and the environment through a dynamic interplay. Let's get into the core aim of this chapter by examining the intersections and potential links between these two areas of study while emphasizing the advantages of fusing knowledge from both disciplines.

9.5.1 *Environment-Health Nexus*

Understanding how environmental elements and human health are connected is one of the most fascinating intersections between smart farming and human bioinformatics. Both agricultural growth and human well-being are substantially impacted by environmental factors. We can decipher complex correlations between air quality, pollen counts, humidity, and respiratory health by fusing data from environmental sensors in agricultural fields with health-related wearable technology. Better disease prevention and management tactics may result from this holistic viewpoint.

9.5.2 Shared Data Analytics

In order to acquire actionable insights, both smart farming and human bioinformatics significantly rely on data analytics. Comprehensive analysis may result from combining data sets from these domains, for instance, uncovering hidden links that may have an effect on public health by comparing crop growth patterns with regional disease outbreaks or changes in air quality. This common data-driven methodology improves the ability to forecast outcomes, allowing for timely interventions in both agriculture and healthcare.

9.5.3 Nutritional Sustainability

We may examine the nutritional implications of food production and human health thanks to the junction of these domains. We may create meals suited to individual health needs while taking environmental sustainability into account by fusing knowledge from smart farming practices, which maximize crop nutrient content and production. This not only fosters human well-being but also responsible agriculture that places a priority on creating foods that are nutrient-rich.

9.5.4 Early Warning Systems

Environmental and health monitoring devices with IoT capabilities can serve as early warning systems for potential dangers. For instance, we can pinpoint locations with increased health risks by examining patterns of environmental pollutants and comparing them with the prevalence of particular health disorders. Through predictive analytics, farmers can also gain the advantage of early crop disease or insect infestation detection, lowering the requirement for excessive chemical use and lowering the risk of ecosystem destruction.

9.6 Benefits of Combining Insights

9.6.1 Holistic Approach

By combining smart farming with human bioinformatics, we can take a holistic approach to well-being that takes into account both human and environmental health. Decisions in both domains are made with greater knowledge thanks to this all-encompassing viewpoint (Shameer et al. 2017).

9.6.2 Cross-Domain Findings

New findings may result from insights gained at the nexus of human bioinformatics and smart farming. Finding out how particular agricultural methods impact human health, or the other way around, brings up new research and innovation opportunities.

9.6.3 Enhanced Resource Management

We can enhance resource management tactics by exchanging insights between various sectors. For instance, information on the best irrigation techniques for crop growth can influence suggestions for how much water to use to stay hydrated and healthy.

Customized health programs are made possible by the integration of agricultural and medical data. Dietary advice can be modified to meet a person's specific nutritional requirements as well as to support organic farming methods.

9.6.4 Impact on Public Health

By addressing both issues of agricultural sustainability and public health, informed policies and actions may be developed, creating healthier environments and communities.

9.7 Challenges, Limitations, and Ethical Considerations

While the fusion of human bioinformatics with smart farming brings exciting possibilities, there are also some difficulties, restrictions, and moral dilemmas that need to be carefully considered (Doukas et al. 2012; Quy et al. 2022).

9.7.1 Data Security and Privacy

Concerns about data privacy and security arise when massive amounts of private information from the healthcare and agricultural industries are combined. The legislative frameworks for personal health data and agriculture data are distinct, necessitating strong steps to ensure data protection, encryption, and secure communication. Such data could be accessed without authorization, which could have serious ethical and legal repercussions.

9.7.2 Data Accuracy and Quality

Accurate data is essential for both smart farming and human bioinformatics. Inaccurate measurements, data transfer issues, or sensor faults might produce inaccurate conclusions. Having accurate data is essential to avoid making the wrong choices, whether they relate to agricultural or medicinal operations.

9.7.3 Interoperability and Standardization

IoT gadgets and sensor systems may be produced by several companies and run on various operating systems. It can be difficult to achieve seamless integration and compatibility between various devices. To ensure data interoperability and effective device connection, standardization activities are required.

9.7.4 Technology Access and Literacy

Not all farmers and people in general have access to cutting-edge IoT technologies or the knowledge required to understand intricate data analyses. Ensuring equitable access and promoting digital literacy among all stakeholders is essential to prevent creating technology-driven disparities.

9.7.5 Environmental Impact

The increased deployment of IoT devices in agriculture can lead to concerns about the environmental impact of manufacturing, energy consumption, and electronic waste. Balancing the benefits of technology adoption with its environmental costs is crucial for sustainable implementation.

9.8 Future Possibilities, Innovations, and Research Areas

There are numerous opportunities for future innovation and research due to the fusion of smart agriculture and human bioinformatics (Wolfert et al. 2017).

9.8.1 Predictive Public Health Models

By combining environmental data from smart farming with health data, predictive models that foresee dangers to public health can be created. Early warnings of illness epidemics associated with environmental factors could help prompt policy choices and responses.

9.8.2 Data-Driven Nutritional Sustainability

Upcoming studies may concentrate on improving crop nutrition in accordance with people's particular health requirements and environmental sustainability. Healthy diets and more sustainable food production would both benefit from this interdisciplinary approach.

9.8.3 Environmentally Friendly Precision Agriculture

New developments in precision agriculture may result in more environmentally friendly methods. Integration with human bioinformatics might enable the development of precision pesticide application strategies that minimize harm to both crops and ecosystems.

9.8.4 Solutions for Personalized Agri-Health

Studies could look into the possibilities for personalized agri-health practices to improve human health. For instance, knowing how particular crops affect people's health profiles could result in tailored dietary suggestions.

9.8.5 Ethical Data Governance

It is crucial to create moral frameworks for sharing and governing data. Future studies should focus on topics like data ownership, informed consent for data usage, and developing open systems for data sharing between the agricultural and healthcare sectors.

9.8.6 Cross-Disciplinary Training

Educational initiatives that close the knowledge gap between the fields of agriculture and medicine could equip professionals with the abilities necessary to fully realize the potential of these interconnected domains.

9.8.7 Climate-Health Resilience

By combining insights from smart farming and human bioinformatics, researchers could develop strategies to enhance climate and health resilience, addressing the growing challenges posed by climate change.

9.9 Conclusion

Through IoT and sensor technology, smart farming and human bioinformatics have come together, creating a paradigm shift that transcends conventional limits. This synergy has the potential to transform agriculture into a data-driven, sustainable industry that boosts sustainable practices, increases illness prevention, and empowers people to take control of their own health. The future promises tremendous potential for the improvement of both people and the environment as technology develops and interdisciplinary collaborations grow.

References

Al-Fuqaha A, Guizani M, Mohammadi M, Aledhari M, Ayyash M (2015) The application of internet of things (IoT) to smart farming: a review. IEEE Internet Things J:857–875

Almalki FA et al (2021) A low-cost platform for environmental smart farming monitoring system based on IoT and UAVs. Sustainability 13(11):5908

Anand KJ et al (2023) Enhancing crop improvement through synergistic integration of advanced plant breeding and proximal remote sensing techniques: a review. Int J Plant Soil Sci 35(19): 121–138

Barbhuiya RK, Ahmad N (2021) IoT applications in translational bioinformatics. In: Translational bioinformatics in healthcare and medicine. Academic Press, pp 69–79

Doukas C et al (2012) Enabling data protection through PKI encryption in IoT m-Health devices. In: 2012 IEEE 12th international conference on bioinformatics & bioengineering (BIBE). IEEE

Khan A, Khan MY, Khan HA, Alam A, Khan MA, Singh RN, Al-Garadi MA, Al-Ghamdi MA (2020) IoT-based smart farming: a comprehensive review. Comput Electron Agric:107041

Quy VK et al (2022) IoT-enabled smart agriculture: architecture, applications, and challenges. Appl Sci 12(7):3396

Ramesh A, Karthikeyan S, Karthik M (2023) Smart farming: a review of internet of things (IoT) applications in agriculture. J Clean Prod:12345–12355

Rezk NG et al (2021) An efficient IoT based smart farming system using machine learning algorithms. Multimed Tools Appl 80:773–797

Routray SK et al (2019) Internet of things based precision agriculture for developing countries. In: 2019 International Conference on Smart Systems and Inventive Technology (ICSSIT). IEEE

Shameer K et al (2017) Translational bioinformatics in the era of real-time biomedical, health care and wellness data streams. Brief Bioinform 18(1):105–124

Singh AK, Singh PK, Singh RK (2019) The role of IoT in smart farming: a comprehensive review. ACM Trans Sensor Netw:1–24

Syed L et al (2019) Data science algorithms and techniques for smart healthcare using IoT and big data analytics. In: Mishra M, Mishra B, Patel Y, Misra R (eds) Smart techniques for a smarter planet: towards smarter algorithms, vol 374. Springer, Cham, pp 211–241

Wolfert S et al (2017) Big data in smart farming—a review. Agric Syst 153:69–80

Yoosefzadeh Najafabadi M, Hesami M, Eskandari M (2023) Machine learning-assisted approaches in modernized plant breeding programs. Genes 14(4):777

Zhang M et al (2021) Wearable Internet of Things enabled precision livestock farming in smart farms: a review of technical solutions for precise perception, biocompatibility, and sustainability monitoring. J Clean Prod 312:127712

Chapter 10
Deep Learning-Assisted Techniques for Detection and Prediction of Colorectal Cancer From Medical Images and Microbial Modality

Ravi Kumar, Amritpal Singh, and Aditya Khamparia

Abstract In the past decade, significant progress has been made in the fields of artificial intelligence, machine learning, and deep learning (DL). These advancements have opened up wide applications and opportunities in the medical field. Colorectal cancer (CRC) has gained substantial interest from researchers due to its ranking as the third most prevalent cancer type after breast and lung cancer, affecting around 10% of all cancer patients globally each year. It is the second leading cause of cancer-related death worldwide, making the development of efficient computer-assisted methods for its detection, prediction, and treatment crucial. There are modalities used for colorectal cancer screening and detection such as colonoscopy images, biopsy samples, biomarker data, blood samples, CT scan, MRI, ultrasound, PET, and microbial data. The advancement of technology has made deep learning an attractive choice for fast and effective detection, segmentation, and prediction of diseases through image analysis. This technology has the potential to assist and empower medical professionals in making timely and informed decisions. Deep learning has proven to be highly effective when ample high-quality features and large datasets are available. However, one of the main challenges in using deep learning for medical image analysis is the limited availability of datasets from medical centers. This chapter provides an overview of DL-based models and their application in detecting and predicting CRC from various modalities. On the

R. Kumar (✉)
Research Scholar, Department of Computer Science Engineering, Lovely Professional University, Punjab, India

Department of Computer Science Engineering, Jawaharlal Nehru Government Engineering College, Sundernagar, Mandi, Himachal Pradesh, India
e-mail: ravi.42100244@lpu.in

A. Singh
Department of Computer Science Engineering, Lovely Professional University, Punjab, India

A. Khamparia
Department of Computer Science, Babasaheb Bhimrao Ambedkar University, Amethi, India

A. Khamparia et al. (eds.), *Microbial Data Intelligence and Computational Techniques for Sustainable Computing*, Microorganisms for Sustainability 47,
https://doi.org/10.1007/978-981-99-9621-6_10

SCPolyps dataset the OEM model achieved training and test accuracy of 98% and 96% respectively.

Keywords Deep learning · Colorectal cancer · Detection · Segmentation · Prediction

10.1 Introduction

Based on the 2020 global cancer statistics, colorectal cancer (CRC) holds the third position among the most prevalent cancer types worldwide, following breast and lung cancer (Sung et al. 2021). It affects approximately 10% of all cancer patients annually, impacting both men and women. CRC, which is regrettably the second leading cause of fatalities due to cancer worldwide, is just behind lung cancer. In the same year, the World Health Organization reported approximately ten million fatalities attributed to cancer alongside an alarming rise of 19.3 million newly diagnosed cancer cases (Sung et al. 2021). People over the age of 50 are particularly vulnerable, with hereditary factors posing the greatest risk at 35%, along with other elements like smoking, unhealthy eating habits, and obesity (Arnold et al. 2017). CRC is a significant worldwide health issue and one of the primary causes of fatalities due to cancer worldwide. Early CRC detection leads to timely treatment and thus improves patient outcome.

There are many tests used for colorectal cancer diagnosis and detection such as colonoscopy, biopsy, biomarker testing, blood tests, CT scan, MRIs, ultrasound, and positron emission tomography (PET). Generally, biopsy is preferred for cancer diagnosis but colonoscopy is the most efficient method for CRC screening (Colorectal Cancer 2022). There are mainly four domains in existing AI applications for CRC as listed below (Qiu et al. 2022). The most common technique used for screening is colonoscopy. It is considered the most reliable technique for CRC screening. These techniques require experienced experts otherwise they may lead to omission and misdiagnosis. Accuracy can be improved using electronic medical records and omic data with these methods (Winawer 2007). The radiography and pathological examination are mainly used for diagnosis and staging of CRC (Goyal et al. 2020). Chemotherapy, surgery, and radiotherapy are the three most popular CRC treatments (Gao et al. 2020).The prediction of CRC recurrence and estimation of the survival period comes under prognostic analysis. Disease progression is estimated using statistical methods, e.g., Cox regression (Zhu et al. 2020).

Recent research has focused on the application of machine learning (ML)- and deep learning (DL)-based technologies in the medical arena, presenting great opportunities for the creation of efficient computer-assisted approaches in cancer diagnosis, prediction, and treatment (Thakur et al. 2023; Srivastava et al. 2022). This chapter provides an introduction to leveraging DL for the detection of CRC. Traditional approaches to colorectal cancer (CRC) detection rely heavily on human expertise, including visual examination of colonoscopy images and analysis of biopsy sample reports by experienced endoscopists. However, this subjective approach can be susceptible to human error, potentially resulting in missed or misdiagnosed cases of CRC. Furthermore, the increasing demand for colonoscopy screenings strains healthcare systems, lead to longer waiting times and delayed diagnoses. Deep learning presents an opportunity to automate and enhance this process, ultimately improving early detection rates. One of the fundamental requirements for effective deep learning models is access to large datasets for training. However, acquiring substantial volumes of medical data can be challenging. Transfer learning, generative adversarial networks (GANs), and data augmentation emerge as a solution to this problem. Transfer learning enables the models to be trained effectively with smaller datasets. This approach involves fine-tuning of pretrained models, which were initially trained on extensive datasets like ImageNet, with minimal adjustments to achieve optimal results in medical image analysis and cancer detection tasks (Thakur et al. 2023).

The remainder of this chapter is organized as follows. The overview of deep neural network-based techniques for medical image analysis is given in Sect. 10.2. The deep learning-based detection of colorectal cancer using various data modalities is presented in Sect. 10.3. The methodology for classification of tumor and non-tumor images is presented in Sect. 10.4. Section 10.5 presents results and discussions. Finally the conclusion is presented in Sect. 10.6.

10.2 Deep Neural Networks for Image Analysis

A crucial area of research in medical imaging is computerized diagnosis. Various imaging modalities are employed in medical applications, including colonoscopy images, whole slide images (WSI), RNA sequences, microbial samples, CT (computed tomography) scan images, magnetic resonance imaging (MRI), X-ray images, and positron emission tomography (PET). In the past, basic techniques like thresholding, region growing, and edge tracing were utilized for medical image processing as shown in Fig. 10.1 (Qiu et al. 2022). However, the increasing volume

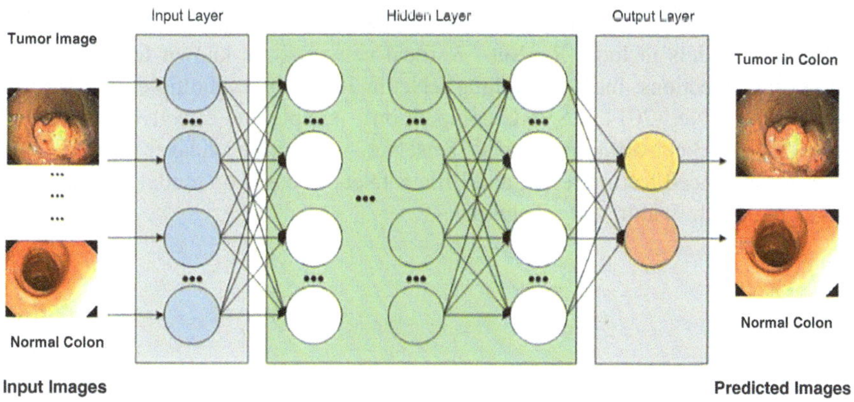

Fig. 10.1 Deep neural network for medical image classification

and complexity of medical imaging data have spurred the adoption of machine learning techniques.

Traditional machine learning approaches, relying on manually engineered features, often necessitate substantial manual effort in algorithm design. This limitation has led to the adoption of neural networks. The widespread adoption of artificial neural networks is enabled by factors such as the accessibility of data and advancements in computational processing capabilities (Winawer 2007). The development of DL techniques, particularly convolutional neural networks (CNNs), has significantly expanded the possibilities for automating and enhancing the processing of medical images. The neural network-based approaches for medical image analysis are described below.

10.2.1 Convolutional Neural Networks

CNN is a special type of DL model that proved to be very efficient and useful for processing and analyzing visual data such as images and videos. Convolution is joining two functions to obtain a third function. It is applied on input images to extract important information. It plays a crucial role in computer vision tasks such as medical image analysis, classification, object detection, and segmentation (Thakur et al. 2023).

10.2.2 Transfer Learning Models

The primary issue with DL is that a lot of data is required for it to perform at its best. But mostly it is very difficult to get or create large datasets, especially in the medical domain. This problem can be resolved to a great extent by transfer learning. With transfer learning it is possible to train models with less data and obtain optimal results. Many deep learning models, pretrained on large datasets like ImageNet, are available for transfer learning across various image analysis tasks. Popular pretrained models include: ResNet—residual networks are known for their depth and skip connections; Inception (GoogLeNet in 2014)—uses multiple filter sizes in parallel; VGGNet (2014 ILSVRC competition)—simple and effective architecture with small 3×3 convolutional filters; DenseNet—connects each layer to every other layer for enhanced feature reuse; and EfficientNet—focuses on model efficiency by balancing depth, width, and resolution.

10.2.3 Ensemble Learning Models

Ensemble learning is a powerful technique in DL and ML where multiple models are combined to improve robustness, predictive performance, and generalization. One of the drawbacks of ensemble models is that they require more resources.

10.2.4 Object Detection Models

Object detection models locate and classify objects within an image. Well-known models include YOLO, faster R-CNN, and SSD (Krenzer et al. 2023).

10.2.5 Image Segmentation Models

Image segmentation models classify each pixel in an image, distinguishing different objects or regions. U-Net, FCN (fully convolutional network), and SegNet are popular choices for semantic and instance segmentation.

10.2.6 Network Pruning

Network pruning or network compression is currently gaining much attention . In network pruning the model is compressed and becomes lightweight while preserving useful features and demonstrating comparable performance as compared to the original model. These lightweight models are very easy to deploy and also utilize much fewer resources than their counterparts. Li-SegPNet is a lightweight encoder-decoder model for colorectal polyps segmentation. Debesh Jha introduced NanoNet, a lightweight model that utilizes YOLOX for real-time polyp detection, group normalization, and video-adjacent frame association algorithm for unstable polyp detection.

10.2.7 Deep Belief Networks (DBNs)

In a variety of domains, including medical image analysis, DBNs have been applied to obtain hierarchical features from images and improve classification and segmentation tasks. They are less frequently utilized for medical image processing tasks than other neural network architectures like CNNs.

10.2.8 Generative Adversarial Networks (GANs)

GANs are used for image generation, style transfer, and super-resolution. Notable GAN architectures include deep convolutional GAN (DCGAN), CycleGAN, and BigGAN. GAN is a powerful ML model developed by Goodfellow, I. et al. in 2014 (Park et al. 2023). Since then, GANs have significantly advanced the field of generative modeling and have found use in a number of fields, including computer vision and natural language processing. A GAN consists of two neural networks that are trained concurrently, one of which is a generator and the other a discriminator. The generator attempts to produce data samples (e.g., images and text) that resemble genuine data using random noise as input, which is often taken from a straightforward distribution like the Gaussian distribution. The generator is trained to reduce the likelihood that the discriminator will identify the samples it generates as fraudulent. A binary classifier called the discriminator seeks to distinguish between authentic and fraudulent data. It has been honed to increase the likelihood that it will accurately identify real data as real and fake data as fake (Park et al. 2023).

10.2.9 Transformers

Transformer-based models have made significant inroads into medical image analysis, offering state-of-the-art performance in various tasks. Dense transformer extends the traditional transformer architecture for dense prediction tasks, making it suitable for tasks like medical image segmentation. It utilizes self-attention mechanisms to capture contextual information. MedT is designed specifically for medical imaging tasks. It combines the transformer architecture with spatial attention mechanisms tailored for the unique characteristics of medical images. TransUNet integrates transformers with a U-Net architecture, which is well suited for medical image segmentation. It combines the strength of transformers in capturing long-range dependencies with the spatial awareness of U-Net. RETINAnet is adapted from object detection models and employs transformers for the detection and localization of retinal lesions, making it valuable for diabetic retinopathy screening. TransV3D extends transformer architectures to three-dimensional medical image data. It is suitable for tasks such as organ or tumor segmentation in 3D medical scans.

10.3 Deep Learning for Classification, Segmentation, Detection, and Prediction of CRC From Different Modalities

Colorectal cancer stands among the most prevalent cancer types, ranking as the second most common in females and the third in males. Detecting and removing CRC can substantially lower the risk of patient fatalities. As a result, researchers are dedicated to exploring deep learning (DL) solutions to automate the detection of CRC using different modalities (Gao et al. 2020). Various imaging modalities can be employed in medical applications, including colonoscopy images, whole slide images (WSI), RNA sequences, microbial samples, CT (computed tomography) scan images, magnetic resonance imaging (MRI), X-ray images, and positron emission tomography (PET). The different modalities considered by various researchers for detection and prediction of CRC using ML and DL are shown in Fig. 10.2.

10.3.1 CRC Detection and Prediction Using Endoscopic Images and Videos

The colonoscopy is the gold standard for colorectal cancer screening. Automatic CRC detection from endoscopic images becomes possible with the advancement in technology and deep learning techniques. One effective approach to early detection is the analysis of polyp images, as polyps are often precursors to colorectal cancer (Goyal et al. 2020). By leveraging their ability to automatically learn complex

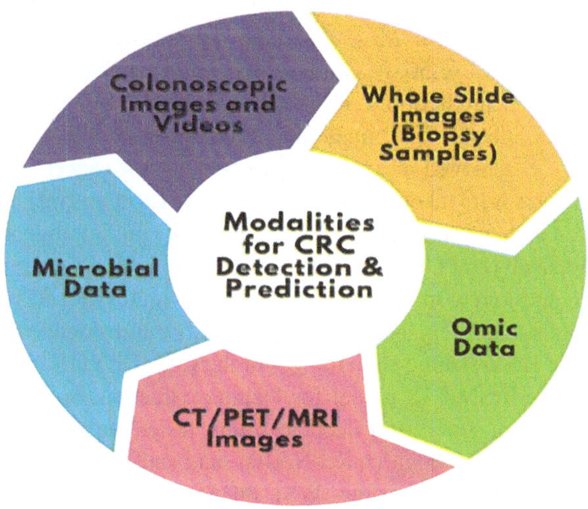

Fig. 10.2 Modalities for detection and prediction of colorectal cancer

patterns and features from large datasets, DL techniques can aid in the early detection of CRC from colon images (Gao et al. 2020). Some of the techniques for CRC classification and segmentation from colonoscopy images and videos are discussed here. Pogorelov et al. (2017) proposed the Kvasir dataset developed from endoscopic images of the gastrointestinal tract and implemented machine learning and deep learning models on this dataset. Choi et al. (2020) applied InceptionV3, ResNet50, and DenseNet161 for detecting CRC using endoscopic images. DenseNet161 achieved the highest accuracy rate of 92.48%. Yao et al. (2021) implemented UNet with pretrained VGG19 and Jha et al. (2021) proposed NanoNet, a lightweight DL-based model for semantic segmentation and categorization of tissue (Polyps) automatically. It has been observed that it performs differently in different datasets but for the Kvasir-SEG dataset, NanoNet-C used much fewer parameters, and had better DSC, mean IoU, and frames/second than the ResUnet model. They concluded that a dataset consisting of small and large sized polyps is required for better and more realistic results. Some authors proposed instance segmentation with DL, an end-to-end learning and attention-based DL model for classification, segmentation, and localization of polyps from colonoscopy images (Nogueira-Rodríguez et al. 2021; Akilandeswari et al. 2022; Wang et al. 2022a; Yang et al. 2022). Colonoscopy is performed for detection of polyps which may lead to cancer in later stages. Cancer confirmation was done by taking tissue samples from polyps and investigating them under a microscope.

10.3.2 CRC Detection and Prediction Using Tissue Images (Biopsy Samples)

Tissue images are very helpful in detecting and classifying CRC. Some of the studies based on biopsy samples for CRC detection and classification are discussed here. Lu, L. et al. and Tsai, M. J. et al. proposed transfer learning-based deep learning models such as ResNet50, AlexNet, VGG19, GoogLeNet, SqueezeNet, and InceptionV3 for the early prediction of metastatic CRC from haematoxylin and eosin (H&E)-stained tissue images. The image augmentation was applied for scaling the dataset. Their model achieved mean accuracy of 94.8% for cancer slides. The H&E-stained tissue images were taken from public datasets (Lu et al. 2021; Tsai and Tao 2021). Another researcher proposed a slide-based artificial intelligence predictor (SBAIP), with ResNet-18, a DL model, to predict lymph node cancer conditions from histological images of colorectal cancer and logistic regression classifier for clinical data such as patient age, sex, T-stage, tumor location, and sidedness as input (Kiehl et al. 2021). The H&E-stained slide images of the CAMELYON16 challenge were given as input to the ResNet-18 model for feature extraction. Some authors proposed supervised and semi-supervised deep learning, faster region-based convolutional neural network (R-CNN) with a ResNet101, and end-to-end learning with EfficientNet-B0 for examining CRC using WSI images on whole slide images. The benign and inflammatory tissues were categorized as low risk and dysplasia and malignant tissues were categorized as high risk. In the future, annotations and effective use of unlabeled data are required to improve efficiency. These models achieved accuracy around 94% (Yu et al. 2021; Ho et al. 2022; Teichmann et al. 2022). The images were divided into patches or tiles, and then the shape and size of the cells were analyzed to detect CRC.

10.3.3 CRC Detection Using Omic Data

Genetic sequence and biomarkers can be used to detect and predict CRC. Lee et al. implemented the InceptionV3 model and a Random Forest classifier for the classification of CRC based upon MSI (microsatellite instability) in 20 mm magnified patches of WSI as an input. The data was taken from The Cancer Genome Atlas (TCGA) dataset and samples were classified as cancerous and non-cancerous. For the TCGA datasets, the model obtained AUROC of 0.892 (Lee et al. 2021). The researcher proposed end-to-end DL utilizing WSI for detecting molecular changes. The DL was implemented using k-Siamese CNN architecture, Efficientnet-B0, and stochastic gradient-descent (SGD) with Adam heuristic. The model achieved AUC scores of up to 94% (Teichmann et al. 2022). Liu et al. implemented ResNet-18 with the attention-based Multiple Instance Learning (MMIL) model for preoperative Lymph Node Metastatis (LNM) assessment to predict CRC. Blood and tumor tissue serum biomarkers were used for prediction. The model attained an AUC of 0.855 for T1, 0.832 for T2, 0.691 for T3, and 0.792 for stage T4 of CRC. In the future, a multiclass classification can be done (Liu et al. 2022). Zhao et al. proposed deep learning techniques using enhanced venous-phase CT and RNA sequencing patterns to predict CRC (Zhao et al. 2022). The RNA sequencing genes of tumor tissues and CT images of CRC were gathered and examined by Spearman's correlation.

10.3.4 CRC Detection From MRI, (FDG)-PET, and CT Scan Data

Vorontsov et al. (2019) applied DL for segmenting liver regions from CT images of patients with CRC liver metastases. They applied the fully connected network (FCN)-based deep learning model for lesion size <10 mm, lesion size 10–20 mm, and lesion size >20 mm (Yang et al. 2022). Yuan et al. implemented the ResNet3D model with the support vector machine (SVM) classifier to detect peritoneal carcinomatosis in colorectal cancer using preoperative high-contrast CT images. Their model achieved accuracy of 94.11% and an AUC of 0.922 (Wang et al. 2022b). Wang et al. proposed a dense residual single-axis super resolution network on abdominal CT images for non-peritonealized CRC diagnosis (Yang and Liu

2020). The SPSS 22.0 software was used for analysis. To diagnose CRC T-staging, the extramural depth and the extramural vascular invasion grading must be applied to the CT image of the tissue. The author concluded that CT can be used for preoperative staging of non-peritonealized colorectal cancer. The author proposed deep learning techniques using enhanced venous-phase CT and RNA sequencing patterns to predict CRC (Zhao et al. 2022). Yang et al. proposed a FDG-PET/CT-based diagnostic model for CRC using regional LNM. The SPSS 20.0 was used for statistical analysis and Medcalc software for receiver operator characteristic (ROC) curves. The model's AUC was 95% but it had high sensitivity (He et al. 2021). He et al. proposed ML-based techniques for the estimation of regional LNM using FDG-PET/CT images (Li et al. 2021). They implemented least absolute shrinkage and selection operator approach and five ML techniques out of which logistic regression achieved an AUC of 0.866 and eXtreme gradient boosting achieved an AUC of 0.903 and outperformed the other models. Li et al. proposed a classification method of LNM in CRC using MRI images. Transfer learning was implemented using pre-trained weights of AlexNet. The model obtained accuracy of 0.8358 and an AUC of 0.8569. Only the LNM $>$ 3 mm was considered for the study (Baxter et al. 2016).

10.3.5 Deep Learning for Detecting Colorectal Cancer Using Microbial Data

In recent studies on detection and screening of CRC it has been observed that microbial data also plays a significant role in detecting CRC and the role of bacteria in increasing cancerous growth in affected areas. Though few studies have used gut microbiota as a diagnostic biomarker for CRC, the evidence discussed next showed an association between gut microbiota and CRC. Collectively, these studies showed how ML and DL are effective in the processing of microbiome data for the identification and categorization of colorectal cancer. In order to increase the sensitivity of fecal immunochemical testing for colonic lesion diagnosis, Baxter et al. (2016) used a Random Forest approach to build a microbiota-based model, which produced a reasonable prediction performance with an AUC of 67.3%. The MicrobiomeHD dataset was used in Namkung's paper from 2020, which investigated a number of machine learning techniques, such as SVM, RF, and ANN. Their models impressively attained an AUC of 90%, demonstrating strong predictive ability (Namkung 2020). Topçuoğlu et al. (2020), using techniques including Decision Trees, SVM, XGBoost, and RF, established a framework to solve microbiome-based categorization problems. The AUROC of their Random Forest model was 0.680 (Topçuoğlu et al. 2020). AUC scores of 0.96 and 0.89 on two different microbiome datasets were achieved in 2021 by Mulenga et al. using CNN to classify colorectal cancer from gut microbiome data, highlighting the possibility for excellent predictive accuracy (Mulenga et al. 2021). A collection of 1056 public fecal samples was used by Wu et al. (2021) to identify microbial markers for early colorectal cancer detection using Random Forest, yielding an AUC of 0.89 and indicating reliable prediction performance (Wu et al. 2021). In their 2023 study, Lu et al. used ML methods such as naive Bayes, RF, and logistic regression to identify colorectal cancer using information on the gut microbiome. With an AUC of 0.926 and an accuracy rate of 91.7%, their model displayed remarkable predictive accuracy (Lu et al. 2023). Yinhang et al. (2023) concentrated on using RF and SVM models to estimate CRC lymph node metastasis, identifying critical microbiological variables such as Lachnospiraceae_FCS020_group and Lachnospiraceae_UCG 004 as significant contributions (Yinhang et al. 2023). These results demonstrate the potential of ML to enhance CRC detection using microbiome data. They demonstrate differing levels of predictive efficacy while utilizing different algorithms, underscoring the field's exciting potential for microbiome-based models.

We have discussed different modalities that can be utilized by different researchers to detect colorectal cancer. In the next section we will implement deep learning techniques for the early detection of CRC. The early detection of CRC plays a vital role in reducing mortality rates and improving patient outcomes. Detecting and removing precancerous polyps during colonoscopy significantly reduces the risk of developing CRC. The next section describes the workflow for early detection of CRC from colonoscopy images.

10.4 Methodology

The classification workflow for colon diseases or colorectal cancer is described as follows. Firstly, the dataset is downloaded from a public source and subsequently unzipped for further analysis. The images within the dataset are resized, normalized, and subjected to data augmentation techniques. The augmentation helps in increasing the size and diversity of the dataset. The second step involves splitting the dataset into two sets: a training set and a test set. This division allows for effective model training, validation, and evaluation. Thirdly, the CNN models, pretrained on the ImageNet, are loaded. The top layers of these models are changed to enable the classification of colon diseases. The model training process begins in the subsequent phase, utilizing the training set data. Fourthly, the models are compiled, setting important hyper-parameters such as the optimization method, learning rate, loss function, dropout rate, and regularization techniques and then validating the trained model using the validation dataset. This step provides an opportunity to evaluate the model's generalization and performance on unseen data. Finally the model's performance is verified using the test dataset. The results are also visually verified by displaying the corresponding images. The overall process of classifying colon illnesses or colorectal cancer is visually depicted in Fig. 10.3, presented as a flowchart.

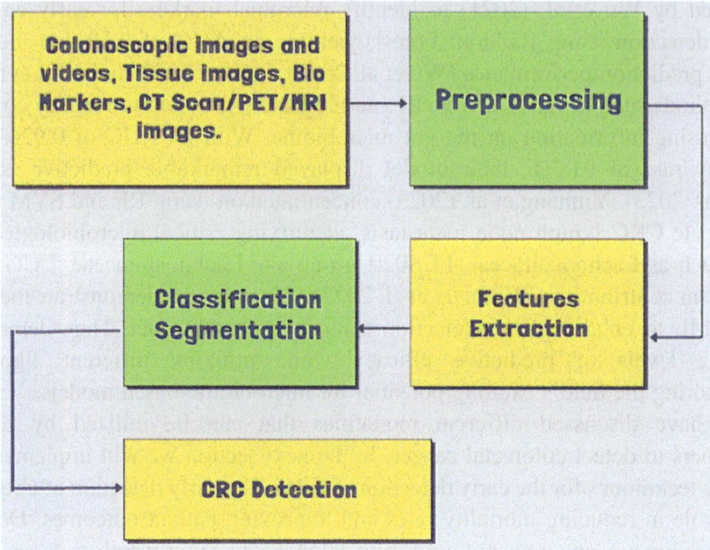

Fig. 10.3 Workflow

10.4.1 Dataset

There are very few datasets available for colorectal cancer detection because the acquisition of medical data is a very tedious task, and low-quality images hinder accurate prediction. The quality of the dataset plays a crucial role in the performance of the model. In this study three datasets with endoscopic images were selected for detection and prediction of CRC. The dataset used for this task was SCPolyps as shown in Fig. 10.4. SCPolyps is a self-created dataset that has 3000 images with four categories, normal, polyp, esophagus, and ulcerative, with 750 images each. In this study we have only used normal and polyp images. The sample polyp and normal colon images are shown in Fig. 10.4.

10.4.2 Model Selection for CRC Classification

The models selected for this study were VGG19 (Simonyan and Zisserman 2014), InceptionV3 (Szegedy et al. 2015), and our ensemble model. The transfer learning was implemented using VGG19, InceptionV3, and OEM. The ensemble model was created by combining two models (EfficientNet (He et al. 2016) and ResNet50 (Tan and Le 2019)) and taking the average. Hyper-parameters used were batch size, epochs, dropout, and regularization (L2). The weights of models pretrained on

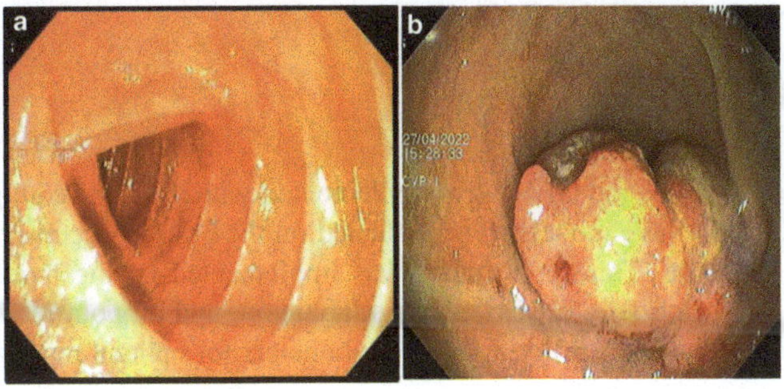

Fig. 10.4 Sample normal colon and polyp images of SCPolyps dataset. (**a**) Normal colon. (**b**) Polyp

ImageNet were used during transfer learning. Then the models were fine-tuned by varying hyper-parameters to obtain optimal results.

10.4.3 Tools

All the models were implemented on Google Colab using Python.

10.4.4 Model Evaluation

In our study, classification of two classes and evaluation were done on the basis of loss, accuracy, precision, recall, and AUC and F1 score. The calculation of accuracy, precision, recall, and F1 score was done as given in Eqs. (10.1)–(10.4), respectively.

$$\text{Accuracy} = \frac{\text{TP} + \text{TN}}{\text{TP} + \text{TN} + \text{FP} + \text{FN}} \tag{10.1}$$

$$\text{Precision} = \frac{\text{TP}}{\text{TP} + \text{FP}} \tag{10.2}$$

$$\text{Recall} = \frac{\text{TP}}{\text{TP} + \text{FN}} \tag{10.3}$$

$$\text{F1 score} = 2 * \frac{\text{Precision} \ * \ \text{Recall}}{\text{Precision} \ + \ \text{Recall}} \tag{10.4}$$

10.5 Results and Discussion

10.5.1 Model Implementation

VGG19, InceptionV3, and OEM were implemented with the same configuration on Google Colab. All three models were evaluated on the basis of accuracy, precision, recall, and F1 score. Transfer learning was implemented on the models. Hyper-parameters like batch size, epochs, L2-regularization, and dropout rate were used. Models were fine-tuned by varying hyper-parameters for optimal results.

10.5.2 Model Performance on Datasets

The performance of these models on the basis of selected parameters on different datasets is presented in tabular as well as graphical form. The results for the SCPolyps training dataset are given in Table 10.1.

The model performance was also evaluated on the SCPolyps test set and the results are shown in Table 10.2.

It has been observed that OEM outperformed the other models in training and test accuracy, precision, recall, AUC, and F1 values. The performance of OEM on the SCPolyps dataset is shown in Fig. 10.5. OEM outperformed the other models in classifying polyps on the SCPolyps dataset.

The OEM model correctly predicted polyp and normal colon images as shown in Figs. 10.6 and 10.7 respectively.

Table 10.1 Performance of models on SCPolyps training set

Model	Loss	Accuracy	Precision	Recall	AUC	F1
InceptionV3	0.8109	0.9262	0.8809	0.9785	0.9405	0.8944
VGG19	0.487	0.9861	0.9861	0.9863	0.9862	0.9862
OEM	0.0012	0.995	0.995	0.996	0.9951	0.9954

Table 10.2 Performance of models on SCPolyps test set

Model	Loss	Accuracy	Precision	Recall	AUC	F1
InceptionV3	0.1338	0.9007	0.8110	0.9433	0.8896	0.8944
VGG19	0.803	0.9858	0.9857	0.9858	0.9858	0.9858
OEM	0.0125	0.993	0.993	0.994	0.9928	0.9929

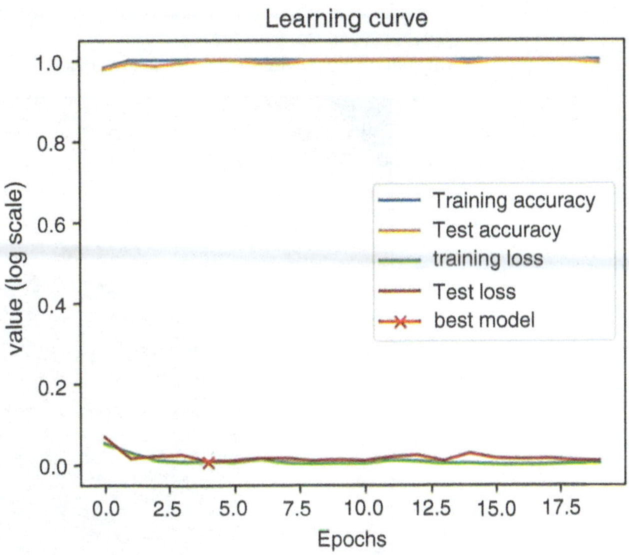

Fig. 10.5 Performance of OEM on SCPolyps training set and test set

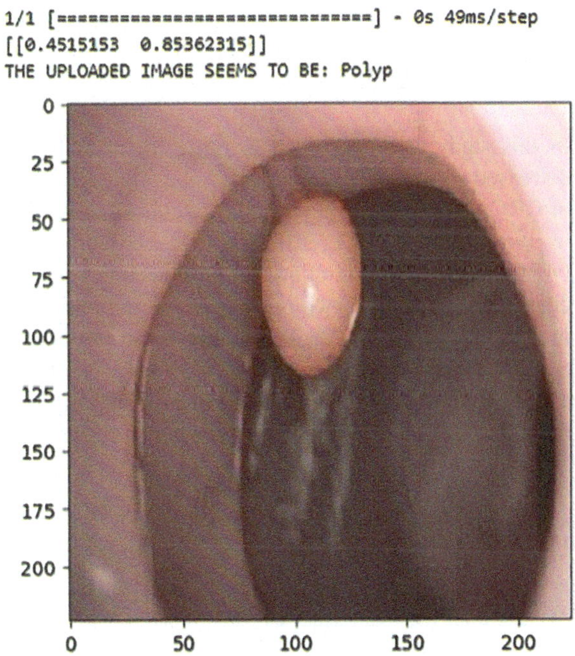

Fig. 10.6 OEM predicting polyp

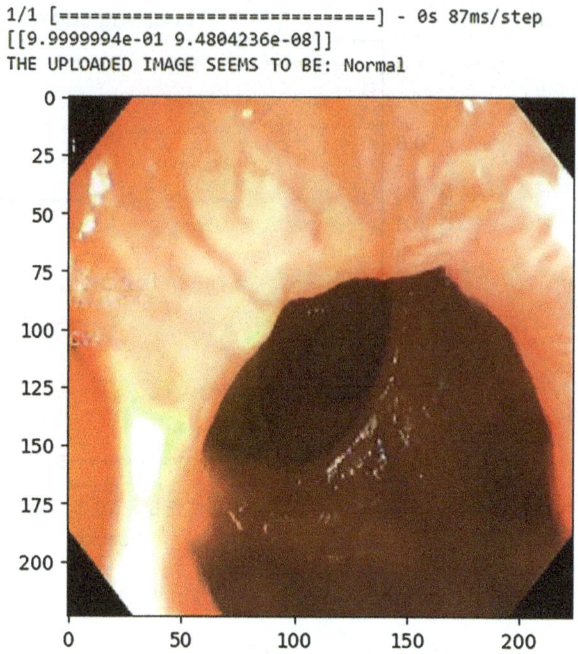

Fig. 10.7 OEM predicting normal colon

10.6 Conclusion

In this study, different deep learning methods for medical image analysis were discussed as well as different modalities that are used for detection and prediction of CRC. The deep learning models were implemented to detect polyps which may lead to CRC in later stages from colonoscopy images. On the SCPolyps dataset, the OEM model achieved training and test accuracy of 98% and 96% respectively. The challenge of identifying colorectal cancer emphasizes the value of precise prediction models. When used in conjunction with the expertise of competent medical practitioners, these models considerably contribute to the formulation of more accurate diagnoses. By applying deep learning, AI-driven systems continue to improve in this field, and there is still plenty of room for further research and improvement.

References

Akilandeswari D, Sungeetha CJ, Thaiyalnayaki K, Baskaran K, Jothi Ramalingam R, Al-Lohedan H, Al-dhayan DM, Karnan M, Hadish KM (2022) Automatic detection and segmentation of colorectal cancer with deep residual convolutional neural network. Evid Based Complement Alternat Med 2022:3415603. https://doi.org/10.1155/2022/3415603

Arnold M, Sierra MS, Laversanne M, Soerjomataram I, Jemal A, Bray F (2017) Global patterns and trends in colorectal cancer incidence and mortality. Gut 66(4):683–691. https://doi.org/10.1136/gutjnl-2015-310912

Baxter NT, Ruffin MT, Rogers MAM et al (2016) Microbiota-based model improves the sensitivity of fecal immunochemical test for detecting colonic lesions. Genome Med 8:37. https://doi.org/10.1186/s13073-016-0290-3

Choi K, Choi SJ, Kim ES (2020) Computer-Aided diagnosis for colorectal cancer using deep learning with visual explanations. In: 42nd annual international conference of the IEEE engineering in medicine and biology society (EMBC). IEEE Publications, pp 1156–1159. https://doi.org/10.1109/EMBC44109.2020.9176653

Colorectal Cancer (2022). https://www.cancer.net/cancer-types/colorectal-cancer/screening, last accessed on August 20, 2022

Gao Y, Zhang XX, Li S, Lu Y (2020) Application of artificial intelligence technology in the diagnosis and treatment of colorectal cancer. Chin J Gastrointest Surg 23:1155–1158

Goyal H, Mann R, Gandhi Z, Perisetti A, Ali A, Aman Ali K, Sharma N, Saligram S, Tharian B, Inamdar S (2020) Scope of artificial intelligence in screening and diagnosis of colorectal cancer. J Clin Med 9(10):3313. https://doi.org/10.3390/jcm9103313

He K, Zhang X, Ren S, Sun J (2016) Deep residual learning for image recognition. In: Proceedings of the IEEE conference on computer vision and pattern recognition, pp 770–778

He J, Wang Q, Zhang Y, Wu H, Zhou Y, Zhao S (2021) Preoperative prediction of regional lymph node metastasis of colorectal cancer based on 18F-FDG PET/CT and machine learning. Ann Nucl Med 35(5):617–627. https://doi.org/10.1007/s12149-021-01605-8

Ho C, Zhao Z, Chen XF, Sauer J, Saraf SA, Jialdasani R, Taghipour K, Sathe A, Khor LY, Lim KH, Leow WQ (2022) A promising deep learning-assistive algorithm for histopathological screening of colorectal cancer. Sci Rep 12(1):2222. https://doi.org/10.1038/s41598-022-06264-x

Jha D, Tomar NK, Ali S, Riegler MA, Johansen HD, Johansen D, de Lange T, Halvorsen P (2021) Nanonet: Real-time polyp segmentation in video capsule endoscopy and colonoscopy. In: 34th International symposium on computer-based medical systems (CBMS). IEEE Publications, pp 37–43

Kiehl L, Kuntz S, Höhn J, Jutzi T, Krieghoff-Henning E, Kather JN, Holland-Letz T, Kopp-Schneider A, Chang-Claude J, Brobeil A, von Kalle C, Fröhling S, Alwers E, Brenner H, Hoffmeister M, Brinker TJ (2021) Deep learning can predict lymph node status directly from histology in colorectal cancer. Eur J Cancer 157:464–473. https://doi.org/10.1016/j.ejca.2021.08.039

Krenzer A, Heil S, Fitting D, Matti S, Zoller WG, Hann A, Puppe F (2023) Automated classification of polyps using deep learning architectures and few-shot learning. BMC Med Imaging 23(1):59

Lee SH, Song IH, Jang H-J (2021) Feasibility of deep learning-based fully automated classification of microsatellite instability in tissue slides of colorectal cancer. Int J Cancer 149(3):728–740. https://doi.org/10.1002/ijc.33599

Li J, Wang P, Zhou Y, Liang H, Lu Y, Luan K (2021) A novel classification method of lymph node metastasis in colorectal cancer. Bioengineered 12(1):2007–2021. https://doi.org/10.1080/21655979.2021.1930333

Liu H, Zhao Y, Yang F, Lou X, Wu F, Li H, Xing X, Peng T, Menze B, Huang J, Zhang S, Han A, Yao J, Fan X (2022) Preoperative prediction of lymph node metastasis in colorectal cancer with deep learning. BME Front 12:1–12, article ID 9860179. https://doi.org/10.34133/2022/9860179

Lu L, Dercle L, Zhao B, Schwartz LH (2021) Deep learning for the prediction of early on-treatment response in metastatic colorectal cancer from serial medical imaging. Nat Commun 12(1):6654. https://doi.org/10.1038/s41467-021-26990-6

Lu F, Lei T, Zhou J, Liang H, Cui P, Zuo T et al (2023) Using gut microbiota as a diagnostic tool for colorectal cancer: machine learning techniques reveal promising results. J Med Microbiol 72(6):001699

Mulenga M et al (2021) Feature extension of gut microbiome data for deep neural network-based colorectal cancer classification. IEEE Access 9:23565–23578. https://doi.org/10.1109/ACCESS.2021.3050838

Namkung J (2020) Machine learning methods for microbiome studies. J Microbiol 58:206–216. https://doi.org/10.1007/s12275-020-0066-8

Nogueira-Rodríguez A, Domínguez-Carbajales R, López-Fernández H, Iglesias Á, Cubiella J, Fdez-Riverola F, Reboiro-Jato M, Glez-Peña D (2021) Deep neural networks approaches for detecting and classifying colorectal polyps. Neurocomputing 423:721–734. https://doi.org/10.1016/j.neucom.2020.02.123

Park HC, Hong IP, Poudel S, Choi C (2023) Data augmentation based on generative adversarial networks for endoscopic image classification. IEEE Access

Pogorelov K, Randel KR, Griwodz C, Eskeland SL, de Lange T, Johansen D, Spampinato C, Dang-Nguyen DT, Lux M, Schmidt PT, Riegler M (2017) Kvasir: a multi-class image dataset for computer aided gastrointestinal disease detection. In: Proceedings of the 8th ACM on Multimedia Systems Conference, pp 164–169

Qiu H, Ding S, Liu J, Wang L, Wang X (2022) Applications of artificial intelligence in screening, diagnosis, treatment, and prognosis of colorectal cancer. Curr Oncol 29(3):1773–1795

Simonyan K, Zisserman A (2014) Very deep convolutional networks for large-scale image recognition. arXiv preprint arXiv:1409.1556

Srivastava A, Tomar NK, Bagci U, Jha D (2022) Video capsule endoscopy classification using focal modulation guided convolutional neural network. In: 2022 IEEE 35th international symposium on computer-based medical systems (CBMS). IEEE, pp 323–328

Sung H, Ferlay J, Siegel RL, Laversanne M, Soerjomataram I, Jemal A et al (2021) Global cancer statistics 2020: GLOBOCAN estimates of incidence and mortality worldwide for 36 cancers in 185 countries. CA Cancer J Clin 71:209–249

Szegedy C, Liu W, Jia Y, Sermanet P, Reed S, Anguelov D, Erhan D, Vanhoucke V, Rabinovich A (2015) Going deeper with convolutions. In: Proceedings of the IEEE conference on computer vision and pattern recognition (CVPR)

Tan M, Le Q (2019) Efficientnet: rethinking model scaling for convolutional neural networks. In: International conference on machine learning. PMLR, pp 6105–6114

Teichmann M, Aichert A, Bohnenberger H, Ströbel P, Heimann T (2022) End-to-end learning for image-based detection of molecular alterations in digital pathology. arXiv preprint arXiv:2207.00095.

Thakur D, Saini JK, Srinivasan S (2023) DeepThink IoT: the strength of deep learning in internet of things. Artif Intell Rev 56:14663–14730

Topçuoğlu BD, Lesniak NA, Ruffin MT IV, Wiens J, Schloss PD (2020) A framework for effective application of machine learning to microbiome-based classification problems. MBio 11(3):10–1128

Tsai MJ, Tao YH (2021) Deep learning techniques for the classification of colorectal cancer tissue. Electronics 10(14):1662. https://doi.org/10.3390/electronics10141662

Vorontsov E, Cerny M, Régnier P, Di Jorio L, Pal CJ, Lapointe R, Vandenbroucke-Menu F, Turcotte S, Kadoury S, Tang A (2019) Deep learning for automated segmentation of liver lesions at CT in patients with colorectal cancer liver metastases. Radiol Artif Intell 1(2):180014. https://doi.org/10.1148/ryai.2019180014

Wang D, Chen S, Sun X, Chen Q, Cao Y, Liu B, Liu X (2022a) AFP-mask: anchor-free polyp instance segmentation in colonoscopy. IEEE J Biomed Health Inform 26(7):2995–3006. https://doi.org/10.1109/JBHI.2022.3147686

Wang X, Guo C, Zha Y, Xu K, Liu X (2022b) Diagnosis of nonperitonealized colorectal cancer with computerized tomography image features under deep learning. Contrast Media Mol Imaging 2022:1886406. https://doi.org/10.1155/2022/1886406

Winawer SJ (2007) The multidisciplinary management of gastrointestinal cancer. Colorectal cancer screening. Best Pract Res Clin Gastroenterol 21(6):1031–1048. https://doi.org/10.1016/j.bpg.2007.09.004

Wu Y, Jiao N, Zhu R, Zhang Y, Wu D, Wang AJ et al (2021) Identification of microbial markers across populations in early detection of colorectal cancer. Nat Commun 12(1):3063

Yang Z, Liu Z (2020) The efficacy of 18F-FDG PET/CT-based diagnostic model in the diagnosis of colorectal cancer regional lymph node metastasis. Saudi J Biol Sci 27(3):805–811. https://doi.org/10.1016/j.sjbs.2019.12.017

Yang K, Chang S, Tian Z, Gao C, Du Y, Zhang X, Liu K, Meng J, Xue L (2022) Automatic polyp detection and segmentation using shuffle efficient channel attention network. Alex Eng J 61(1): 917–926. https://doi.org/10.1016/j.aej.2021.04.072

Yao Y, Gou S, Tian R, Zhang X, He S (2021) Automated classification and segmentation in colorectal images based on self-paced transfer network. Biomed Res Int 2021:6683931. https://doi.org/10.1155/2021/6683931

Yinhang W, Jing Z, Jie Z, Yin J, Xinyue W, Yifei S et al (2023) Prediction model of colorectal cancer (CRC) lymph node metastasis based on intestinal bacteria. Clin Transl Oncol 25(6):1–12

Yu G, Sun K, Xu C, Shi XH, Wu C, Xie T, Meng R-Q, Meng XH, Wang KS, Xiao HM, Deng HW (2021) Accurate recognition of colorectal cancer with semi-supervised deep learning on pathological images. Nat Commun 12(1):6311. https://doi.org/10.1038/s41467-021-26643-8

Zhao J, Wang H, Zhang Y, Wang R, Liu Q, Li J, Li X, Huang H, Zhang J, Zeng Z, Zhang J, Yi Z, Zeng F, Zeng F (2022) Deep learning radiomics model related with genomics phenotypes for lymph node metastasis prediction in colorectal cancer. Radiother Oncol 167:195–202. https://doi.org/10.1016/j.radonc.2021.12.031

Zhu W, Xie L, Han J, Guo X (2020) The application of deep learning in cancer prognosis prediction. Cancers 12(3):603. https://doi.org/10.3390/cancers12030603

Chapter 11
Smart Farming and Human Bioinformatics System Based on Context-Aware Computing Systems

Sini Anna Alex, T. P. Pallavi, and G. C. Akshatha

Abstract Fertilizer is an important product that contributes to the growth of crops. As soil nutrients decrease, exotic and special fertilizers such as nitrogen, phosphorus, potassium, calcium, magnesium, soy milk, and sulfur are replenished in the soil. Anyhow the use of chemical fertilizers affects the lifestyle of farmers as well as the health of their crops. This chapter addresses the health of farmers and the health of crops by analyzing the characteristics of soil, environmental characteristics, and characteristics of healthy farmers. The main health problems of farmers include skin problems, lung problems, heart diseases, and cancer. This advice can suggest to farmers the best fertilizer to use to increase future crops. Hadoop Distributed File System (HDFS) has four levels of processing, like data polishing, extraction of features and matching similarity, binary analysis, and data clustering. The first stage cleans the data, removes missing values, and then performs data normalization and component decomposition. In the second stage, soil, environmental, and farmer health characteristics are extracted. The similarity is then evaluated based on the construction of ontology-supported grid reduction (OMR) to predict farmers' health problems. In the third stage, the FP-growth algorithm and densely connected recurrent neural network (DC-RNN) are used to classify healthy farmers and healthy crops. In the fourth stage, the last group of farmers is presented with product health information from the self-planning map and, accordingly, product and fertilizer recommendations that will reduce health risks. Recommendations were made by HDFS, and performance was evaluated concerning parameters like precision, recall, F measurement, and accuracy on papaya, banana, and leafy vegetables.

S. A. Alex (✉) · G. C. Akshatha
Department of CSE (AI & ML), Ramaiah Institute of Technology, Bangalore, Karnataka, India
e-mail: sinialex@msrit.edu; akshatha.gc@msrit.edu

T. P. Pallavi
Department of CSE (Cyber Security), Ramaiah Institute of Technology, Bangalore, Karnataka, India
e-mail: pallavitp@msrit.edu

A. Khamparia et al. (eds.), *Microbial Data Intelligence and Computational Techniques for Sustainable Computing*, Microorganisms for Sustainability 47,
https://doi.org/10.1007/978-981-99-9621-6_11

Keywords Context-aware computing · Clustering · Crop health · Health risks · Recommendation

11.1 Introduction

Agriculture is the main source of food production in different climates, fertilizers, pesticides, and soil types (Shirsath et al. 2017). The use of chemical fertilizers can be harmful to health. In general, excessive use of chemical fertilizers leads to reduced crop yields and affects the health of farmers. On the other hand, inadequate use of fertilizer will not support crop growth. Therefore, the best fertilizer should be used in the soil to increase productivity every year. Using faster learning machines such as Adaptive Neuro-Fuzzy Interference System (ANFIS), Support Vector Machine (SVM) (Karandish et al. 2017) to predict fertilizer concentration. Nitrogen (N) is an essential fertilizer that helps the growth of new fruits such as apples, peaches, peaches, oranges, walnuts, olives, and kiwis (Carranca et al. 2018). Like nitrogen, other commonly used nutrients are phosphorus (P), potassium (K), calcium (Ca), magnesium (Mg), sulfur (S), iron (Fe), manganese (MN), copper (Cu), and zinc (Zn) (Moreno et al. 2018). Nutrients in the form of a person's manure should enrich the consistency of the soil to promote the growth of crops.

Fertilizer recommendation systems specifically take into account soil parameters to obtain the required results (Jethva et al. 2018). The simple approval process is shown in Fig. 11.1. In agriculture, farmers have to suffer from deadly diseases. Cultivation is often done by farmers without gloves and dust covers, causing health hazards. Farmers who are not protected and use fertilizer unbalanced will experience health problems (Nguyen Viet et al. 2019). Determine the root cause of pesticides by modeling machine learning algorithms such as artificial neural network (ANN), K-nearest neighbor (K-NN), deep learning (DL), ensemble learning (EL), and SVM (Tomiazzi et al. 2019; Liakos et al. 2018). Use classification techniques to predict farmers' health problems. Health risks include health effects such as sore throat, headache, fatigue, skin irritation, eye irritation, and difficulty breathing

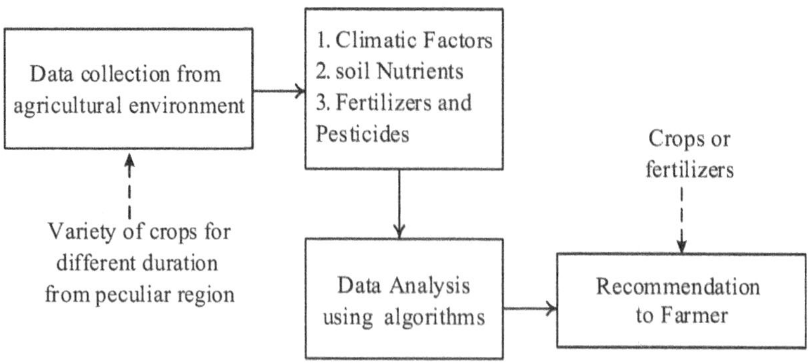

Fig. 11.1 Agriculture recommendation system

(Sonchieu et al. 2018). Due to these health effects, food availability in agricultural lands is determined by yield, soil content, and nutritional status (Raut et al. 2018).

Get more out of big data by using cluster algorithms to predict crops. Use clustering algorithms based on classification around central point (PAM), clustering of uppercase letters (CLARA), and speed-based matching of applications with noise (DBSCAN) (Majumdar et al. 2017). Crop yield also depends on fertilizer use. In the analysis, it is tested for quality and product density (Senthil Vadivu et al. 2017). Nitrogenous fertilizer will be given, while the split plant trial is carried out. Therefore, the nutrients nitrogen, phosphorus, and potassium are all depleted, and rice is very useful. Climate change is also having a greater impact on growth and quality.

In addition to crops, other health risks posed by fertilizers to farmers include heart disease, immune system disorders, skin problems, and cancer (Zhang et al. 2019). Fertilizer use also affects crop growth and farmers' health.

11.1.1 Motivation

Ecological impacts related to food agriculture; but it also depends on how the fertilizer is used. The two main reasons for using fertilizer in agriculture are profitability and health risks. Many studies have examined farming techniques based on crop health analysis and farmer health analysis, focusing on agricultural characteristics and health characteristics (genetic problems and symptoms), respectively. The main purpose of this research is to determine crop health and farmers' health, depending on fertilizer use.

Based on this motivation, two principal objectives are defined:

- To develop a larger-scale system for evaluating both crop health and farmer health risks simultaneously with higher accuracy.
- To deliver appropriate recommendations of fertilizers for farmers that help to increase future yield from the farm by reducing health risks.

11.1.2 Contributions of This Chapter

The main conclusions of this chapter are as follows:

- Dual analysis of farmer health and crop health to recommend better fertilizers to reduce health risks and increase yields. Hadoop Distributed File System (HDFS) has been adopted to analyze larger data. The importance of milk consumption for different crops and its impact on cancer risk was evaluated.
- HDFS works in four stages: data polishing, feature extraction, binary analysis, and data clustering. In order to ensure the accuracy of the analysis, the first two stages were completed, and in the last stage, it was suggested that appropriate suggestions be made.

- The first stage is data cleaning based on expectation maximization (EM), noise distance, normalization using *Z*-score, data normalization, and component decomposition.
- The second stage is responsible for revealing three different elements such as land, environment, and farmer health. Farmers' health characteristics may have clinical content, so they are extracted from ontology-assisted map reduction (OMR).
- The third stage plays an important role in identifying the poor, average, and good for crop growth and farmers' health. This farmer health classification system is based on FP-growth algorithm and densely connected recurrent neural network (DC-RNN) for crop identification.
- The fourth objective is to provide appropriate advice to farmers and ranchers. The analyzed data was divided into three categories using the self-reporting system (SOM) to obtain the best recommendations to improve farmers' agricultural practices.
- The impact of using different fertilizers on banana, papaya, and vegetable farms containing lead on health risks for farmers and crops; accuracy was evaluated based on regression, F-measure, and precision.

11.2 State-of-the-Art

This section is composed of details about previous research works carried out over the analysis of crops, soil, fertilizer, and recommendations in the field of agriculture for different crop types.

Management of crop health in agriculture is an important factor limiting productivity increase. Higher yield means healthier crop. Using multi-model ensemble technique to predict performance (Wallach et al. 2018). Climate change in the environment often affects crops. This was done by collecting sufficient climate change models for 25 crops. Since crop growth in recent days depends on the characteristics of the soil, the limitation of nutrients in the soil causes crop failure. In this case, fertilizer is the best medicine to stimulate crop growth. Dynamic Land Ecosystem Model (DLEM) is used to model crop growth and yield (Xhang et al. 2018). Data were collected and tested for three crops between 1980 and 2012: wheat, corn, and rice. Due to the large amount of agricultural data, research efforts on machine learning and artificial intelligence algorithms are still ongoing. Deep neural networks (DNN) have been used to evaluate maize and bean yields (Kim et al. 2019; Khaki et al. 2019). The data includes cultivated land, weather data, hydrological data, and crop yield statistics. DNN generator is equipped with a hidden algorithm, loss function, optimization, activation function, and output value. Each function contains an estimate of the sum of the squared error, mean square error, and cross-entropy.

In Jeong et al. (2016), the Random Forest (RF) algorithm was studied to accurately predict crop yield. Weather conditions and nitrogen fertilizer were taken into account when verifying the data. However, the dataset contains some of the fertilizers, and the RF algorithm takes more time to build the decision tree, and the complexity increases as the depth of the tree increases. It needs more help handling small files. The fuzzy logic system uses two factors, namely rainfall and temperature, to determine crop yield (Bang et al. 2019). Precipitation characteristics were modeled using autoregressive moving average (ARMA) and externally variable autoregressive moving average (ARMAX), then temperature was modeled for ARMA and seasonal autoregressive integrated moving average (SARIMA). These two factors are used in the fuzzy system to classify crop yields as good, very good, medium, bad, and very poor.

Clustering is a solution to predict crop yield by considering multiple characteristics. Clustering algorithms such as k-means, distribution around medoids (PAM), clustering of uppercase letters (CLARA), and density-based spatial clustering of noisy applications (DBSCAN) have been proposed in Afrin et al. (2018). After collection, crop yield is estimated by linear regression method. The main characteristics required for separation are soil properties and weather conditions. The clusters were formed based on similarities in soil nutrients and climate conditions. The three climate parameters used in this project are temperature, humidity, and precipitation. Among the four integration methods, the results of the DBSCAN algorithm have higher accuracy and lower cost than the good one. However, other important factors such as water level and fertilizer are ignored in this study. In Suresh et al. (2018), clustering and classification algorithms, such as K-means and improved K-nearest neighbor algorithms, have been proposed. In this study, we discussed the classification as rainwater, groundwater, and cultivated land. The choice of distribution k value is complex.

Classify soil properties to predict fertilizer application needs. Suchithra and Pai proposed a machine learning technique (ELM) that can operate using five different forces (Gaussian radial basis, sine square, hyperbolic tangent, triangle, and constant limit) (Suchithra and Pai 2020). This function calculates the nutrient fertility index and pH of the soil based on the soil parameters. The effect of fertilizers on the soil creates a nutrient imbalance and reduces productivity. ELM cannot provide accurate results with higher accuracy than real machine learning algorithms. Appropriate fertilizer can be recommended by distributing nutrients into the soil. Prepare a recommended fertilizer (Suchithra and Pai 2018). This study used the sigmoid kernel in a multi-class SVM to provide product recommendations. Parameters in SVM are optimized by genetic algorithm (GA) and particle swarm optimization (PSO) algorithm. Tuning with GA takes more time, and traditional SVM requires large learning time and memory consumption, so it is not suitable for larger datasets.

The effects of chemical fertilizers and pesticides pose risks to the environment (Rahman and Zhang 2018). A binary logistic regression model was applied to the data collected from farmers in 2016. Fertilizer levels for specific crops were determined, but health recommendations and risks were not disclosed. Recommendations to prevent harm to farmers were examined in Mubushar et al. (2019), Ichami et al. (2019). Fertilizer recommendations are not made using fast machine learning algorithms. A fertilizer recommendation letter was previously developed by volunteers who collected information from farmers who also needed to understand the health issues of fertilizer use and lighting. Most agricultural activities involve growing crops or approving fertilizers, but both are equally important to increase production and reduce the health risks of fertilizers to farmers. The main issues identified in this research area will be continued in the next section.

11.3 Data Analytics in Agriculture

The two main points in agriculture in this research are evaluating the health of farmers and making recommendations. Health risks to farmers in agriculture arise from the use of chemical fertilizers and mental and physical stress. The health risks of neurobehavioral and musculoskeletal problems have been studied (Khan et al. 2019; Sang et al. 2018). Snowball and saturation methods were used in this study, but the results were incorrect due to a lack of information. To improve the accuracy of the results, it is suggested to use genetic algorithms to develop fuzzy failure mode and effects analysis (FMEA). Most genetic algorithm methods consume a lot of time, and the fuzzy concept has 125 rules that must always work. But this is a better algorithm and cannot support big data in a short time. Fertilizer-based farmer health assessment was conducted in Mabe et al. (2017). In this study, farmers asked many questions about pesticide and fertilizer use. Multiple linear regression models were used to determine health outcomes. This regression model is capable of handling linear effects and survey-based statistical errors to predict farmer health.

An agricultural consensus has been formed to measure soil toxicity and inform farmers. Here, the J48 decision tree algorithm is used for classification; as the height of the tree increases, more storage space is needed (Pawar and Chillarge 2018). Similarly, Bodake et al. (2018) proposed the Naive Bayes algorithm for classification, but its accuracy is lower. Additionally, using pH to predict soil quality is not enough because soil consistency also depends on the surrounding air.

The problems mentioned in estimating the welfare of farmers without and with fertilizer make fertilizer lower. After that, the limit could not be determined in the agreement, and no information regarding fertilizer was produced. These issues are well addressed by the solutions proposed in this study and detailed in the next section.

11.4 Proposed HDFS Recommendation System

11.4.1 System Model

The proposed HDFS recommendation system is modeled by analyzing crop health and farmer health using multiple features. The recommendation system aims to provide proper guidance for the use of fertilizer that helps to improve crop growth as well as mitigate farmer's health risks. Multi-feature-based dual analysis in the HDFS recommendation system is employed to function with four sequential phases as follows:

- Phase 1: Data polishing.
- Phase 2: Feature extraction and similarity matching.
- Phase 3: Dual analysis.
- Phase 4: Data clustering.

Figure 11.2 depicts the overall working processes handled in the proposed HDFS recommendation system. The main goal of this system is to analyze different fertilizers and the risks caused to crop and farmer health due to the utilization of fertilizer. The result to be obtained from this HDFS recommendation is helpful for future crop selection and fertilizer selection.

11.4.2 Phase 1: Data Polishing

In this phase, the collected dataset is polished by data cleaning, noise elimination, normalization, data canonicalization, and component breakdown. This phase is performed for improving accuracy. The dataset is first processed with a data cleaning step which re-fills the missing data in the dataset. The expectation maximization (EM) for data cleaning consists of the expectation (E) step and the maximization (M) step; the missing value is updated only if convergence is satisfied. Based on E, missing values are re-filled, and the parameter quality is maximized by M. The mathematical expressions for E and M are given as:

$$E \Longrightarrow Q\left(\theta|\theta^{(t)}\right) = E_{z|X,\theta^{(t)}}[logL(\theta;X,Z)] \tag{11.1}$$

$$M \Longrightarrow \theta^{(t+1)} = \underset{\theta}{agr\ max}\ Q\left(\theta|\theta^{(t)}\right) \tag{11.2}$$

Let Z be the missing values in the dataset, θ are the unknown parameters, X is the currently determined parameter of Z, $logL(\theta; X, Z)$ is the likelihood function, and $Q(\theta| \theta^{(t)})$ is the expected value. After re-filling the missed values in the dataset noise elimination is handled by the estimation of cosine similarity. This cosine similarity is applied to identify nearest neighbors, i.e., values very close to the values that are eliminated. The cosine similarity for two dataset attributes is given as:

$$Sim = \frac{\sum_i^n A_i B_i}{\sqrt{\sum_i^n A_i^2}\sqrt{\sum_i^n B_i^2}} \tag{11.3}$$

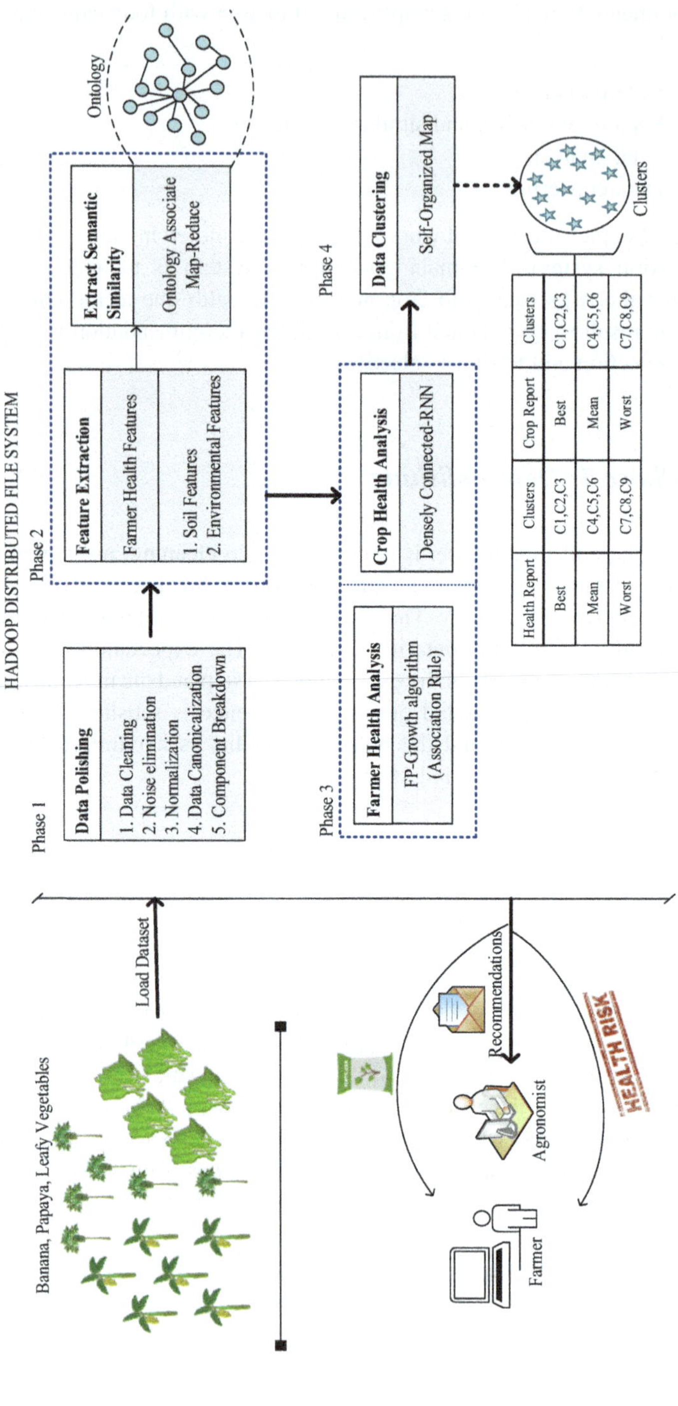

Fig. 11.2 Proposed HDFS recommendation system

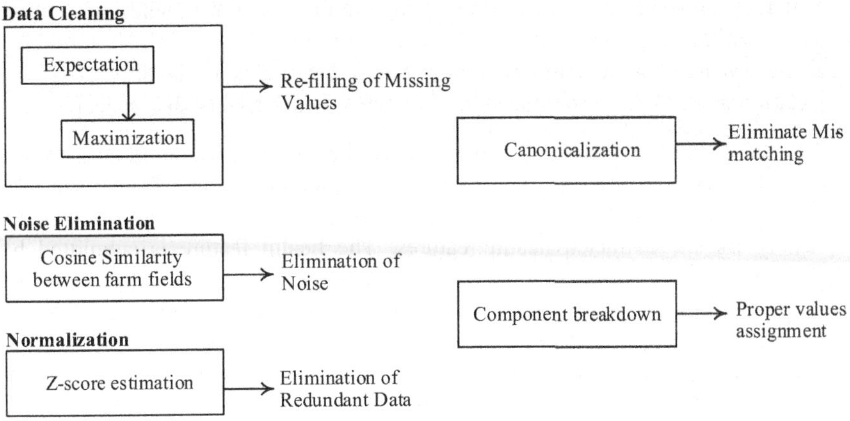

Fig. 11.3 Purpose of individual process in data polishing

The terms A_i and B_i are considered features from two specific farm fields; only the nearest values are eliminated. Then the dataset is normalized using Z-score for eliminating redundant data. This Z-score is expressed as follows:

$$z = \frac{x - \mu}{\sigma} \tag{11.4}$$

Let x be the raw score, and μ and σ are the mean and standard deviation of the population. Hereby, the dataset is completely removed with noise and redundant data.

Then data canonicalization is handled for transforming the data into standard form. A fertilizer's representation can be given in more than one form that corresponds to a similar fertilizer. Whatever form is provided, it is transformed into standard form; by doing canonicalization unwanted mismatching of fertilizers is mitigated, which enables to increase accuracy in analysis. This process is also effective in reducing incorrectly represented features. The last step in data polishing is component breakdown, which enables to split off the improperly specified soil nutrients. For example, "25–4-2" denotes 25% of nitrogen, 4% of phosphorous, and 2% of potassium. The sequential process handled in data polishing is depicted in Fig. 11.3 along with the major attainment of each process. On completion of data polishing, the dataset is taken over for the next phase of processing.

11.4.2.1 Phase 2: Feature Extraction and Similarity Matching

In this phase features are extracted from the dataset; the three different key features that are extracted are:

- Soil feature—Nitrogen, phosphorous, potassium, calcium, magnesium, sulfur, and soymilk.
- Environmental feature—Temperature, rainfall, pH, and water level.
- Health feature—Skin problems, lung problems, heart defects, and cancer.

The soil features are extracted based on the utilization of the amount of fertilizers in the soil for crop growth. On extraction of these three key features from the dataset they are processed, before which ontology-based map-reduce process is executed, especially for the extracted health features. The health feature is submitted by farmers based on their symptoms. The medical terms are generally scientific, which requires more details for accurate analysis. So the extracted health features are processed into the OMR process.

Ontology is constructed based on the relationship between the domain concept and the topic. In this work the domain concept denotes health risk and the topic denotes the corresponding symptoms of the farmer. The associated relationship between symptoms and health risks enables to create ontology. Figure 11.4 demonstrates the constructed ontology on the left side, where the key, value (k, v) pairs are generated, and on the right side, the map-reduce processing is carried over.

Split: Initially, the given input symptoms of a farmer are split into individuals.

Map: Mapping functions are enabled to map the occurrences of each symptom in the ontology using (k, v) pair.

Combine: The mapped (k, v) pairs are consolidated and sorted in order for the next process.

Reduce: Lastly reducer function is performed to reduce the given inputs. Finally the reducer supports to identify particular health risks.

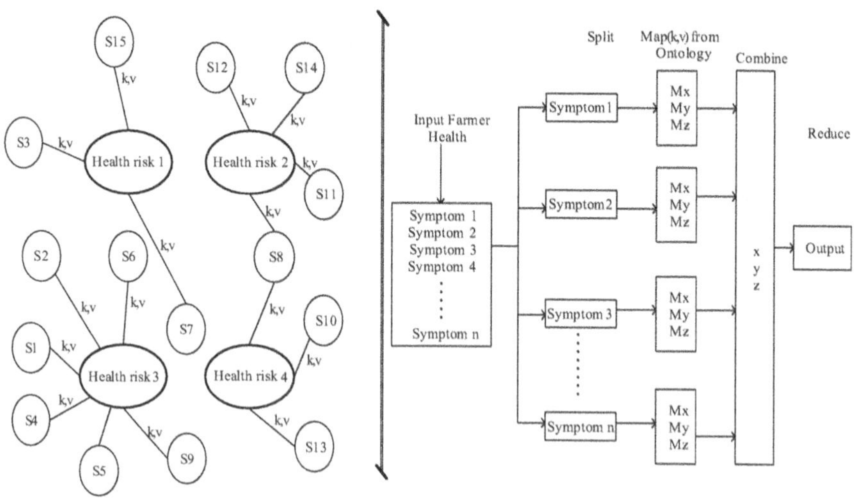

Fig. 11.4 OMR working model

11.4.3 Phase 3: Dual Analysis

In this phase, farmer health and crop health are analyzed, and so this phase is named as dual analysis. Farmer health analysis is handled by FP-growth algorithm, and crop health analysis is handled by densely connected RNN.

Farmer health analysis—The FP-growth is proposed to generate association rules. Two processes performed in farmer growth analysis are the construction of FP-tree and the extraction of association rules. The tree is built from the transactional data that includes farmer's health constraints.

Initially the FP-tree is generated in the form of an acyclic graph $G = (N, E)$ in which N and E denote nodes and edges respectively. Then by traversing through the path, frequent patterns are extracted, which is followed by the generation of association rules.

For instance, Table 11.1 enlists a set of ten transition identities as {T(ID1), T (ID2), T(ID3), T(ID4), T(ID5), T(ID6), T(ID7), (ID8), T(ID9), T(ID10)} with their corresponding itemset {a, b, c, d, x, y}. In this proposed work, the itemset denotes farmer's health symptoms, diseases, and fertilizers using which the frequent health risks are extracted. After constructing F-Tree the support count for each transaction ID is determined, then the larger support count itemsets are filtered out, and the frequent pattern from all the transactions is extracted. From the frequent itemset, strong association rules are built.

Let the frequent pattern mined from the FP-growth be $p = \{A, B\ldots\}$. In association rules, support and confidence are two main properties that are given as:

$$S(A \Rightarrow B) = P(P \cup B) \tag{11.5}$$

$$C(A \Rightarrow B) = P(B|A) \tag{11.6}$$

Association rules are defined from frequent patterns. Here A and B represent disjoint patterns, and S and C are support and confidence respectively. For generating strong association rules, the individual frequent pattern from p creates nonempty subsets Ns_b. On taking each Ns_b the rule is constructed as follows:

Table 11.1 Transaction data

Transaction ID	Itemset
T(ID1)	a, b, c, d, x, y
T(ID2)	d, x, y
T(ID3)	a, b, c
T(ID4)	c, x, y
T(ID5)	a, b
T(ID6)	c, d, x, y
T(ID7)	b, c, d
T(ID8)	x, y
T(ID9)	c, d, x, y
T(ID10)	b, c, d

$$Ns_b \Rightarrow (p - Ns_b) \text{ if } C = S(p)/S(Ns_b) \geq C_{\min} \tag{11.7}$$

$C_{\min}$ denotes minimum confidence value. FP-growth algorithm performs faster in extracting frequent patterns. The better performance of FP-growth than the conventional a priori algorithm presents absolute farmer health analysis.

Crop health analysis—The crop health is analyzed by using DC-RNN in which the output determined from the previous layer will be processed as one of the inputs into subsequent layers. This DC-RNN is chosen since it increases performance by reprocessing the features from previous layers. The first layer receives input from the dataset which includes crop features, i.e., soil features and environmental features. The consecutive layers are connected in the feed-forward model. The potential benefits of DC-RNN are feature reuse, stronger feature propagation, and others. The first layer extracts features from the dataset. In this network the features are processed and concatenated for better results. Let l be a layer in DC-RNN which receives the following feature maps as $(x_0, x_1, x_2, \ldots, x_{l-1})$, so the input of the previous layer is given as:

$$x_1 = H_l([x_0, x_1, x_2, \ldots, x_{l-1}]) \tag{11.8}$$

From the above, $[x_0, x_1, x_2, \ldots, x_{l-1}]$ are the concatenated features between 0, 1, ..., $1 - l$ layers, and H_l is the input sample present in l^{th} DC-RNN blocks. The feature maps are fed into RNN layers. The three operations performed in this proposed DC-RNN are batch normalization, rectified linear unit (ReLU), and 3×3 convolution. Let the output from DC-RNN be $H_{lk}(t)$ for l^{th} layer has k^{th} feature of $H_{lk}(\cdot)$with t time. Hereby the output from the layer is mathematically expressed as shown below:

$$H_{lk}(t) = \left(w^f_{(l,k)}\right)^T * H^{f(i,j)}_{(l,k)}(t) + \left(w^r_{(l,k)}\right)^T * H^{f(i,j)}_{(l,k)}(t-1) + b_{(l,k)} \tag{11.9}$$

The inputs to the convolution layer and l^{th} recurrent convolution layer are represented as $H^{f(i,j)}_{(l,k)}(t)$ and $H^{f(i,j)}_{(l,k)}(t-1)$. Further the weighted values of the convolution neural layer and the recurrent layer are $w^f_{(l,k)}$ and $w^r_{(l,k)}$, respectively, that denote l^{th} layer and k^{th} feature map, and $b_{(l,k)}$ is the bias. As a result, crop health is divided into three classes: poor, average, and best.

11.4.4 Phase 4: Data Clustering

SOM is presented in this phase for clustering crop health and farmer health individually. The SOM clustering is performed by initializing weights, mapping, and matching units. The weighted values for each input data are initialized, and the clustering operations are performed after the construction of the map. Then the matching unit is predicted using L2 distance. Let $X = (x_1, x_2)$ and $Y = (y_1, y_2)$ be two points for which the L2 distance is predicted as follows:

$$d = \sqrt{(y_1 - x_1)^2 + (y_2 - x_2)^2} \tag{11.10}$$

After identifying the best matching unit by using the distance measurement, it is then updated with weighted values. The weight update is handled by:

$$W_v(s + 1) = W_v(s) + \theta(u, v, s).\alpha(s).D((t) - W_v(s)) \tag{11.11}$$

From the above equation, W_v denotes the weight vector of node v, s is iteration, $\theta(u, v, s)$ is neighborhood function, $\alpha(s)$ is learning restraint, u is the best matching unit that is present in the map, and $D(t)$ represents the target data vector. On measuring the distance, similarity between points is predicted, and clusters are constructed for both health records and crop records.

Based on the selected target vector, the clusters are constructed. Here, we construct clusters in three categories: best, mean, and worst. Each category is composed of clusters which enable the prediction of the accurate status of the crop and farmer health. This clustering is the key to providing a recommendation system. The health report of best, mean, and worst denotes that farmers are caused with lesser health risk of cancer, either caused by some other risk and cancer respectively. Similarly, the crop reports are also clustered in three such categories. The following pseudo-code illustrates the procedure followed for clustering analyzed data.

The procedure of the SOM Clustering algorithm

Step 1: Initialize weight vectors for nodes.
Step 2: Randomize weight vectors into the map.
Step 3: Pick target input vector//Select best, mean, and worst values.
Step 4: Estimate L2 distance//Predict similarity between the values.
Step 5: Pick the best matching unit.
Step 6: Update weight vectors.
Step 7: Construct a cluster based on the similarity.

The obtained clusters are helpful in forwarding recommendations for farmers. Hereby the key goal of analyzing farmer health and crop health is attained, and the recommendation is delivered to farmers regarding the health risks and crop yield. Understanding the health risk posed by the utilized fertilizer is essential for the farmer to predict the severity of a particular fertilizer in soil for crop growth.

11.5 Conclusion

The optimal farm plan in the HDFS environment is designed to analyze the health of crops and farmers. Although the use of fertilizer in agriculture plays an important role in crop growth, it can also cause serious harm to farmers. The standard HDFS environment is completed in four stages: data polishing, feature extraction, binary analysis, and clustering. To improve the accuracy of our analysis, we perform data polishing to eliminate unnecessary variables that reduce accuracy. Features were extracted after data polishing was completed following data cleaning, normalization, and noise removal. By analyzing the crops' and farmers' health, three types of features were extracted: soil, environment, and farmer. Based on the results obtained, the farmer's health conditions were specially processed in the OMR to show health-related symptoms.

Analysis of two main roles in predicting crop weight and farmers' welfare using DC-RNN and FP breeding, respectively. These two methods are preferred because they are faster and provide more accurate analysis. Finally, the SOM aggregation is based on best, average, and worst crops and farmer health to provide reasonable recommendations. The causes of fertilizer deficiency in crops and farmers' health are analyzed, and recommendations on fertilizer are given to farmers and agronomists.

Future Work: This HDFS recommendation system is planned to be extended with the following directions in the future:

- Analysis of different crops that cause farmer health risks under varying concentrations of fertilizers from agricultural fields.
- Build an authentication scheme for enabling secure participation of farmers, since the health details are sensitive.

In the future, integrated deep learning methods can be studied for improving the performances of HDFS systems in agricultural environments.

References

Afrin S, Khan AT, Mahia M, Ahsan R, Mishal MR, Ahmed W, Rahman RM (2018) Analysis of soil properties and climatic data to predict crop yields and cluster different agricultural regions of Bangladesh. In: 2018 IEEE/ACIS 17th international conference on computer and information science (ICIS) 2018 Jun 6. IEEE, pp 80–85

Bang S et al (2019) Fuzzy logic based crop yield prediction using temperature and rainfall parameters predicted through ARMA, SARIMA, and ARMAX model. In: 2019 Twelfth international conference on contemporary computing (IC3). IEEE, pp 1–6

Bodake K, Ghate R, Doshi H, Jadhav P, Tarle B (2018) Soil based fertilizer recommendation system using Internet of Things. MVP J Eng Sci 1(1):13–19

Carranca C, Brunetto G, Tagliavini M (2018) Nitrogen nutrition of fruit trees to reconcile productivity and environmental concerns. Plan Theory 7(1):4

Ichami S, Shepherd K, Sila A, Stoorvogel J, Hoffland E (2019) Fertilizer response and nitrogen use efficiency in African smallholder maize farms. Nutr Cycl Agroecosyst 113(1):1–19

Jeong J, Resop JP, Mueller N, Fleisher D, Yun K, Butler E, Timlin D, Shim K, Gerber J, Reddy V, Kim S (2016) Random forests for global and regional crop yield predictions. PLoS One 11(6): e0156571

Jethva JM et al (2018) A review on data mining techniques for fertilizer recommendation. Int J Sci Res Comput Sci Eng Inform Technol 3(1):1386–1390

Karandish F, Darzi-Naftchali A, Asgari A (2017) Application of ML models for diagnosing health hazard of nitrate toxicity in shallow aquifers. Paddy Water Environ 15(1):201–215

Khaki S et al (2019) Crop yield prediction using deep neural networks. J Front Plant Sci 10:621

Khan N, Kennedy A, Cotton J, Brumby S (2019) A pest to mental health, exploring the link between exposure to agrichemicals in farmers & mental health. Int J Environ Res Public Health 16(8):1327

Kim N, Ha KJ, Park NW, Cho J, Hong S, Lee YW (2019) A comparison between major artificial intelligence models for crop yield prediction: case study of the Midwestern United States, 2006–2015. Int J Geoinform 8(5):240

Liakos K et al (2018) Machine learning in agriculture: a review. Sensors 18(8):2674

Mabe FN, Talabi K, Danso-Abbeam G (2017) Awareness of health implications of agrochemical use: effects on maize production in Ejura-Sekyedumase municipality, Ghana. Adv Agric 2017:7960964. https://doi.org/10.1155/2017/7960964

Majumdar J et al (2017) Analysis of agriculture data using data mining techniques: application of big data. J Bid Data 4:20

Moreno RH et al (2018) Model of neural networks for fertilizer recommendation and amendments in pasture crops. In: 2018 ICAI workshops (ICAIW), pp 1–5

Mubushar M et al (2019) Assessment of farmers on their knowledge regarding pesticide usage and biosafety. Saudi J Biol Sci 26(7):1903–1910

Nguyen Viet H, Grace D, McDermott J (2019) Integrated approaches to tackling health issues-related to agri-food systems. Int J Public Health 64(1):5–6

Pawar M, Chillarge G (2018) Soil toxicity prediction and recommendation system using data mining in precision agriculture. In: 2018 3rd international conference for convergence in technology (I2CT) 2018 Apr 6. IEEE, pp 1–5

Rahman K, Zhang D (2018) Effects of fertilizer broadcasting on the excessive use of inorganic fertilizers and environmental sustainability. Sustainability 10(3):759

Raut R et al (2018) Soil monitoring, fertigation, and irrigation system using IoT for agricultural application. In: Intelligent communication and computational technologies: proceedings of Internet of Things for Technological development, IoT4TD 2017. Springer, Singapore, pp 67–73

Sang AJ, Tay KM, Lim CP, Nahavandi S (2018) Application of a genetic-fuzzy FMEA to rainfed lowland rice production in sarawak: environmental, health, and safety perspectives. IEEE Access 6:74628–74647

Senthil Vadivu S et al (2017) Modelling a predictive analytics methodology for forecasting rice variety and quality on yield on farm and farming attributes using Bigdata. Int J Pure Appl Math 116(5):61–65

Shirsath R et al (2017) Agriculture decision support system using data mining. In: International conference on intelligent computing and control (I2C2). IEEE

Sonchieu J et al (2018) Health risk among pesticide sellers in Bamenda (Cameroon) and peripheral areas. Environ Sci Pollut Res 25(10):9454–9460

Suchithra M, Pai M (2018) Improving the performance of sigmoid kernels in multiclass SVM using optimization techniques for agricultural fertilizer recommendation system. In: Soft computing systems: second international conference, ICSCS 2018, Kollam, India, April 19–20, 2018, revised selected papers 2. Springer, Singapore, pp 857–868

Suchithra M, Pai M (2020) Improving the prediction accuracy of soil nutrient classification by optimizing extreme learning machine parameters. Inform Process Agric 7(1):72–82

Suresh A, Ganesh Kumar P, Ramalatha M (2018) Prediction of major crop yields of Tamilnadu using K-means and modified KNN. In: 2018 3rd International conference on communication and electronics systems. IEEE

Tomiazzi JS et al (2019) Performance of machine-learning algorithms to patterns recognition and classification of hearing impairment in Brazilian farmers exposed to pesticide and/or cigarette smoke. Environ Sci Pollut Res 26(7):6481–6491

Wallach D et al (2018) Multimodel ensembles improve predictions of crop-environment-management interactions. Glob Change Biol 24(11):5072–5083

Xhang J et al (2018) Improving representation of crop growth and yield in the dynamic land ecosystem model and its application to China. J Adv Model Earth Syst 10(7):1680–1707

Zhang J et al (2019) Farm machine use and pesticide expenditure in maize production: health and environment implications. Int J Environ Res Public Health 16(10):1808. https://doi.org/10.3390/ijerph16101808

Chapter 12
Plant Diseases Diagnosis with Artificial Intelligence (AI)

Syed Muzammil Munawar, Dhandayuthabani Rajendiran, and Khaleel Basha Sabjan

Abstract India's agriculture is significant because of the country's expanding population and rising food demands. Therefore, it is necessary to increase crop productivity. One of these significant factors contributing to reduced agricultural yields is the prevalence of bacterial, fungal, and viral illnesses. Applying techniques for plant disease identification helps stop and manage this. Machine learning techniques will be used in the process of identifying plant illnesses since they apply information most frequently and provide excellent methods for disease diagnosis. Machine learning-based techniques can be used to identify diseases because they focus mostly on data superiority outcomes for a certain goal. In this method, machine learning and deep learning based on artificial intelligence (AI) have been used to conduct a thorough assessment of the numerous methodologies used in plant disease diagnosis. In the realm of computer vision, deep learning has also become increasingly important for providing improved performance results for identifying plant diseases. There has been significant progress in the machine learning and computer vision fields as a result of the application of deep learning improvements to a variety of disciplines. In order to demonstrate the superiority of the deep learning model over the machine learning model, a comparison of the two techniques' performances and applications in numerous research articles has been made. The deep learning technology can be used to identify leaf diseases from collected photos in order to prevent significant crop losses.

Keywords Agriculture · Neural Networks · Plant diseases detection · Machine learning methods · Artificial intelligence and deep learning · Supervised learning

---

S. M. Munawar (✉) · D. Rajendiran · K. B. Sabjan
Department of Biochemistry, C. Abdul Hakeem College (Autonomous), Melvisharam, Vellore, Tamil Nadu, India

A. Khamparia et al. (eds.), *Microbial Data Intelligence and Computational Techniques for Sustainable Computing*, Microorganisms for Sustainability 47,
https://doi.org/10.1007/978-981-99-9621-6_12

12.1 Introduction

A thorough study on several machine and deep learning algorithms for plant disease recognition and classification has been published. After that, different machine learning categorization algorithms are useful to perceive plant diseases in an effort to assist farmers with automatic disease detection of all types of agricultural diseases to be establish. The study goes over diverse deep learning methods to detecting plant diseases. Additionally, a number of methods/mappings for identifying the disease symptoms were summarized. Here, the newest advancements in deep learning technology in analysis of plant leaf disease the effort determination is valuable resource to researchers attempting to identify plant diseases. Additionally, a comparison flanked by deep learning, machine learning methods is done. Despite the fact that there has been a lot of notable progress within modern years, nearby a quantity of research gaps in need to be filled in order so to put into practice efficient strategies for plant disease identification. Agricultural areas can detect plant leaf illnesses and accurately report them to the right parties thanks to the integration of IoT, AI, and unmanned aerial vehicles. No one is interested in farming or agriculture because of the challenges farmers confront every day in contemporary society so that all members of the younger generation move to modern cities in order to live safely and steer clear of such agricultural challenges. The issue of effectively preventing plant diseases is closely tied to climate and agricultural change (Harakannanavara et al. 2022). The physiological nature of host-pathogen interactions may vary as a result of climate change, which may also alter host resistance and the stages and rates of pathogen development (Das and Sengupta 2020). The issue is made worse by the ease with which illnesses are disseminated globally today. New diseases may develop in regions where they haven't yet been identified and, naturally, where there isn't any local expertise to treat them. Careless pesticide use can cause long-term diseases to develop resistance, which makes it much harder for people to battle them. One of the fundamental tenets of precision farming is the rapid and precise identification of plant diseases (Sujatha et al. 2021). In order to address the problem of long-lasting pathogenic resistance and lessen the negative effects of climate change, it is essential that no unnecessary expenditures of money or other resources be made and that the results are of a high quality. The significance of precise and quick disease identification, especially early impediment, has never been higher in this changing environment. Plant diseases can be detected using a variety of techniques. A more complete assessment is required when there are no visible signs or when it is too late to respond. However, because most diseases result in some sort of outward expression, a skilled professional examination is the primary way for the detection of plants. A plant pathologist must become increasingly adept at identifying distinguishing symptoms in order to correctly diagnose plant diseases (Bhagat et al. 2020). The signs of sick plants may be harder for amateurs and enthusiasts to identify than they are for a trained pathologist, which could result in a wrong diagnosis. Both inexperienced gardeners and seasoned professionals can benefit greatly from an automated method created to detect plant illnesses utilizing the appearance and visual symptoms of the plant as validate of disease diagnosis.

Advances in computer vision have the potential to strengthen and expand the practice of precise plant protection, as well as the market for particular agricultural computer vision applications. Plant diseases were identified and categorized using industry-standard digital image processing technologies like color detection and threshold (Cui et al. 2020). The most popular deep learning method currently being used for plant disease diagnosis is Convolutional Neural Network (CNN). Deep learning is a novel development in machine learning, with ground-breaking results in a number of study areas, including computer vision, pharmaceuticals, and bioinformatics. Deep learning benefits from being able to use raw data without directly utilizing manual labor (Ananthi 2020). For two main reasons, deep learning has recently produced favorable results in both academia and industry (Baidar 2020). First of all, a large amount of data is generated daily. Therefore, utilizing this information, a comprehensive model might be developed. Second, the processing power of the Graphics Processing Unit allows deep models to be developed and applied to boost compute parallelism. Machine learning has made it possible for PCs to learn without intentional customization, which is fundamentally analogous to how people learn. The computer is using information about a few classes of errands to learn from previous experiences if the presentation of the task improves as more understanding is gained. A supervised activity is learning. unsupervised, semi-supervised, and reinforcement.

12.1 Supervised Learning

Specified datasets with input and output boundaries are referred to as supervised learning for the purposes of developing the models (Panigrahi et al. 2020). An 80:20 split is maintained between data collection and model testing when building a model. Regression and classification are other subcategories of supervised learning. The arrangement is an example of a supervised learning task, which produces a discrete value. This discrete value may have many classes or run concurrently. Relapse is a supervised learning strategy that yields long-term value as opposed to reach. To expect a worth that is more in step with production esteem is the aim of the relapse. Several supervised learning methods include Nearest Neighbor, Gaussian Naive Bayes, Decision Trees, Support Vector Machine (SVM), and Random Forest.

12.1 Unsupervised Learning

In unsupervised learning, targets are not provided for the model to create displays for; instead, merely input boundaries are provided. Unsupervised learning can be classified as either bundling or association (Majeed et al. 2020). Information that has been organized into groups by various examples and distinguished by a machine-learning model is clustered, while the term "Association" refers to a procedure based on standards for classifying relationships between the boundaries of a sizable informational collection. K-Means Clustering, BIRCH—Balanced Iterative

Reducing, and Clustering Using Hierarchies are examples of unsupervised learning models.

12.1 Semi-supervised Learning

Semi-supervised learning functions similarly to the methods already stated. This kind of learning is used when dealing with information, both named and unnamed information. Marks are computed using an unsupervised method, and the determined attributes are then dealt with using supervised learning techniques. This method is more well known in image databases, where many of the images are unidentified.

12.1 Reinforcement Learning

Every time information is handled, it is discovered and added to the information that is being prepared, and the model's execution keeps improving with criticism to learn from examples. As a result, it gets more skilled and experienced the more it learns (Feng and Tian 2020). Algorithms for reinforcement learning include Temporal Difference, Q-Learning, and Deep Adversarial Networks (Figs. 12.1 and 12.2).

12.1 Techniques and Tools

Pathogens such as bacteria, fungus, nematodes, viruses, pests, weeds, insects, photoplasma, and other organisms can cause plant illnesses. Based on routine inspections, ranchers can recognize the symptoms and indicators of a plant's condition. Possible warning signals include overflow, a cottony mass, or an apparent mass on the plant. Among the symptoms are galls, wilt, rots, cankers, necrosis, chlorosis, as well as underdevelopment and overdevelopment.

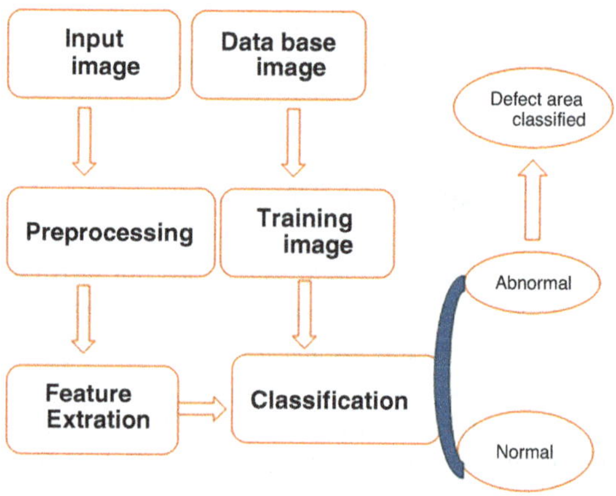

Fig. 12.1 Common stepladder for detection of crops

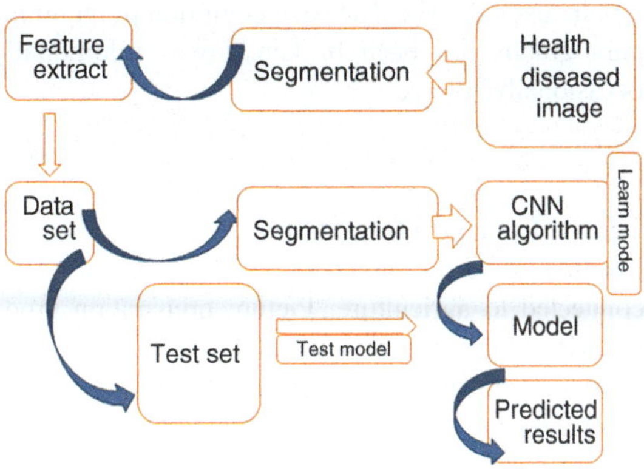

Fig. 12.2 System design for detecting plant leaf disease

12.1 Crop Diagnosis

The CropDiagnosis mobile app's major objective is to provide users with solutions for executives who are causing users concern with accurate yield diagnosis and customized application aid. In order to conduct a full survey, yield specifics, including type, area, soil, and other factors, are gathered and presented (Fegade and Pawar 2020). Different aspects, like the nature, appearance, and development of the harvest, are also taken into account, and maybe a diagnosis is made. The program requires details on finer aspects of development parameters, such as kind, area, soil, and attributes like look, kind, and development of a plant, in order to make decisions.

12.1 Plantix App

A major aspect of the Plantix app, in addition to some other functions, is the capacity to identify plant illnesses. The Plantix app was created by the horticultural IT Company PEAT in Berlin. It is employed to identify soil inadequacies and faults. Images of plants are used by the software to detect diseases. A few of these photos are saved in an advanced cell and synced with the image in the worker for diagnosis (Fegade and Pawar 2020). The Plantix app's robotized crop disease distinguishing proof is a key element. Based on images of the hazardous plants that ranchers have supplied the app's analysis. The app not only highlights disease symptoms but also offers advice on how to lessen disease activity as well as useful details on how to avoid harvest illness the next season. Additionally, the software keeps a database of diseases so that farmers without internet connection can easily consult it.

12.1 Saillog Agrio

Saillog, an AI tool, can help farmers spot and take care of pests and diseases that affect their crops. Saillog comes with Agrio, a user-friendly mobile app that is free to use. Customers of this software send photos of dangerous plants using sophisticated mobile devices. The disease recognition programming is completed when this set of photographs has been broken down. Additionally, a temporary arrangement is occasionally offered.

12.1 Future Directions

Machine learning and deep learning are increasingly being used in applications connected to agriculture. Picture preparation procedures are used for the precise finding and grouping of harvest illness as well as the precise location and order of the plant diseases significant for the productive development of the crop. For their ability to identify plant diseases and treatment options as well as for assisting farmers in raising the profitability of their crop yields, a variety of commercially available products are quickly gaining popularity.

Acknowledgments We are thankful to the management and principal of C. Abdul Hakeem College (Autonomous) Melvisharam, Tamil Nadu, India; to Dr. Deepak Gupta, Maharaja Agrasen Institute of Technology, India; and Dr. Aditya Khamparia, BabaSaheb Bhimrao Ambedkar University, Lucknow, India, for their encouragement, providing the necessary facilities and support in carrying out the work.

Conflict of Interest The authors declare that there are no conflicts of interest.

References

Ananthi V (2020) Fused segmentation algorithm for the detection of nutrient deficiency in crops using SAR images. In: Artificial intelligence techniques for satellite image analysis. Springer, pp 137–159

Baidar T (2020) Rice crop classification and yield estimation using multi-temporal sentinel-2 data: a case study of Terrain Districts of Nepal.

Bhagat M, Kumar D, Haque I, Munda HS, Bhagat R (2020) Plant leaf disease classification using grid search based SVM. In: 2nd International conference on data, engineering and applications (IDEA), pp 1–6

Cui J, Zhang X, Wang W, Wang L (2020) Integration of optical and SAR remote sensing images for crop-type mapping based on a novel object-oriented feature selection method. Int J Agric Biol Eng 13:178–190

Das S, Sengupta S (2020) Feature extraction and disease prediction from paddy crops using data mining techniques. In: Computational intelligence in pattern recognition. Springer, pp 155–163

Fegade TK, Pawar B (2020) Crop prediction using artificial neural network and support vector machine. In: Data management, analytics and innovation. Springer, pp 311–324

Feng K, Tian RS (2020) Forecasting reference evapotranspiration using data mining and limited climatic data 54. Taylor & Francis, pp 363–371

Harakannanavara SS, Rudagi JM, Puranikmath VI, Ayesha Siddiqua R, Pramodhini R (2022) Plant leaf disease detection using computer vision and machine learning algorithms. Global Trans Proc 3(1):305–310

Majeed Y, Zhang J, Zhang X, Fu L, Karkee M, Zhang Q et al (2020) Deep learning based segmentation for automated training of apple trees on trellis wires. Comput Electron Agric 170:105277

Panigrahi KP, Das H, Sahoo AK, Moharana SC (2020) Maize leaf disease detection and classification using machine learning algorithms. In: Progress in computing, analytics and networking. Springer, pp 659–669

Sujatha R, Chatterjee JM, Jhanjhi N, Brohi SN (2021) Performance of deep learning vs machine learning in plant leaf disease detection. Microprocess Microsyst 80:103615

Chapter 13
Analyzing the Frontier of AI-Based Plant Disease Detection: Insights and Perspectives

Mridula Dwivedi, Babita Pandey, and Vipin Saxena

Abstract Plant diseases (PDs) are a significant risk to agriculture all over the world and constitute a threat not only to economic security but also to nutritional safety. The utilization of artificial intelligence (AI), to be more precise, machine learning (ML) and computer vision, has recently emerged as a potentially useful technique for the early and accurate detection of a PD. The prime objective of this survey is to provide readers with an in-depth look at the cutting edge on AI-based plant disease detection (PDD). In this chapter, we explore a number of AI- and ML-based approaches that have the potential to assist with PDD. In addition, we have shed light on the potential for AI-driven solutions to be utilized in agricultural contexts, and we have identified research gaps and difficulties. This study tries to fulfill the expectation that it will be of use to other researchers, agricultural professionals, and policymakers in their search for disease-control strategies that are both more successful and more permanent. This study not only identifies the AI methods that hold the most promise for PDD but also brings to light some of the challenges and problems that remain unresolved that could lead to additional developments in the field.

Keywords Plant disease detection · AI · Crop · Machine learning

13.1 Introduction

Farming is essential for the global availability of food and economic prosperity. However, this critical sector faces numerous challenges, with one of the most formidable being the prevalence of PDs. These diseases not only lead to substantial crop yield losses but also pose a threat to the livelihoods of millions of farmers worldwide. Authentic and efficient diagnosis of PDs is imperative for effective disease management, reducing crop losses, and optimizing resource utilization.

M. Dwivedi (✉) · B. Pandey · V. Saxena
Department of Computer Science, Babasaheb Bhimrao Ambedkar University, Lucknow, Uttar Pradesh, India

A. Khamparia et al. (eds.), *Microbial Data Intelligence and Computational Techniques for Sustainable Computing*, Microorganisms for Sustainability 47,
https://doi.org/10.1007/978-981-99-9621-6_13

Historically, the identification of PDs was predominantly carried out by visual inspections performed by farmers or farm professionals. Nevertheless, these techniques were plagued by several constraints. The tasks were seen to be significantly time consuming, requiring a major allocation of workers, and their outcomes were limited to a rather restricted geographical scope.

Technology plays an essential part in the farming industry, exerting a significant impact on various agricultural enterprises and emerging farms. The involvement of technology in agriculture, in recent times, has presented the potential to cultivate crops in arid desert environments. Automation techniques are highly coveted instruments in the agricultural business. Numerous studies support the claim that the adoption of automation technologies in agricultural settings has the potential to enhance crop productivity, leading to an increase in farmers' annual earnings (Bagde et al. 2015).

The emergence of PDs is dependent upon a simultaneous presence of three contributing aspects: the host, the pathogen, and the environmental circumstances that are favorable for the proliferation of the disease. The term "host" refers to the particular crop that is being examined for disease detection. On the other hand, "pathogens" cover the various agents that cause diseases, including fungi, viruses, and bacteria. Additionally, environmental factors play a crucial role in determining whether the disease will flourish or decline (Patil and Kumar 2020).

In the event that any of these components are lacking, the disease will not exhibit any symptoms. PDs originate from two main sources: biotic and abiotic causes. Biotic factors are derived from the impact of creatures that are alive, such as fungi, bacteria, viruses, and nematodes. On the other hand, abiotic factors stem from ecological elements such as temperature, humidity, soil moisture, and the general ambient conditions.

Lately, the convergence of AI and agriculture has opened up new avenues for addressing the challenges of PDD. AI, particularly ML and computer vision, has demonstrated remarkable potential in automating the detection and identification of PDs. By harnessing the power of AI, farmers and agronomists can not only identify diseases in their early stages but also make data-driven decisions regarding treatments and resource allocation, thereby reducing the environmental footprint of agriculture.

The primary aim of this study is to offer a generalized idea to the novice researchers about PDD with the help of AI technology. We have provided the steps that are required for PDD. We have even discussed the best suitable existing AI methods for PDD, depending upon the type of crop along with the accuracy of AI method. The impact of PD on human life and environment is provided in detail along with the challenges and open issues of PDD.

The following sections will delve into the methodologies employed in AI-based PDD, showcase exemplary applications, discuss challenges, and offer insights into the trajectory of this field. The remaining sections of this chapter are organized as follows: Section 13.2 will provide a summary of related works, Sect. 13.3 generalizes the idea of steps involved in plant disease detection, and Sect. 13.4 discusses the AI methods best suitable for plant type with its accuracies. Sections 13.5, 13.6, and

13.7 discuss the impact of plant diseases on human life and environment, challenges, and open issues respectively. At last, the chapter concludes in Sect. 13.8.

13.2 Related Work

Extensive literatures have been conducted on PDD by various researchers depicting comprehensive analysis and proposing novel methodologies which have left behind some drawbacks as well. Plant illnesses have been seen as a substantial hazard to global food security due to the lack of essential infrastructure in several regions worldwide. Histogram of Oriented Gradients (HOG) and Support Vector Machines (SVM) are employed for the purpose of predicting the presence of disease in certain crops.

Gokulnath and Usha Devi (2020) have highlighted a significant issue characterized by the fast growth of agricultural goods. However, this growth is hindered by the negative impact of diseases, pests, and weeds, which are responsible for causing a decline in productivity levels of these items. ML techniques are employed to identify the symptoms it produces. One shortcoming of this approach is the restricted access to just three spectral groups, which consequently imposes limitations on the investigation of plant state.

Thangavel et al. (2022) presented many methodologies for predicting plant illnesses and proposed an effective approach for detecting and diagnosing infections in plants. This approach involved the use of datasets and subsequent training modules to enable the recognition of PD, achieving a commendable 96.5% accuracy. A total of 32 distinct plants and illnesses were employed along with convolutional neural network (CNN) as shown in Table 13.1.

13.3 Steps in the Detection of Plant Diseases

Due to the fact that manual diagnosis of PDs is a time-consuming process that is also prone to errors, it is most useful in the setting of more modest-sized agricultural operations. On the other side, one could argue that autonomous diagnosis demonstrates a greater level of accuracy and efficiency, using less time and labor resources than traditional methods do. As a direct result of this, a significant amount of research has been carried out within the same framework. The phases of the PDD process are illustrated in Fig. 13.1.

Input Image: The input image is provided to the model for detection if it is a healthy plant or an unhealthy or a diseased one.

Image Preprocessing: The input image consists of noise which makes it inappropriate to work with. Preprocessing step is performed for the removal of such noise.

Table 13.1 Exhaustive literature survey

Author	Method	Drawbacks
G. Dhingra (Dhingra et al. 2018)	Image Processing (IP)	Less accurate
V. Singh (Singh and Misra 2017)	Image Segmentation: Genetic Algorithm	Algorithm selection is difficult
G. Shrestha (Shrestha et al. 2020)	CNN	Accuracy might be improvised
R. Sujatha (Sujatha et al. 2021)	Clustering and Classification	Using IoT might be much beneficial
J. Liu (Liu and Wang 2021)	Supervised and Unsupervised Learning	Fails to maintain a balance between accuracy and complexity
K Lin (Lin et al. 2019)	VGG16	Comparison is not made properly
S. Zhang (Zhang et al. 2019)	Clustering	Requires clusters of same dimensions
A. Badage (Badage 2018)	Edge Detection	Problem choosing edge in multi-edge image
J. Singh (Singh and Kaur 2019)	Regional	High computation time and memory needed

Feature Extraction: Feature extraction is carried out for choosing parameters like shape, texture, color for using them as parameters to provide details and information of the image. These parameters are fed into the classifier as input features.

Database: The database consists of a variety of plant images that are both healthy and unhealthy. These images help the classifier train.

Training: The images from the database are trained in the training step for better and much accurate classification.

Classification: The trained classifier classifies the inputted, preprocessed image as a healthy or unhealthy one based on the image's features.

13.4 Disease Detection AI Methods and Type of Crops

The technique known as PDD requires the investigation of a huge number of plant species that come from a variety of locations all over the world. It is important to note that India and China are currently serving as the primary locations where a significant portion of the research endeavors in the subject of PDD are being undertaken. This endeavor is being carried out on a global scale, and it involves the cultivation of a broad variety of crops, including soybeans, apples, bananas, avocados, citrus fruits, coffee, maize, cucumbers, millet, oil palms, rice, potatoes, berries, wheat, and tomatoes, each of which is subjected to in-depth testing for the detection of disease.

In the subject of PDD, a great number of in-depth research have been carried out, the majority of which have concentrated on crops that are typically cultivated. This is

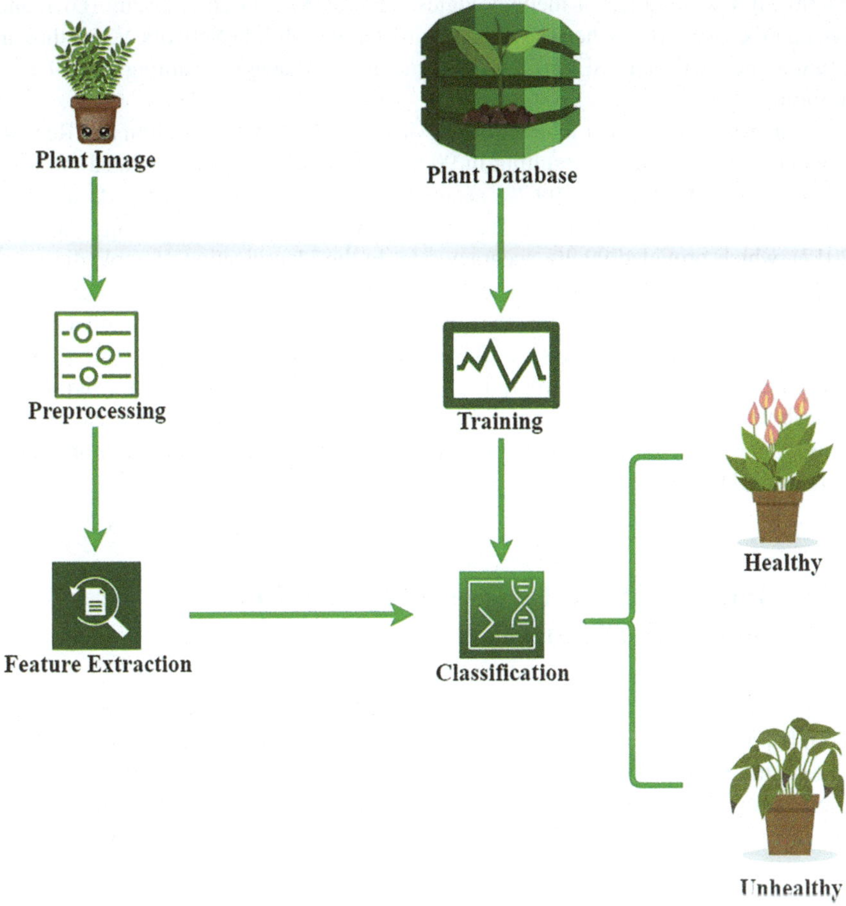

Fig. 13.1 Plant disease detection steps

Table 13.2 Analysis of AI method depending upon crop

Author	Plant	Method	Accuracy
M. Ji (Ji and Wu 2022)	Grape	DeepLabV3+	97.75%
J. Ma (Ma et al. 2019)	Cucumber	CART	90.67%
G. Yang (Yang et al. 2020)	Tomato	LFC-Net	99.7%
R. Sreevallabhadev (Sreevallabhadev 2020)	Rice	CNN + SVM	96.8%
H. Ali (Ali et al. 2017)	Citrus fruits	Bagging tree	99.9%
W. Haider (Haider et al. 2021)	Wheat	CNN + DT	97.2%

shown in Table 13.2, where we have analyzed the results obtained for each type of crop and described the visual methods that were utilized to obtain these conclusions. For example, rice has been the focal point of several research investigations that employed a wide range of methodologies, such as ML, Deep Learning (DL), and Image Processing (IP). Chen et al. (2021) utilized the Mobile Network V2 method in conjunction with the SE block-based attention strategy, resulting in 99.33% accuracy.

Similarly for cucumber investigation (Ma et al. 2019), Classification and Regression Trees were adopted, resulting in 90.67% accuracy. A tomato-related research (Yang et al. 2020) yielded 99.7% accuracy, adopting feedback network and area suggestion network. The Bagged Tree classifier was utilized in citrus crops (Ali et al. 2017), which provided 99.9% accuracy rate. In the identification of apple diseases, DenseNet121 model attained an accuracy of 93.71% (Zhong and Zhao 2020).

It is essential to recognize that the quality of the data has a significant influence on the efficacy of these models, as this is a factor that cannot be ignored. As a consequence of this, the incorporation of attention processes has been a significant contributor to the advancement in the diagnosis of diseases affecting rice and grapes. This demonstrates how important thorough data management and analysis are in the field of PDD.

13.5 Impact of Plant Diseases on Human Life and Environment

Diseases have a big effect on a plant's health because they stop important things from happening, like absorbing water and nutrients, making food, growing healthy, and splitting cells. How much damage is done depends on a lot of different things, such as how dangerous the bacteria are, how strong the host plants are, the weather, how long the infection has been going on, and a lot more. Because of this, these diseases can cause a wide range of signs, from mild problems to complete plant death.

These sneaky pathogens, which are often carried by polluted plant parts, cause a wide range of problems in plant populations. Several manifestations of the disease include the development of lesions on both fruits and leaves, rotting of the roots and fruits, occurrence of leaf blights, wilting, and ultimately, the demise of the plant.

Notably, the effects of plant illnesses are not limited to the world of plants. Take the mild mottle virus, which is known to affect the immune system of humans and cause clinical signs. Also, some PDs can reduce the amount of food available or add harmful chemicals to the food chain, which can be bad for people's health. Also, putting bacteria in the soil that fight PDs, like herbicides and insecticides, which are meant to kill bugs and diseases, can accidentally hurt people.

Due to the fact that different plants are more likely to have different bugs and diseases, traditional farming methods often use a lot of pesticides and herbicides, which unfortunately add to long-term pollution of the land. Even though most plant pathogens can't make people sick, it's best not to eat fruits and veggies that contain mold or are rotten from farms where herbicides are used. Also, you need to be very careful with food that has been affected by fungus that make toxins. Both of these types of food pose serious health risks.

Taking out the sick parts of veggies could reduce the number of disease-causing agents and rotting food. But it's important to know that this method doesn't always get rid of all possible sources of pollution. Some fungi and poisons like to grow and spread, so it's important to deal with them all at once, especially in places where there are a lot of farms with natural plant foods.

13.6 Challenges

PDD has a number of difficult obstacles, each of which calls for innovative solutions. Identification of illnesses in the field in real time presents a significant obstacle. It is vital for there to be a rapid and accurate diagnosis in the field in order to facilitate early action and treatment of sickness. Another challenge is the impact of climate change. As the state of our environment worsens, diseases that were once confined to specific regions may begin to spread. It is essential to modify diagnostic procedures in response to shifting illness patterns. It can be challenging to preserve analytical images of a consistent and good quality. In addition, the limited availability of various datasets for the training of AI models is a factor that delays progress. Large and varied datasets are necessary for the development of robust and effective algorithms for disease detection. Both scalability and computational complexity provide significant challenges. In order to diagnose diseases in real time, AI systems need to be able to effectively process huge amounts of data. Environmental factors also contribute to a reduction in accuracy. In order to avoid producing false positives and negatives as a result of illumination, humidity, and other conditions, more robust AI models are required. The use of Unmanned Aerial Vehicles (UAVs) and the Internet of Things (IoT) for the diagnosis of illnesses has both potential and limitations. Integration of these technologies into agricultural practices will require extensive planning as well as the creation of supporting infrastructure. In conclusion, the identification of diseases by crowdsourcing and public engagement is an innovative approach; nonetheless, it is fraught with difficulties in terms of data quality, privacy, and scalability. These challenges bring into focus the essential role that teamwork and creativity play in PD control.

13.7 Open Issues

The potential breakthroughs in the field of PD identification through the utilization of AI are highly promising and set for substantial progress. One of the primary areas of emphasis lies in the ongoing improvement of AI algorithms in order to augment the precision and dependability of illness detection systems. The attainment of this objective necessitates the optimization of algorithms and the augmentation of varied and comprehensive datasets, hence guaranteeing the accurate identification of diseases.

An additional promising area of exploration involves the implementation of real-time monitoring techniques. In this context, AI-driven systems that are equipped with various sensors, drones, and cameras are employed to actively and attentively examine agricultural crops. This approach facilitates the early detection of diseases, hence enabling prompt intervention measures to be taken. These technologies have the capacity to significantly transform the field of agriculture through their ability to proactively mitigate the occurrence of disease outbreaks.

Moreover, the incorporation of many sensing modalities, including hyperspectral imaging, thermal imaging, and chemical analysis, into AI systems is imminent. The utilization of a multi-modal approach holds the potential to enhance disease detection by including a wider range of factors and providing more precise results, hence addressing the specific requirements of different crops and settings.

Within the domain of edge computing, AI algorithms are being deployed on edge devices, including IoT sensors and UAVs, in order to minimize latency and decrease reliance on uninterrupted internet connectivity. This advancement enhances the feasibility of AI-driven disease detection, particularly in underserved regions characterized by restricted availability of high-speed internet connectivity.

The idea of automating therapy is a promising development, wherein AI systems not only detect problems but also provide recommendations and occasionally independently administer therapies, such as precise application of pesticides or herbicides. The use of this approach has the potential to result in a substantial decrease in the amount of labor and resources needed for the management of diseases.

The possibility for early disease forecasting in the future is anticipated, wherein AI will be utilized to predict disease outbreaks. This prediction will be based on historical data, weather patterns, and environmental factors. Consequently, farmers will be empowered to proactively protect their crops.

The significance of data integration and sharing should be underscored as the domain of PDD employing AI progresses. The utilization of platforms that enable the efficient sharing of PD data among researchers, farmers, and agricultural organizations has the potential to enhance the collective knowledge and disease management endeavors.

Additional avenues encompass the amalgamation of AI with robotics to facilitate autonomous farming, the augmentation of ML explainability to foster user trust, the use of blockchain technology for the purpose of agricultural traceability, and the promotion of international cooperation to address global PDs. In addition, educational and training programs strive to provide farmers and agricultural professionals with the necessary knowledge and resources to proficiently utilize AI systems for the purpose of detecting and managing diseases. Furthermore, it is imperative to address ethical issues, environmental implications, and responsible deployment in order to guarantee the sustainable and ethical integration of AI in the field of agriculture.

The field of PDD with AI is characterized by its dynamic nature and quick evolution. It has significant promise for transforming the agricultural sector, as it has the capacity to enhance crop productivity, mitigate environmental consequences, and safeguard global food supplies. Nevertheless, the realization of this transformational potential necessitates collaborative endeavors from researchers, policymakers, and agricultural stakeholders in order to surmount obstacles.

13.8 Conclusion

PDD is a substantial barrier to overcome for those working in the agricultural sector. As a result, it is of the utmost importance to detect plant diseases as quickly as possible in order to lessen the impact of these problems and limit their spread within agricultural settings. Lately, there has been a substantial number of studies that has been carried out on a variety of approaches to PD detection, and many of these studies have made important and ground-breaking discoveries. This research was conducted with the intention of introducing various AI technologies and approaches to aid farmers in achieving higher yields. In addition to this, it places an emphasis on PDD methodologies throughout numerous literatures. In this study chapter, the important connection between AI and PDD is investigated, and insights into its significance, methodology, challenges, and potential future routes of investigation are provided. With the help of this work, we have endeavored to present a fundamental concept for utilizing AI in the detection of PDs. This chapter details the processes involved in PDD as well as the AI methods that are utilized for PDD. A crucial component of this presentation is a discussion of the difficulties and opportunities that lie ahead for PDD. The impact of PDs on human life and the environment is also explored, which is something that is quite helpful for beginning researchers.

References

Ali H, Lali MI, Nawaz MZ, Sharif M, Saleem BA (2017) Symptom based automated detection of citrus diseases using color histogram and textural descriptors. Comput Electron Agric 138:92–104

Badage A (2018) Crop disease detection using machine learning: Indian agriculture. Int Res J Eng Technol 5(9):866–869

Bagde S, Patil S, Patil S, Patil P (2015) Artificial neural network based plant leaf disease detection. Int J Comput Sci Mobile Comput 4:900–905

Chen J, Chen J, Zhang D, Nanehkaran YA, Sun Y (2021) A cognitive vision method for the detection of plant disease images. Mach Vis Appl 32:1–18

Dhingra G, Kumar V, Joshi HD (2018) Study of digital image processing techniques for leaf disease detection and classification. Multimed Tools Appl 77:19951–20000

Gokulnath BV, Usha Devi G (2020) A survey on plant disease prediction using machine learning and deep learning techniques. Intel Artif 23:136–154

Haider W, Rehman AU, Durrani NM, Rehman SU (2021) A generic approach for wheat disease classification and verification using expert opinion for knowledge-based decisions. IEEE Access 9:31104–31129

Ji M, Wu Z (2022) Automatic detection and severity analysis of grape black measles disease based on deep learning and fuzzy logic. Comput Electron Agric 193:106718

Lin K, Gong L, Huang Y, Liu C, Pan J (2019) Deep learning-based segmentation and quantification of cucumber powdery mildew using convolutional neural network. Front Plant Sci 10:422622

Liu J, Wang X (2021) Plant diseases and pests detection based on deep learning: a review. Plant Methods 17:1–18

Ma J, Du K, Zheng F, Zhang L, Sun Z (2019) A segmentation method for processing greenhouse vegetable foliar disease symptom images. Inform Process Agric 6:216–223

Patil RR, Kumar S (2020) A bibliometric survey on the diagnosis of plant leaf diseases using artificial intelligence. Libr Philos Pract 2020:3987

Shrestha G, Deepsikha, Das M, Dey N (2020) Plant disease detection using CNN. In: Proceedings of 2020 IEEE applied signal processing conference, ASPCON. IEEE, pp 109–113

Singh J, Kaur H (2019) Plant disease detection based on region-based segmentation and KNN classifier. In: Proceedings of the international conference on ISMAC in computational vision and bio-engineering 2018 (ISMAC-CVB), vol 30. Springer International Publishing, pp 1667–1675

Singh V, Misra AK (2017) Detection of plant leaf diseases using image segmentation and soft computing techniques. Inform Process Agric 4:41–49

Sreevallabhadev R (2020) An improved machine learning algorithm for predicting blast disease in paddy crop. Mater Today Proc 33:682–686

Sujatha R, Chatterjee JM, Jhanjhi NZ, Brohi SN (2021) Performance of deep learning vs machine learning in plant leaf disease detection. Microprocess Microsyst 80:103615

Thangavel M, Gayathri PK, Sabari KR (2022) Plant leaf disease detection using deep learning. Int J Eng Res Technol 10:3599–3605

Yang G et al (2020) Self-supervised collaborative multi-network for fine-grained visual categorization of tomato diseases. IEEE Access 8:211912–211923

Zhang S, You Z, Wu X (2019) Plant disease leaf image segmentation based on superpixel clustering and EM algorithm. Neural Comput & Applic 31:1225–1232

Zhong Y, Zhao M (2020) Research on deep learning in apple leaf disease recognition. Comput Electron Agric 168:105146

Chapter 14
Fuzzy and Data Mining Methods for Enhancing Plant Productivity and Sustainability

Khalil Ahmed, Mithilesh Kumar Dubey, Devendra Kumar Pandey, and Sartaj Singh

Abstract The agriculture sector plays a pivotal role in addressing global food security challenges while also grappling with the imperative of sustainability. In recent years, the use of data mining and fuzzy logic together has emerged as a potent toolset for optimizing plant productivity and fostering sustainable agricultural practices. This chapter explores the intersection of Fuzzified reasoning and Knowledge discovery in data in the context of plant productivity and sustainability. We delve into the principles, methodologies, and real-world applications of these technologies to empower agricultural stakeholders with the knowledge and tools needed to advance sustainable farming practices.

Keywords Sustainable agriculture · Fuzzy logic · Data mining · Fuzzy rule

14.1 Introduction

The need for food and agricultural goods is increasing as the world's population escalates. Simultaneously, environmental concerns, resource constraints, and climate change impose unprecedented challenges on the agriculture sector (Deshpande 2017). To address these challenges, it becomes essential to optimize plant productivity while minimizing resource use and environmental impact. Fuzzy logic and data mining are two prominent computational techniques that, when combined, offer innovative solutions to enhance plant productivity and sustainability (Gharde et al. 2018).

Agriculture is at the nexus of global challenges in the twenty-first century, confronted with the formidable task of simultaneously feeding an ever-expanding global population while mitigating the profound environmental and resource sustainability challenges it faces. There are still people in the globe to burgeon, the

K. Ahmed · M. K. Dubey (✉) · D. K. Pandey · S. Singh
School of Computer Application, Lovely Professional University, Phagwara, Punjab, India
e-mail: mithilesh.21436@lpu.co.in; devendra.15673@lpu.co.in

A. Khamparia et al. (eds.), *Microbial Data Intelligence and Computational Techniques for Sustainable Computing*, Microorganisms for Sustainability 47,
https://doi.org/10.1007/978-981-99-9621-6_14

demand for food, Fiber, and bioenergy intensifies, putting immense pressure on agricultural systems (Antia 2023). This necessitates a fundamental transformation in the way we approach agriculture—a transformation that not only amplifies plant productivity but also safeguards the planet's delicate ecological balance.

Central to this agricultural revolution is the integration of cutting-edge technologies, including fuzzy logic and data mining, which have emerged as a powerful synergistic duo for optimizing plant productivity while championing sustainability (Misra 2014).

14.2 The Imperative of Sustainable Agriculture

Sustainable agriculture stands as a pivotal pillar of our future as shown in Fig. 14.1. It encapsulates a holistic approach that seeks to meet the burgeoning demand for agricultural products while minimizing the ecological footprint (Sharma et al. 2022a).

Key concerns in sustainable agriculture encompass the judicious use of resources such as water and soil emissions of carbon dioxide are being reduced, the conservation of biodiversity, and the promotion of resilient farming systems that can withstand climate variability (Kumar et al. 2018). However, achieving these multifaceted goals is a Herculean task. It requires not only technological innovation but also a deep understanding of the complex, interrelated factors that affect plant growth and overall farm performance. This is where fuzzy logic and data mining step into the limelight, offering advanced computational techniques to unravel the intricacies of agricultural systems as shown in Table 14.1.

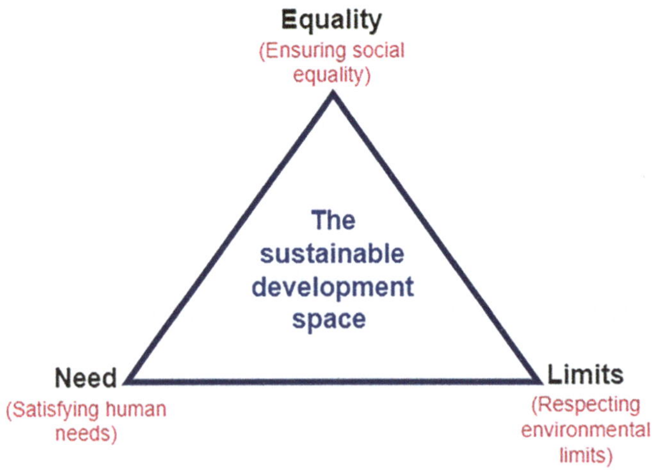

Fig. 14.1 Normative imperatives model for sustainable development strategy

Table 14.1 Agriculture that is both customary and sustainable

Customary	Sustainable
Growing regions and using creativity and science to control nature	Procedures for least impedance with attributes
Extreme soil breakdown and technological cultivation	No-till or definitely less culturing (organic culturing)
Windy conditions and deteriorating soil	Slow winds and eroded soil
Low water intrusion	Water invades at a rapid rate

Table 14.2 Primary difficulties facing sustainable agriculture

Parameters	Obstacles
Areas	Ways to stop soil erosion, stop ecosystem dissolution, and halt habitat loss without lowering the profitability of crops
Management	Efficiently use H_2O
Low inputs	Efficiently use nutrients
	Pest control
Human inference	Increase yield productivity Reduce the negative impact on human health
Environment	Lessen the effect of environmental degradation Catastrophe in habitat

14.2.1 *Fuzzy Logic*

Taming Agricultural Uncertainty Fuzzy logic, initially formulated by Lotfi Zadeh in the 1960s, is a mathematical framework that excels in handling uncertainty, imprecision, and vagueness—an inherent feature of agricultural data. In agriculture, concepts like "optimal soil moisture," "ideal temperature range," or "high pest susceptibility" often elude precise numerical definitions. Fuzzy logic introduces the notion of fuzzy sets, which enable the representation of such ambiguous concepts using membership functions (Kumar et al. 2018). This enables agricultural stakeholders to work with imprecise data in a structured manner.

14.2.2 *Data Mining: Unearthing Insights from Agricultural Data*

In tandem with fuzzy logic, data mining techniques play an instrumental role in modern agriculture. The agriculture sector generates a staggering volume of data, encompassing information on climate, soil properties, crop characteristics, and more. Data mining techniques, such as classification, clustering, and regression, serve as a treasure trove of tools for extracting valuable knowledge from this data deluge (Naseem et al. 2022). They enable researchers and practitioners to identify patterns, make predictions, and optimize agricultural processes as shown in Table 14.2.

Data mining techniques: Data mining techniques in agriculture involve the application of various methods and algorithms to extract valuable insights and patterns from large agricultural datasets. These techniques are crucial for making informed decisions, optimizing resource allocation, and improving crop yields in the farming sector. Some key data mining techniques in agriculture.

14.2.3 The Synergy: Fuzzy Logic and Data Mining.

However, the true power of these technologies comes to the fore when they are integrated. The synergy between fuzzy logic and data mining forms a formidable partnership that empowers agricultural stakeholders (Mabhaudhi et al. 2019). Fuzzy logic allows for the representation of im-precise knowledge, while data mining techniques help unearth hidden patterns and relationships in vast datasets (Klir and Yuan 1995). Together, they pave the way for intelligent decision support systems that can enhance crop yields, reduce resource consumption, and promote sustainability.

14.2.4 Fuzzy Logic: A Foundation for Uncertainty Handling.

Fuzzy logic provides a mathematical framework for handling uncertainty, imprecision, and vagueness in agricultural data as shown in Table 14.3 (Raorane and Kulkarni 2012). This section introduces the fundamentals of fuzzy logic, highlighting its key components:

Fuzzy sets: Fuzzy sets allow the representation of vague concepts, such as "high temperature" or "adequate soil moisture," using membership functions (Majumdar et al. 2017).

Fuzzy rules: Fuzzy rules translate expert knowledge into a formal structure that guides decision-making processes (Doğan et al. 2015).

Inference systems: Fuzzy inference systems process fuzzy inputs, apply fuzzy rules, and generate crisp or fuzzy outputs (Naseem et al. n.d.). Fuzzy rules in agriculture, when combined with technology and data mining, offer a powerful

Table 14.3 Fuzzy logic techniques in the agriculture domain

SDM domain	Usage	Technique/method
Agriculture	Precision agriculture	Cross-validation techniques
	Crop yield prediction	ML, DL, and artificial intelligence
	Crop disease detection and classification	ML, DL, IoT, and Artificial intelligence
	Recommendation system for crop	CNN, ANN

approach to decision-making and optimization in farming practices. Fuzzy logic is a mathematical framework that deals with uncertainty and imprecision, making it well suited for addressing the complex and uncertain nature of agricultural systems. When integrated with technology and data mining, it enables farmers and agricultural experts to make more informed and adaptive decisions. Let's explore this concept in detail: Fuzzy logic extends traditional binary (true/false) logic by allowing for degrees of truth between 0 and 1, representing the level of membership or certainty. Fuzzy sets describe the degree of membership of an element to a set. Fuzzy rules consist of "if-then" statements that use fuzzy linguistic variables to express relationships. Fuzzy rules allow for adaptive decision-making, considering the uncertainty and variability in agricultural systems. Fuzzy logic provides transparent and interpretable rules that farmers and experts can understand and trust. By optimizing resource usage (e.g., water, pesticides, fertilizers), fuzzy logic can improve crop yields while minimizing environmental impact. Fuzzy logic helps mitigate risks associated with unpredictable factors like weather and pests, leading to more reliable outcomes. Fuzzy rules in agriculture, when integrated with technology and data mining, enable data-driven, adaptive, and efficient decision-making (Melin and Castillo 2014). This approach leverages the power of fuzzy logic to handle uncertainty and imprecision inherent in agricultural systems, ultimately improving crop yields, resource management, and sustainability in farming practices.

14.3 Data Mining Techniques for Agricultural Insights

Data mining techniques extract knowledge from vast datasets, enabling data-driven decision-making in agriculture as shown in Fig. 14.2 (RURAL DO 2014). This section presents an overview of data mining techniques applicable to plant productivity and sustainability.

Classification: Classify crops based on yield potential, disease susceptibility, or other factors.

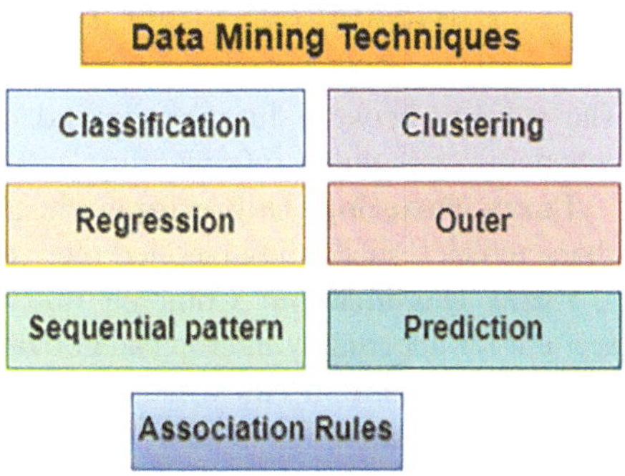

Fig. 14.2 Data mining techniques

14.3.1 *Classification and Clustering*

Group similar crops or regions to identify patterns and trends (Jamshidi et al. 2016). Classification algorithms are employed to categorize data into distinct groups or classes. In agriculture, this can be highly valuable: Crop Disease Identification. Using historical data on crop health, machine learning classification models can identify diseases or pests affecting crops and suggest appropriate interventions. Soil Type Classification: classifying soil types based on their characteristics (e.g., texture, pH levels) helps farmers make informed decisions about crop selection and irrigation methods.

14.3.2 *Regression*

Regression analysis is a statistical way to understand the relationships between variables. In agriculture, it can be applied to predict outcomes based on various factors. Predict crop yields or resource requirements based on historical data (Pantazi et al. 2019). By analyzing historical data on crop yields and factors like weather conditions, soil quality, and fertilizer usage, regression models can predict future crop yields. This helps farmers plan their harvest and resources effectively. Regression can be used to predict the growth rate of livestock based on factors like diet, environment, and genetics.

Time series analysis: Studies of data series are used to understand data collected over time, making it crucial for agriculture. *Weather Forecasting:* analyzing historical weather data through time series analysis helps in forecasting future weather patterns, which is vital for crop management. Crop Growth Monitoring: continuous monitoring of crop growth over time can help detect anomalies or trends that may require action, such as adjusting irrigation schedules.

Association rule mining: Discover relationships between variables, such as environmental factors and crop performance.

14.4 Integration of Fuzzy Logic and Data Mining

The synergy between fuzzy logic and data mining offers a potent approach to enhancing agricultural sustainability:

Fuzzy clustering: Utilize fuzzy clustering to group crops based on imprecise characteristics, facilitating targeted interventions.

Fuzzy classification: Combine fuzzy logic with classification algorithms to account for uncertainty in crop categorization.

Fuzzy association rules: Develop fuzzy association rules to uncover nuanced relationships between agricultural variables (Zhai et al. 2020).

14.5 Case Studies

This section presents real-world case studies that demonstrate the practical application of fuzzy logic and data mining in agriculture.

Precision agriculture: Explain how to optimize resource allocation in precision agriculture, reducing input waste and environmental impact as shown in Fig. 14.3.

Crop Disease Management: Investigate the use of fuzzy classification and data mining to predict and manage crop diseases, minimizing yield loss (Gavioli et al. 2019).

This form of leaf staging, like the leaf collar approach, starts with a short initial leaf. Leaf counting then changes, concluding with the leaf that is at least 40–50% exposed from the whorl, rather than the highest leaf with a visible collar as shown in Fig. 14.4. The tip of this "indicator" leaf generally "droops" or hangs down in knee-high corn or older, thus I refer to this as the "droopy" leaf approach (Arumugam et al. 2022).

Climate adaptation: Examine the integration of fuzzy logic and data mining in adapting agricultural practices to changing climate conditions.

Fig. 14.3 Precision agriculture life cycle

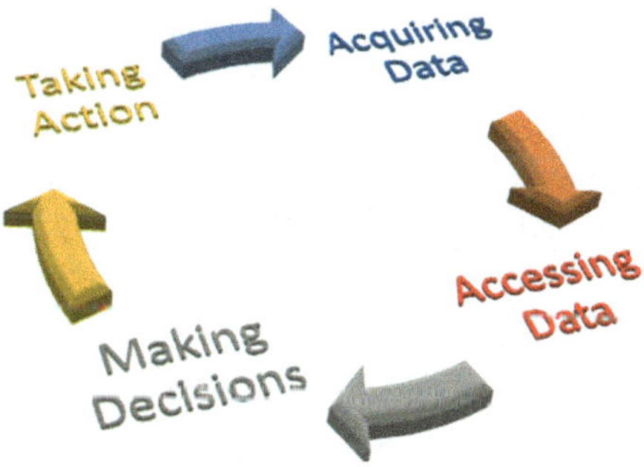

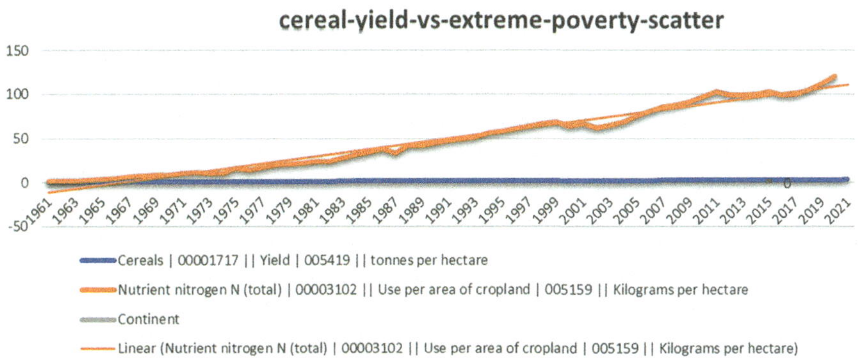

Fig. 14.4 Cereal-yield-vs.-extreme-poverty-scatter due to proper crop management in India (1961–2021) according to Our World Data report

14.6 Challenges and Future Directions

While the integration of fuzzy logic and data mining shows great promise, it also presents challenges such as data quality, model interpretability, and scalability (Arumugam et al. 2022). This section discusses these challenges and outlines potential future directions for research and application in the field of agricultural sustainability.

14.6.1 Challenges

Data Quality and Availability: One of the foremost challenges in applying data mining and fuzzy logic to agriculture is the quality and availability of data. In many regions, data on soil properties, weather, and crop performance may be limited or inconsistent. Ensuring data quality and accessibility remains a critical hurdle (Narmadha et al. 2022).

Interpretability: Complex data mining models, such as deep learning algorithms, often lack interpretability. For stakeholders in agriculture, understanding the rationale behind a decision or prediction is crucial. Balancing model complexity with interpretability is an ongoing challenge.

Scale and Scalability: Agriculture spans a wide range of scales, from small family farms to large agribusinesses. Scaling data mining and fuzzy logic methods to accommodate this diversity is challenging (Sharma et al. 2022b). Methods that work for one scale may not be directly applicable to another.

Integration with Expert Knowledge: Fuzzy logic relies on expert knowledge to define fuzzy sets and rules. Integrating this expert knowledge seamlessly into data mining models can be complex. Ensuring that the models respect domain expertise is an ongoing challenge.

Resource Constraints: Sustainable agriculture often requires the optimization of resource use, such as water, fertilizers, and pesticides. Agricultural land is finite, and in many regions, it's under pressure due to urbanization and degradation. As the global population grows, the need for agricultural land increases, making efficient land use critical. Access to water is a significant constraint, particularly in arid and semi-arid regions. Efficient irrigation practices are essential to conserve water resources. Labor shortages are common in agriculture, especially during peak seasons. Finding and retaining skilled labor can be challenging. Farmers often face financial constraints, limiting their ability to invest in modern machinery and technologies. Access to agricultural knowledge and best practices can be limited in some regions, hindering productivity. Developing models that account for these constraints while maximizing productivity poses a challenge, especially when considering dynamic environmental conditions. Technology such as GPS, sensors, and drones enables precision agriculture. Farmers can target resources like water, fertilizer, and pesticides precisely where they are needed, minimizing waste. Smart irrigation systems monitor soil moisture levels and weather conditions to optimize water usage, reducing water waste and energy costs. Farm machinery, like tractors and harvesters, reduces labor demands, making farming more efficient. Advances in robotics also offer potential solutions to labor shortages. Satellite and drone imagery provide real-time data on crop health, enabling early detection of issues like pests or diseases. This helps in timely interventions, reducing crop losses. Genetically modified crops can be engineered to be more resilient to boosting yields and lowering the demand for chemical pesticides, illnesses, and surrounding inputs.

14.7 Role of Data Mining in Addressing Resource Constraints

Data mining techniques can be applied to various agricultural datasets to extract valuable insights such as. Historical weather data can be analyzed to predict future weather patterns, helping farmers plan planting and harvesting times, and optimize resource allocation as shown in Table 14.4.

Soil data, including texture, pH levels, and nutrient content, can be analyzed to determine optimal crop choices and fertilizer requirements. Analyzing historical crop yield data in conjunction with factors like weather, soil, and farming practices can provide insights into yield optimization. Data mining can identify patterns in pest and disease outbreaks, allowing for early intervention and reduced chemical use. Analyzing market trends and price data can assist farmers in making choices in crop selection and when to sell their produce for maximum profit. Data mining can optimize the supply chain, reducing post-harvest losses and ensuring that crops reach the market efficiently. Machine learning models can predict resource requirements,

Table 14.4 Data analysis tasks and techniques

Data analysis techniques	Data summarization	Segmentation	Classification	Prediction	Dependency analysis
Descriptive and visualization	✓	✓			✓
Correlation analysis					✓
Cluster analysis		✓	✓		
Discriminant analysis				✓	✓
Regression analysis			✓	✓	
Case-base reasoning					✓
Decision tree			✓	✓	
Association rule					✓

such as water and fertilizer, based on historical data, helping farmers plan and budget more effectively.

14.8 Challenges and Considerations

The accuracy and quality of data are critical for meaningful analysis. Data collected from various sources may require cleaning and validation. Not all farmers have access to the latest technology, and investments can be costly. Addressing the digital divide is crucial. Farmers must be cautious about sharing sensitive data. Proper data security measures must be in place. Farmers and agricultural professionals need training to effectively use technology and interpret data-driven insights. Technology and data mining are powerful tools for addressing resource constraints in agriculture. They enable farmers to make more informed decisions, optimize resource usage, and increase overall productivity while considering environmental sustainability. However, it's essential to ensure that these tools are accessible to all farmers and that data privacy and security concerns are addressed.

14.9 Future Directions

Enhanced Data Collection and Integration: Future research should focus on improving data collection techniques, including the use of remote sensing, IoT devices, and drones, to gather high-resolution and real-time data. Additionally, efforts to integrate diverse data sources seamlessly will be crucial for comprehensive analysis.

Explainable AI in agriculture: The development of explainable AI models that combine the strengths of data mining and fuzzy logic can bridge the interpretability gap. This will make it easier for agricultural stakeholders to trust and act upon model recommendations.

Machine learning for precision agriculture: Continued research into machine learning techniques tailored to precision agriculture can lead to more efficient resource management. Algorithms that adapt to changing conditions and provide real-time recommendations will be invaluable.

Big data and cloud computing: Harnessing the power of big data analytics and cloud computing can address scalability issues. Cloud-based platforms can provide accessible and scalable solutions for agricultural stakeholders, regardless of their scale of operation.

Robotic and autonomous farming: Integrating fuzzy logic and data mining into robotic and autonomous farming systems can revolutionize agriculture. These technologies can enable automated decision-making and precision application of resources.

Climate resilience: Developing models that help agriculture adapt to changing climate conditions will be critical. Predictive models that consider long-term climate trends and their impact on crop yields can aid in climate-resilient farming practices.

Decision support systems: Building user-friendly decision support systems that incorporate fuzzy logic and data mining can empower farmers and policymakers to make informed choices. These systems should consider local conditions and constraints. The challenges and future directions in applying fuzzy logic and data mining methods to enhance plant productivity and sustainability are intertwined. Overcoming challenges while exploring these future directions will require interdisciplinary collaboration among researchers, farmers, policymakers, and technology developers. By addressing these challenges and embracing innovative directions, we can propel agriculture toward a more sustainable and productive future.

14.10 Conclusion

The fusion of fuzzy logic and data mining holds tremendous potential for enhancing plant productivity and promoting sustainable agricultural practices. By harnessing the power of these computational techniques, agriculture can become more resilient, efficient, and environmentally friendly. This chapter has provided a comprehensive overview of the principles, methodologies, and real-world applications of fuzzy logic and data mining in agriculture, paving the way for continued innovation in this critical domain. In this chapter, we embark on a comprehensive exploration of the fusion of fuzzy logic and data mining in the context of plant productivity and sustainability. We delve into the underlying principles, methodologies, and real-world applications of these technologies, providing a roadmap for agricultural stakeholders to harness the potential of these cutting-edge tools. Through this synergy, agriculture can not only meet the demands of a growing world but also do so in a way that respects the delicate balance of our planet's ecosystems, ensuring a sustainable future for generations to come.

References

Antia DD (2023) Desalination of saline irrigation water using hydrophobic, metal–polymer hydrogels. Sustainability 15(9):7063

Arumugam K, Swathi Y, Sanchez DT, Mustafa M, Phoemchalard C, Phasinam K, Okoronkwo E (2022) Towards applicability of machine learning techniques in agriculture and energy sector. Mater Today: Proc 51:2260–2263

Deshpande T (2017) State of agriculture in India. PRS Legisl Res 53(8):6–7

Doğan O, Aşan H, Ayç E (2015) Use of data mining techniques in advance decision making processes in a local firm. Eur J Business Econ 10(2):6821. https://doi.org/10.12955/ejbe.v10i2.682

Gavioli A, de Souza EG, Bazzi CL, Schenatto K, Betzek NM (2019) Identification of management zones in precision agriculture: an evaluation of alternative cluster analysis methods. Biosyst Eng 181:86–102
Gharde Y, Singh PK, Dubey RP, Gupta PK (2018) Assessment of yield and economic losses in agriculture due to weeds in India. Crop Prot 107:12–18
Jamshidi P, Sharifloo A, Pahl C, Arabnejad H, Metzger A, Estrada G (2016) Fuzzy self-learning controllers for elasticity management in dynamic cloud architectures. In: 2016 12th International ACM SIGSOFT conference on quality of software architectures (QoSA). IEEE, pp 70–79
Klir G, Yuan B (1995) Fuzzy sets and fuzzy logic, vol 4. Prentice Hall, New Jersey, pp 1–12
Kumar V, Jat HS, Sharma PC, Gathala MK, Malik RK, Kamboj BR, Yadav AK, Ladha JK, Raman A, Sharma DK, McDonald A (2018) Can productivity and profitability be enhanced in intensively managed cereal systems while reducing the environmental footprint of production? Assessing sustainable intensification options in the breadbasket of India. Agric Ecosyst Environ 252:132–147
Mabhaudhi T, Chimonyo VGP, Hlahla S, Massawe F, Mayes S, Nhamo L, Modi AT (2019) Prospects of or-phan crops in climate change. Planta 250:695–708
Majumdar J, Naraseeyappa S, Ankalaki S (2017) Analysis of agriculture data using data mining techniques: application of big data. J Big Data 4(1):20
Melin P, Castillo O (2014) A review on type-2 fuzzy logic applications in clustering, classification and pattern recognition. Appl Soft Comput 21:568–577
Misra AK (2014) Climate change and challenges of water and food security. Int J Sustain Built Environ 3(1):153–165
Narmadha R, Latchoumi TP, Jayanthiladevi A, Yookesh TL, Mary SP (2022) A fuzzy-based framework for an agriculture recommender system using membership function. Appl Soft Comput: Tech Appl:207–223
Naseem M, Singh V, Ahmed K, Mahroof M, Ahamad G, Abbasi E (2022) Architecture of automatic irrigation system in Hilly area using wireless sensor net-work: a review. In: 2022 2nd International conference on emerging frontiers in electrical and electronic technologies (ICEFEET). IEEE, pp 1–6
Naseem M, Alam M, Ahmad K, Singh V, Mahroof M, Ahamad G (n.d.) Machine learning approaches for automatic irrigation system in Hilly areas using wireless sensor networks
Pantazi XE, Moshou D, Bochtis D (2019) Intelligent data mining and fusion systems in agriculture. Academic Press
Raorane AA, Kulkarni RV (2012) Data mining: an effective tool for yield estimation in the agricultural sector. Int J Emerg Trends Technol Comput Sci 1(2):1–4
RURAL DO (2014) IRLA2014
Sharma V, Tripathi AK, Mittal H (2022a) Technological revolutions in smart farming: current trends, challenges and future directions. Comput Electron Agric:107217
Sharma A, Abrahamian P, Carvalho R, Choudhary M, Paret ML, Vallad GE, Jones JB (2022b) Future of bacterial disease management in crop production. Annu Rev Phytopathol 60:259–282
Zhai Z, Martínez JF, Beltran V, Martínez NL (2020) Decision support systems for agriculture 4.0: survey and challenges. Comput Electron Agric 170:105256

Chapter 15
Plant Disease Diagnosis with Artificial Intelligence (AI)

Muhammad Naveed, Muhammad Majeed, Khizra Jabeen, Nimra Hanif, Rida Naveed, Sania Saleem, and Nida Khan

Abstract Plant diseases, which are unseen but deadly, endanger our crops and the food security of nations. However, optimism stems from the convergence of powerful artificial intelligence (AI) approaches, each of which plays a distinct role in the protection of our domains. A change in the way farming is practiced at the moment could be embodied by an unwavering application of artificial intelligence and its subsets in agriculture. A farmer may accomplish more with fewer resources thanks to AI-powered farming solutions, which also improve quality and ensure speedy GTM (go-to-market) strategies for crops. Direct use of AI (artificial intelligence) or machine intelligence in the agricultural industry could represent a paradigm shift in the way farming is now carried out. Deep learning, driven by neural networks, has transformed how we perceive and diagnose many diseases. Deep learning overcomes the constraints of existing approaches by autonomously extracting detailed visual information, providing greater precision and efficiency in recognizing plant diseases. Convolutional neural networks (CNNs), a subset of deep learning, have emerged as powerful tools, with elaborate network structures and local receptive fields that enable them to interpret complex visual input, making them indispensable in the field of image recognition. Machine learning approaches, such as support vector machine (SVM) and artificial neural network (ANN) classifiers, have also

M. Naveed · K. Jabeen · N. Hanif · R. Naveed
Department of Biotechnology, Faculty of Science and Technology, University of Central Punjab, Lahore, Punjab, Pakistan
e-mail: dr.naveed@ucp.edu.pk

M. Majeed (✉)
Department of Botany, University of Gujrat, Gujrat, Pakistan
e-mail: m.majeed@uog.edu.pk

S. Saleem
Department of Plant Sciences, Quaid-i-Azam University, Islamabad, Pakistan
e-mail: saniasalem@bs.qau.edu.pk

N. Khan
Department of Botany, University of Science and Technology Bannu, Bannu, Khyber Pakhtunkhwa, Pakistan

A. Khamparia et al. (eds.), *Microbial Data Intelligence and Computational Techniques for Sustainable Computing*, Microorganisms for Sustainability 47,
https://doi.org/10.1007/978-981-99-9621-6_15

stepped up to the plate, automating the diagnosis of plant diseases with remarkable precision. Deep learning-capable robotics and machine intelligence have had a profoundly disruptive and enabling impact on industry, governments, and society. They are also having an impact on more general trends in international sustainability. Weather patterns, soil composition, and disease trends all tell their own tales, providing forecast insights and personalized preventative steps to safeguard the harvest. A symphony of IoT devices orchestrates vigilance across smart farms. By capturing the afflicted plant sections, farmers may quickly and correctly identify illnesses and find remedies using a mobile app through AI advancements. The most recent artificial intelligence (AI) algorithms for cloud-based image processing enable real-time diagnosis. Artificial intelligence and its thorough learning capabilities have developed into a crucial strategy for addressing a range of farming-related difficulties.

Keywords Artificial intelligence · Plant diseases · Remarkable precision

15.1 Introduction

Agriculture has always been considered the most essential need of livelihoods throughout the ecosystem. It plays a vital role in supporting every type of organism in establishing habitats, producing foods, or utilizing the raw materials to produce good-quality industrial products, etc. Agriculture not only defines its significance in natural shelters and foods but has a direct link with the income of a human being and also of its country (Majeed et al. 2020a). Building a strong economy of a country requires agricultural development followed by the establishment of industrial processing to increase the GDP value of a country. It makes up half of the employed labor force, accounts for around 24% of the GDP, and is the main source of foreign exchange profits (Haq et al. 2022b). The entire rural and urban population's feed depends on agriculture. As agriculture is linked to the quality, production, and growth of plants, it is necessary to simply protect and increase their production, so as to live (Talaviya et al. 2020).

Plant diseases affect the quality of vegetables, organic goods, vegetables, and cereals. Organisms which cause infectious diseases in plants, referred to as the "plant pathogens," include fungi, viroid, protozoa, oomycetes, parasitic weeds and plants, nematodes, phytoplasmas, bacteria (Aziz et al. 2023a), virus-like organisms, and viruses (Naveed et al. 2022a; Omara et al. 2023). In less developed nations, where access to disease-control measures is limited, they are subjected to a yearly loss of 30–50% of vital crops which often leads to hunger and starvation (Aggarwal et al. 2022).

Plant disease identification or diagnosis is one of the major challenges in agriculture and a key component affecting the outcome of cultivating crops (Arif et al. 2021). If not detected before, plant pathogens can decrease the crop production rate from 10 to 95%. Identification through the naked eye can be a tedious task for local farmers as it requires professional observatory skills and expertise in plant pathology

(Talaviya et al. 2020). Additionally, if a rare disease attacks a field, farmers seek specialist help to make a precise and timely diagnosis, which inevitably results in higher treatment expenses (Arshad et al. 2022). To avoid such complications, early diagnosis of various plant diseases is one of the many farming aspects where computerization in agriculture has made tremendous strides. Almost every country has shifted its emphasis toward mechanizing agriculture in order to achieve accuracy and precision in identifying plant diseases (Orchi et al. 2021). There is a vast array of technology available to us now, including high-quality digital cameras with excellent focus and resolution as well as cameras built into even our mobile phones. These technologies are currently employed to collect datasets about plants (Chithambarathanu and Jeyakumar 2023).

The development of artificial intelligence (AI) is based on the idea of prior learning experiences. Artificial intelligence (AI) is playing a crucial role in replacing traditional methods with quick, accurate, computerized, cost-effective, and, most importantly, precise ways of identifying plant sickness (Majeed et al. 2021a). The use of totally computer-based image managing innovation in rural designing research produces amazing results about the capability of a programmed method for identifying diseases and their treatments in plants so that we could protect them at a very early stage before the destruction is done (de Oliveira and de Silva 2023). Through early identification, prompt treatment, and reduced disease occurrence, automatic detection techniques help farmers improve crop quality. Image processing is utilized to quantify the size of the diseased area and identify color differences in the affected area (Kamdar et al. 2021).

Machine learning (ML) techniques like backpropagation, artificial neural networks (ANN), and convolutional neural networks (CNNs) are automating the use of machines and the development of cutting-edge inventions (Nawaz et al. 2020). The ML-based techniques are easier to implement and do not need a large amount of training data, but they are slow because of complicated preprocessing and rely on the expertise of experienced human professionals for feature extraction and optimal feature selection needed to accomplish classification (Tariq et al. 2023). ML-based models were also introduced for the classification and identification of plant diseases such as decision tree (DT), random forest (RF), K-nearest neighbors (KNN), and support vector machine (SVM) (Kukadiya and Meva 2023).

The convolutional neural network (CNN) is a well-known DL model that demonstrated efficient pattern recognition performance and is frequently used for detecting early plant leaf diseases. In recent studies, CNN has mostly been employed for diagnosing and classifying crop plant diseases (Hassan et al. 2022a). It has convolutional layers, which are sets of image channels that have been integrated to create images or spotlight maps (Kothari 2018). CNN optimizes the weights and channel parameters in the protected-up layers to create highlights that are becoming for solving problems in contrast to conventional device/algorithm learning methodologies that depend on hand-crafted highlights (Shoaib et al. 2023). Due to the efficient feature representation used in these methods, crop-related classification tasks have demonstrated encouraging results. Computer vision's mature CNN architectures, such as AlexNet, GoogLeNet, VGGNet, ResNet, EfficientNet, MobileNet, and DenseNet, are extensively used in existing plant disease categorization methods (Albahli and Masood 2022).

Approaches based on deep learning (DL) and AI are widely used in many industries, including agriculture (Ahmad et al. 2023). These methods avoid laborious image preprocessing and have a smaller memory footprint since they automatically synthesize discriminative features from the input samples (Albattah et al. 2022). Therefore, we must use technology for great use and should keep progressing in these fields of research; otherwise, if we do not focus on these problems today, it will not be good for the future of humanity.

15.1.1 AI Technologies in Plant Diseases Diagnosis

15.1.1.1 Deep Learning and Neural Network

The first concept of the deep learning method originated from a paper by Hinton et al. published in the *Journal of Science*. Deep learning uses neural networks to analyze and learn from data, and several hidden layers act like artificial neurons to extract data features. The artificial neuron is employed to first capture basic features and subsequently merge these basic features to derive more advanced, abstract features; this approach has the potential to significantly mitigate the basic problems. A growing number of scientists are becoming interested in deep learning since it addresses the problem that traditional methods rely on artificially manufactured characteristics. Systems of recommendation, pattern recognition, speech recognition, and natural language processing now use it successfully (Ghosh and Roy 2021).

Conventional image classification and recognition techniques that depend on manual design components make it difficult to obtain deep and complex picture feature information. And the deep learning technique has the potential to eliminate this barrier (Albahli and Masood 2022). It can do autonomous learning on the original image right away to learn about numerous image feature stages, such as low- to high-level semantic elements. Conventional plant disease and pathogen identification algorithms mostly use manually generated image recognition characteristics, which are complex and rely on luck and expertise; the original image's features cannot be automatically learned and extracted by these algorithms (Mahlein 2016). On the other hand, this approach is able to automatically extract features from huge amounts of data without the need for any operatic system (Majeed et al. 2022a). It consists of numerous coatings and has high unsupervised learning and feature expression abilities. It can also autonomously extract picture features for visual categorization and identification. Hence, deep learning can make a significant contribution to the field of recognizing horticulture disorders in images (Mohanty et al. 2016).

Many renowned deep neural network models have been produced as a result of deep learning techniques like deep convolutional neural networks (CNNs), which is the most commonly employed methodology today. Compared to earlier manual design feature extraction methods in picture recognition, using these deep neural network models to automate feature extraction from high-dimensional feature databases offers several advantages (Waheed et al. 2022). Moreover, with an expansion of developmental data as well as advancements in computational capabilities, deep neural networks continue to enhance their ability to characterize and understand patterns. Currently, the surge in this approach is impacting both research and manufacturing departments, with deep neural network models consistently outperforming their traditional counterparts; over the past few years, the dominant deep learning framework has been the deep convolutional neural network (Khoja et al. 2022).

15.1.2 Convolutional Neural Network (CNN)

CNNs are experts at conducting convolution operations and have complex network architecture. Neural network model, as illustrated in Fig. 15.1, is made up of a convolution layer, an input layer, a pooling layer, an output layer, and a full connection layer. There is no requirement for a full connection in one model, where the convolution layer and the pooling layer alternate several times when the neurons of the convolution layer are linked to the neurons of the pooling layer. A model that's frequently used in the DL field is CNN. This is due to CNN's large model capacity and ability to process complicated information as a result of its fundamental structural features, which gives CNN an edge in picture identification (Guerrero-Ibañez and Reyes-Muñoz 2023). Meanwhile, as a result of CNN's success in visual perception-related tasks, deep learning is growing in popularity. The fundamental advantage of the convolution neural network is its local receptive field, often known as the convolution core. The very initial step in this network is to obtain some essential features information then after it, the neurons are sent into the pooling layer, which retrieves the details twice. Now these pooling strategies have been utilized to perform different computed functions (Biswal and Chestnut 2022). The data moves on to the fully connected layer, where the neurons create

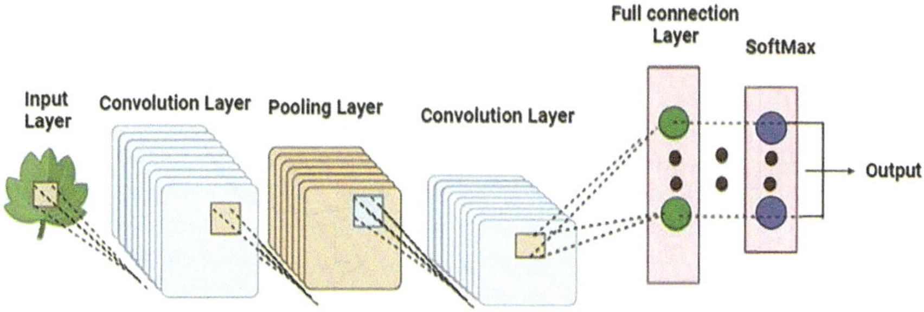

Fig. 15.1 The structure of CNN

complete connections with the neurons in the highest layer, after passing through many convolutional and pooling layers. In the final least, the output results from classifying the full-connection layer data using the SoftMax technique are given to the output layer as output results (Albattah et al. 2022).

15.2 Machine Learning Method

Identification of plant diseases is crucial to a successful agricultural system. The majority of the time, a farmer can see disease symptoms in plants with the naked eye, but this requires regular observation. However, this procedure is costlier and occasionally less precise in big plantations. Farmers in other countries, like India, may be required to show the specimen to specialists, which adds time and costs (Kukadiya and Meva 2023; Thakur and Mittal 2020).

15.2.1 Disease Detection System for General Plant

Plant illnesses can be discovered by inspecting the plant's leaf, stem, and root. Digital image processing may be used to identify damaged leaves, stems, fruits, and flowers, as well as the form and color of the afflicted region (Majeed et al. 2020b). It mainly consists of four basic steps elaborated in Fig. 15.2.

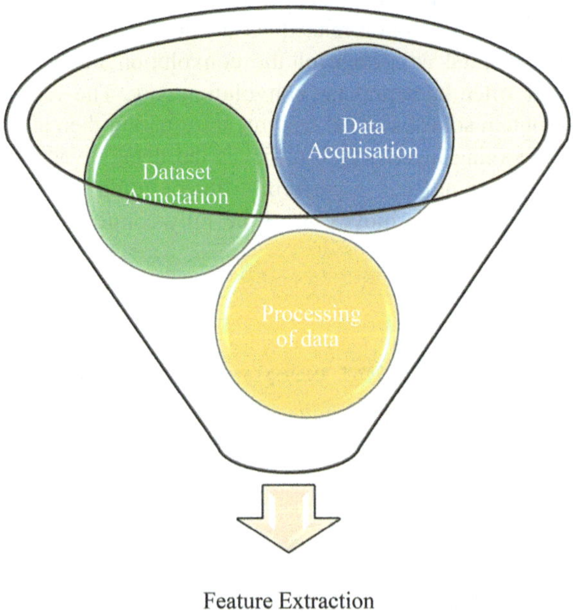

Fig. 15.2 Technique for detecting plant diseases

15.2.2 Data Acquisition

Different types of automated cameras are used to capture images to recognize any illness in plants or crops.

15.2.3 Dataset Annotation

To construct a domain-specific dataset for taking images of various classifications.

15.2.4 Data Processing

Prior to further processing, acquired images must go through data-scrubbing phases to improve certain picture properties. The segmentation process is employed to divide the plant image into distinct segments, facilitating the extraction of disease-affected regions in the plant's stems, fruits, or leaves from the background.

15.2.5 Feature Extraction

In this step, plant physiology is determined by shape, color, height, and texture by using various methods like gray-level co-occurrence matrix (GLCM), Blendvision, and machine intelligence. In the end, plant disease can be classified via any technique (Khan et al. 2022).

15.3 Classification Techniques

15.3.1 Support Vector Machine (SVM) Classifier

The SVM classifier is the most commonly employed method of machine learning technologies to classify and detect any plant pathogen and plant infection. The detection of leaf damage produced by canker and anthracnose infections is required for the diagnosis of illnesses affecting citrus trees, such as grapefruit, lemons, limes, and oranges (Tassadduq et al. 2022). The experimental outcome produced a true acceptance rate of 95%. Detection of grape plant infections achieved an average accuracy rate of 88.89%. The identification of chimera and anthracnose infections in oil palm leaves reached 97% and 95% accuracy rates, respectively (Singh et al. 2021). The identification of potato crop infections, specifically late blight and early blight, was conducted using over 300 publicly accessible images, achieving an accuracy rate of 95% (Liu and Wang 2021).

Created a supervised method for identifiying ailments in tea plants. Using SVM classifiers, three distinct diseases characterized by fewer features were successfully identified, achieving an accuracy rate of 90% through the developed method. A substantial dataset was employed in soybean cultivation to detect three distinct diseases, and the reported mean classification reliability for these diseases stood at approximately 90% (Kumar et al. 2021).

15.3.2 *Artificial Neural Network Classifier*

The use of artificial neural networks (ANN), to diagnose patterns, has been explored in previous studies concerning plant disease detection. An evaluation of a proposed method for recognizing plant diseases was conducted, employing the feed-forward backpropagation algorithm, which demonstrated strong performance with a precision rate of approximately 93%. They examined the solution's effectiveness in addressing diseases that impact plants (Vasudevan and Karthick 2023). To improve the precision of recognizing fungal-induced illnesses in cucumber plants, a model was created. A study conducted on identifying the groundnut plant disease commonly known as leaf spot using neural network backpropagation is proposed (Majeed et al. 2021b). The empirical data demonstrate that they diagnosed four categories of illnesses from 100 sample diseased leaf photos with 97.41% accuracy.

15.3.3 *Fizzy Classifier*

The method to determine the presence of infection in wheat crop photos using Fuzzy Classifier in a related study on Fuzzy Classifier in plant disease detection. This method has been tested on a dataset of healthy and unhealthy leaves. The categorization of healthy and unhealthy leaves was discovered with an accuracy of 88%, and illness detection with an accuracy of 56% (Lee et al. 2017). Table 15.1 summarizes and provides the comparison of several machine learning classifiers employed in identifying plant diseases.

Table 15.1 Summarizes and provides the comparison of several machine learning classifiers employed in identifying plant diseases.

15.4 Advantages and Challenges

AI-powered plant disease diagnosis is transforming agriculture by providing accurate, timely, and sustainable solutions to handle plant diseases.

Table 15.1 Comparative analysis of classification techniques

Techniques classification	Crops culture	Diseases	Accuracy rate
SVM classifier	Citrus	2	95%
	Grape	2	88.89%
	Oil palm	2	97% and 95%
	Potato	2	90%
	Tea	3	93%
ANN classifier	Cucumber	2	Improved accuracy
	Pomegranate	4	90%
	Groundnut	4	97.41%
Fizzy classifier	Wheat	1	88%

15.4.1 Early Detection of Plant Diseases

AI-based systems are able to identify plant diseases in their very earliest stages, normally before any outward signs show up. In most viral attacks, the symptoms appear when either it's too late or it has become difficult to save the crops from spreading the disease (Ahmad et al. 2023). Early detection reduces the spread of illnesses and enables prompt action. The segmentation of the plant leaf images into surface areas like background, diseased area, and non-diseased area of the leaf is ensured by image sensing and analysis. The sick or infected area is then removed and sent to the lab for further analysis.

15.4.2 Rapid Diagnosis

AI systems are able to quickly process enormous amounts of data and images to produce results. This speed is essential for making decisions quickly and taking the appropriate steps to stop the spread of illnesses. The detection of diseases through the naked eye and by observation requires a great level of skills and professionalism, which can only be done by hiring a plant pathologist or expert. This process usually takes up a lot of time. By using deep learning through image-based plant disease techniques, expeditious diagnosis results can be obtained, and the spreading of disease can be ceased (Shoaib et al. 2023).

15.4.3 Accurate Estimations

Large datasets can be analyzed by AI algorithms, which can spot tiny trends that can be challenging for human experts to notice. As a result, disease diagnosis has high rates of accuracy, which lowers the possibility of misdiagnosis. Real-time estimations can be made during the cultivation phase by creating a field map and identifying the places where the crops need water, fertilizer, and pesticides using high-detail photos from drone and copter systems (Waheed et al. 2023).

15.4.4 Cost-Effectiveness

In the long run, using AI for disease diagnosis may be cost effective. It saves time and money by reducing the need for manual labor and the reliance on specialists for each diagnosis. Hiring a plant expert or pathologist is not preferred as it requires exclusive equipment and techniques, which are mostly not affordable (Das et al. 2022).

15.4.5 Global Collaboration and Accessibility

As more data becomes accessible, AI models can continuously learn and increase their diagnostic precision. The system will continue to be successful in recognizing new and emerging diseases thanks to its versatility. AI-based timely disease management can boost crop production. Farmers can grow healthier crops and maximize their harvests by preventing or controlling disease outbreaks. Even small-scale and resource-constrained farmers can access these technologies to enhance their crop management practices as AI systems become more popular and user-friendly. AI-based disease diagnosis can promote international cooperation in plant health management and research. Researchers and specialists from many fields can exchange information and insights to produce more successful disease management measures (Aziz et al. 2023b).

15.4.6 Continuous Monitoring

AI-powered systems can be set up in the field and used for remote plant health monitoring. This makes it possible for farmers to monitor their crops and get instant notifications when a disease outbreak occurs. Remote sensing (RS) techniques along with hyperspectral imaging and 3D laser scanning are crucial to constructing crop metrics over thousands of acres of cultivable land (Majeed et al. 2023).

15.4.7 Improved Yield and Crop Quality

AI can assist farmers in selectively applying remedies to damaged plants by precisely recognizing them, hence minimizing the overall use of pesticides and fungicides. This encourages the use of more environmentally responsible and sustainable farming techniques. AI-based timely disease management can boost crop production. Farmers can grow healthier crops and maximize their harvests by preventing or controlling disease outbreaks (Majeed et al. 2022b). AI-capable equipment that can automate irrigation and increase overall production are aware of historical climate patterns, soil condition, and the kind of crops that should be planted. Irrigation uses up close to 70% of the world's freshwater supply; automation can help farmers manage their water problems while also saving water (Singh et al. 2021).

15.4.8 Data-Driven Insights

AI systems produce useful data that can be utilized to track disease dynamics, comprehend disease dynamics, and guide data-driven crop management decisions. Drone-based photos can assist with field scanning, crop monitoring, and other tasks. Farmers can connect them with IOT and PC vision technology to ensure rapid actions. These feeds may continuously alarm farmers about the climate.

15.4.9 Robotics in Agriculture

Agribot, often known as Agbot, is a farming robot. It helps the farmer boost the crop's productivity and lessens the amount of manual labor he or she needs to do. We may anticipate that these agricultural robots will perform tilling, sowing, harvesting, and many other farm tasks on their own in the following generations. Indeed, these agricultural robots will take care of weeding, pest control, and disease management (Yağ and Altan 2022).

15.4.10 Integration with IoT

A wide range of sectors and businesses, including manufacturing, health, communications, energy, and agriculture, will be impacted by the "Internet of Things (IoT)." The goal of the Internet of Things (IoT) use in agriculture is to equip farmers with decision-making instruments and automation technologies that seamlessly combine goods, information, and services for improved productivity, quality, and profit (Thakur and Mittal 2020). A range of plant types and infections can be easily accommodated by scaling AI models. They are adaptable to various climates and farming methods, making them useful agricultural tools (Bashir et al. 2023). Drone-based photos can assist with field scanning, crop monitoring, and other tasks.

Farmers can connect them with IoT and PC vision technology to ensure rapid actions. These feeds may continuously alarm farmers about the climate.

15.4.11 Challenges and Limitations

AI-driven plant disease diagnosis has almost revolutionized the agriculture industry, but there are some challenges and limitations which need to be addressed for the future betterment and help mankind. These challenges are mentioned below along with their reason for limitations (Salman et al. 2023).

15.4.12 Data Quality and Quantity

To enhance AI accuracy, we need extensive and high-quality datasets of diseased and healthy plants. It required acquiring comprehensive datasets of various plant diseases, which can be challenging, leading to potential biases in AI models. Accurate labeling of images and data is time consuming and requires expertise, making it a bottleneck in model training (Hussain et al. 2023).

15.4.13 Scalability

Scaling AI solutions to accommodate the needs of large agricultural operations is a challenge that needs to be addressed. Implementing AI solutions in large agricultural settings may require significant infrastructure upgrades and investment. Scaling AI systems can be expensive, limiting access for smaller farmers or regions with limited resources.

15.4.14 Ethical Considerations

AI adoption in agriculture must consider ethical concerns, such as data privacy and AI bias. Collecting and sharing agricultural data for AI analysis raise concerns about data privacy and ownership. AI algorithms may exhibit bias if training data is not diverse, leading to unfair recommendations or decisions (Naveed et al. 2022b).

15.4.15 Complexity of Plant Diseases

The complexity of plant disease is a major challenge researchers are facing these days. Plant diseases can manifest differently based on factors like location, climate, and plant variety, making it challenging to create universally applicable models. AI models may struggle to identify diseases when multiple infections occur simultaneously.

15.5 Future Prospects and Implications

AI is revolutionizing many industries like agriculture, food, and textiles. In recent years, AI has been making significant trends in the field of plant disease diagnosis. This chapter explores the current landscape of AI-driven plant disease diagnosis, its future prospects, and the implications it holds for agriculture. Accurate and timely diagnosis of plant diseases are crucial for ensuring food security and optimizing crop yields (Majeed et al. 2022c). Traditional methods of disease detection are often time consuming and less precise. Here, AI steps in to bridge the gap by offering more efficient and accurate solutions. AI is being used for tasks like crop monitoring, soil analysis, and even autonomous farming equipment. We will delve into the fusion of AI techniques and the role of robotics and drones.

15.5.1 Fusion of AI Techniques

AI-powered cameras and sensors can identify disease symptoms by analyzing plant images. This technology aids in early detection. Consider a farmer in the middle of a big cornfield examining if any of the plants have a fungal infection that isn't yet obvious to the human eye simply by using their smartphone, which is loaded with an AI-powered plant disease detection software. They take a snapshot of the corn leaves, and the app analyzes them in seconds, finding minor discolorations and patterns that indicate illness. The farmer receives quick diagnostic and treatment suggestions (Naveed et al. 2022c), allowing them to act and avoid future spread.

Data analytics tools process vast amounts of agricultural data, identifying disease trends and providing insights for preventive measures. Farmers may forecast disease outbreaks and offer particular preventative actions by examining past data. For example, they may recommend modifying irrigation schedules based on previous disease trends to reduce the risk of fungal infections.

Internet of Things (IoT) devices can collect real-time data from the field, allowing for immediate action when disease symptoms are detected. A network of IoT sensors continually monitors the health of crops on a smart farm. Throughout the fields, soil moisture monitors, meteorological stations, and disease detection sensors are carefully positioned. These devices continually gather and transmit real-time data to a centralized control system. An alarm is promptly delivered to the farmer's smartphone when a sensor detects unexpected moisture levels or disease-related characteristics. This timely warning enables the farmer to analyze the problem and take appropriate measures (Hassan et al. 2022b).

15.5.2 Robotics and Drones

Drones equipped with high-resolution cameras can monitor large fields efficiently, detecting early signs of disease and nutrient deficiencies (Ahmed and Yadav 2023). A high-tech drone equipped with a high-resolution camera takes flight in a big vineyard. It flies across rows of grapevines, photographing the leaves and vines in great detail. The photos are sent to a computer that is loaded with AI algorithms. The AI examines the photos for tiny discolorations or anomalies that might indicate grapevine disease. If any problems are discovered, the technology generates a map identifying the afflicted regions, allowing the vineyard manager to precisely focus treatment and avoid additional damage (Haq et al. 2022).

Robots equipped with AI algorithms can autonomously perform tasks like weeding, planting, and harvesting, ensuring healthier crops (Yağ and Altan 2022). Autonomous robots armed with AI algorithms patrol the fields of a vegetable farm. These robots are taught to recognize weeds and differentiate them from cultivated plants (Albattah et al. 2022). They employ computer vision to detect the shape and features of the crops they meet as they travel through the rows. When a weed is detected, the robot accurately administers herbicide to the undesirable plants while leaving the crops alone. Not only does this automated crop management lessen the need for pesticides, but it also encourages healthier, weed-free crops.

15.6 Conclusion

Agriculture, the heartbeat of our world, sustains life in more ways than one. Beyond providing sustenance and raw materials, it shapes economies, livelihoods, and ecosystems. However, the very foundation of agriculture, our plants, faces an unrelenting challenge—disease. Plant diseases jeopardize the quality and quantity of our crops, and in regions with limited resources, they often lead to hunger and hardship. Timely and accurate diagnosis of these diseases is paramount, and this is where technology, particularly artificial intelligence (AI) and machine learning, has emerged as a beacon of hope. The chapter, realm AI-powered plant disease diagnosis, exploring its transformative potential, methodologies, and implications for agriculture and the world at large.

In the grand tapestry of agriculture's future, AI stands as a pivotal thread, weaving together innovation, efficiency, and sustainability. While we celebrate its transformative potential in the diagnosis of plant diseases and its promise to shape a brighter future for farming, we must also acknowledge the tapestry's intricate patterns, marked by the challenges and limitations that need our attention.

References

Aggarwal S, Suchithra M, Chandramouli N, Sarada M, Verma A, Vetrithangam D, Pant B, Ambachew Adugna B (2022) Rice disease detection using artificial intelligence and machine learning techniques to improvise agro-business. Sci Program 2022:1757888. https://doi.org/10.1155/2022/1757888

Ahmad A, Saraswat D, El Gamal A (2023) A survey on using deep learning techniques for plant disease diagnosis and recommendations for development of appropriate tools. Smart Agric Technol 3:100083. https://doi.org/10.1016/j.atech.2022.100083

Ahmed I, Yadav PK (2023) Plant disease detection using machine learning approaches. Expert Syst 40(5):e13136. https://doi.org/10.1111/exsy.13136

Albahli S, Masood M (2022) Efficient attention-based CNN network (EANet) for multi-class maize crop disease classification [Original Research]. Front Plant Sci 13:1003152. https://doi.org/10.3389/fpls.2022.1003152

Albattah W, Javed A, Nawaz M, Masood M, Albahli S (2022) Artificial intelligence-based drone system for multiclass plant disease detection using an improved efficient convolutional neural network [Original Research]. Front Plant Sci 13:808380. https://doi.org/10.3389/fpls.2022.808380

Arif U, Bhatti KH, Ajaib M, Wagay NA, Majeed M, Zeb J, Hameed A, Kiani J (2021) Ethnobotanical indigenous knowledge of Tehsil Charhoi, District Kotli, Azad Jammu and Kashmir, Pakistan. Ethnobot Res Appl 22:1–24. https://doi.org/10.32859/ERA.22.50.1-24

Arshad F, Waheed M, Harun N, Fatima K, Khan BA, Fatima K, Abbas Z, Jabeen S, Majeed M (2022) Indigenous farmer's perception about fodder and foraging species of semi-arid lowlands of Pakistan: a case study of District Kasur, Pakistan. Taiwania 67(4):510–523

Aziz A, Anwar MM, Majeed M, Fatima S, Mehdi SS, Mangrio WM, Elbouzidi A, Abdullah M, Shaukat S, Zahid N, Mahmoud EA, Casini R, Yessoufou K, Elansary HO (2023a) Quantifying landscape and social amenities as ecosystem services in rapidly changing peri urban landscape. Land 12(2):477

Aziz T, Naveed M, Jabeen K, Shabbir MA, Sarwar A, Zhennai Y, Alharbi M, Alshammari A, Alasmari AF (2023b) Integrated genome based evaluation of safety and probiotic characteristics of *Lactiplantibacillus plantarum* YW11 isolated from Tibetan kefir. Front Microbiol 14:1157615. https://doi.org/10.3389/fmicb.2023.1157615

Bashir SM, Altaf M, Hussain T, Umair M, Majeed M, Mangrio WM, Khan AM, Gulshan AB, Hamed MH, Ashraf S, Amjad MS, Bussmann RW, Abbasi AM, Casini R, Alataway A, Dewidar AZ, Al-Yafrsi M, Amin MH, Elansary HO (2023) Vernacular taxonomy, cultural and ethnopharmacological applications of avian and mammalian species in the vicinity of Ayubia National Park, Himalayan Region. Biology 12:4

Biswal B, Chestnut S (2022) DeepTrac: applying artificial intelligence in plant disease detection. SoutheastCon 2022

Chithambarathanu M, Jeyakumar MK (2023) Survey on crop pest detection using deep learning and machine learning approaches. Multimed Tools Appl 82:42277. https://doi.org/10.1007/s11042-023-15221-3

Das S, Pattanayak S, Behera PR (2022) Application of machine learning: a recent advancement in plant diseases detection. J Plant Prot Res 62(2):122–135. https://doi.org/10.24425/jppr.2022.141360

de Oliveira RC, de Silva RDS (2023) Artificial intelligence in agriculture: benefits, challenges, and trends. Appl Sci 13(13):7405. https://www.mdpi.com/2076-3417/13/13/7405
Ghosh A, Roy P (2021) AI based automated model for plant disease detection, a deep learning approach. In: Dutta P, Mandal JK, Mukhopadhyay S (eds) Computational intelligence in communications and business analytics. Springer, Cham
Guerrero-Ibañez A, Reyes-Muñoz A (2023) Monitoring tomato leaf disease through convolutional neural networks. Electronics 12(1):229. https://www.mdpi.com/2079-9292/12/1/229
Haq SM, Tariq A, Li Q, Yaqoob U, Majeed M, Hassan M, Fatima S, Kumar M, Bussmann RW, Moazzam MFU, Aslam M (2022) Influence of edaphic properties in determining forest community patterns of the Zabarwan Mountain range in the Kashmir Himalayas. Forests 13:8
Haq SM, Yaqoob U, Majeed M, Amjad MS, Hassan M, Ahmad R, Morales-de la Nuez A (2022b) Quantitative ethnoveterinary study on plant resource utilization by indigenous communities in high-altitude regions. Front Vet Sci 9:94404
Hassan M, Haq SM, Ahmad R, Majeed M, Sahito HA, Shirani M, Mubeen I, Aziz MA, Pieroni A, Bussmann RW, Alataway A, Dewidar AZ, Al-Yafrsi M, Elansary HO, Yessoufou K (2022a) Traditional use of wild and domestic fauna among different ethnic groups in the Western Himalayas? Cross cultural analysis. Animals 12:17
Hassan M, Haq SM, Majeed M, Umair M, Sahito HA, Shirani M, Waheed M, Aziz R, Ahmad R, Bussmann RW, Alataway A, Dewidar AZ, El-Abedin TKZ, Al-Yafrsi M, Elansary HO, Yessoufou K (2022b) Traditional food and medicine: ethno-traditional usage of fish fauna across the valley of Kashmir: a Western Himalayan region. Diversity 14:6
Hussain S, Raza A, Abdo HG, Mubeen M, Tariq A, Nasim W, Majeed M, Almohamad H, Al Dughairi AA (2023) Relation of land surface temperature with different vegetation indices using multi-temporal remote sensing data in Sahiwal region, Pakistan. Geosci Lett 10(1):33
Kamdar JH, Jasani MD, Jasani JD, Praba J, Georrge JJ (2021) Chapter 11. Artificial intelligence for plant disease detection: past, present, and future. In: Jyotir Moy C, Abhishek K, Pramod Singh R, Vishal J (eds) Internet of things and machine learning in agriculture. De Gruyter, pp 223–238. https://doi.org/10.1515/9783110691276-011
Khan AM, Li Q, Saqib Z, Khan N, Habib T, Khalid N, Majeed M, Tariq A (2022) MaxEnt modelling and impact of climate change on habitat suitability variations of economically important Chilgoza pine (*Pinus gerardiana* Wall.) in South Asia. Forests 13:5
Khoja AA, Haq SM, Majeed M, Hassan M, Waheed M, Yaqoob U, Bussmann RW, Alataway A, Dewidar AZ, Al-Yafrsi M, Elansary HO, Yessoufou K, Zaman W (2022) Diversity, ecological and traditional knowledge of pteridophytes in the Western Himalayas. Diversity 14:8
Kothari J (2018) Plant disease identification using artificial intelligence: machine learning approach. Int J Innov Res Comput Commun Eng 7:11082–11085
Kukadiya H, Meva DD (2023) Machine learning in agriculture for crop diseases identification: a survey. Int J Res GRANTHAALAYAH 11:5099. https://doi.org/10.29121/granthaalayah.v11.i3.2023.5099
Kumar S, Jain A, Shukla AP, Singh S, Raja R, Rani S, Harshitha G, AlZain MA, Masud M (2021) A comparative analysis of machine learning algorithms for detection of organic and nonorganic cotton diseases. Math Probl Eng 2021:1790171. https://doi.org/10.1155/2021/1790171
Lee SH, Chan CS, Mayo SJ, Remagnino P (2017) How deep learning extracts and learns leaf features for plant classification. Pattern Recogn 71:1–13. https://doi.org/10.1016/j.patcog.2017.05.015
Liu J, Wang X (2021) Plant diseases and pests detection based on deep learning: a review. Plant Methods 17(1):22. https://doi.org/10.1186/s13007-021-00722-9
Mahlein A-K (2016) Plant disease detection by imaging sensors—parallels and specific demands for precision agriculture and plant phenotyping. Plant Dis 100(2):241–251. https://doi.org/10.1094/pdis-03-15-0340-fe
Majeed M, Bhatti KH, Amjad MS, Abbasi AM, Rashid A, Nawaz F, Ahmad KS (2020a) Ethnoveterinary practices of Poaceae taxa in Punjab, Pakistan

Majeed M, Bhatti KH, Amjad MS, Abbasi M, Id RWB, Nawaz F, Rashid A, Mehmood A, Id MM, Khan WM, Id SA (2020b) Ethno-veterinary uses of Poaceae in Punjab, Pakistan. PLoS One 15(11):e0241705. https://doi.org/10.1371/journal.pone.0241705

Majeed M, Bhatti KH, Pieroni A, Sõukand R, Bussmann RW, Khan AM, Chaudhari SK, Aziz MA, Amjad MS (2021a) Gathered wild food plants among diverse religious groups in Jhelum District, Punjab, Pakistan. Foods 10:3

Majeed M, Tariq A, Anwar MM, Khan AM, Arshad F, Mumtaz F, Farhan M, Zhang L, Zafar A, Aziz M, Abbasi S, Rahman G, Hussain S, Waheed M, Fatima K, Shaukat S (2021b) Monitoring of land use? And cover change and potential causal factors of climate change in Jhelum District, Punjab, Pakistan, through GIS and multi-temporal satellite data. Land 10:10

Majeed M, Khan AM, Habib T, Anwar MM, Sahito HA, Khan N, Ali K (2022a) Vegetation analysis and environmental indicators of an arid tropical forest ecosystem of Pakistan. Ecol Indic 142:109291. https://doi.org/10.1016/j.ecolind.2022.109291

Majeed M, Lu L, Haq SM, Waheed M, Sahito HA, Fatima S, Aziz R, Bussmann RW, Tariq A, Ullah I, Aslam M (2022b) Spatiotemporal distribution patterns of climbers along an abiotic gradient in Jhelum District, Punjab, Pakistan. Forests 13:8

Majeed M, Tariq A, Haq SM, Waheed M, Anwar MM, Li Q, Aslam M, Abbasi S, Mousa BG, Jamil A (2022c) A detailed ecological exploration of the distribution patterns of wild Poaceae from the Jhelum District (Punjab), Pakistan. Sustainability (Switzerland) 14:7

Majeed M, Lu L, Anwar MM, Tariq A, Qin S, El-Hefnawy ME, El-Sharnouby M, Li Q, Alasmari A (2023) Prediction of flash flood susceptibility using integrating analytic hierarchy process (AHP) and frequency ratio (FR) algorithms. Front Environ Sci 10:1037547. https://doi.org/10.3389/fenvs.2022.1037547

Mohanty SP, Hughes DP, Salathé M (2016) Using deep learning for image-based plant disease detection [methods]. Front Plant Sci 7:01419. https://doi.org/10.3389/fpls.2016.01419

Naveed M, Jabeen K, Naz R, Mughal MS, Rabaan AA, Bakhrebah MA, Alhoshani FM, Aljeldah M, Shammari BRA, Alissa M, Sabour AA, Alaeq RA, Alshiekheid MA, Garout M, Almogbel MS, Halwani MA, Turkistani SA, Ahmed N (2022a) Regulation of host immune response against i proteins via computational mRNA vaccine design through transcriptional modification. Microorganisms 10(8):10081621. https://doi.org/10.3390/microorganisms10081621

Naveed M, Makhdoom SI, Ali U, Jabeen K, Aziz T, Khan AA, Jamil S, Shahzad M, Alharbi M, Alshammari A (2022b) Immunoinformatics approach to design multi-epitope-based vaccine against Machupo virus taking viral nucleocapsid as a potential candidate. Vaccines (Basel) 10(10):1732. https://doi.org/10.3390/vaccines10101732

Naveed M, Mughal MS, Jabeen K, Aziz T, Naz S, Nazir N, Shahzad M, Alharbi M, Alshammari A, Sadhu SS (2022c) Evaluation of the whole proteome to design a novel mRNA-based vaccine against multidrug-resistant *Serratia marcescens*. Front Microbiol 13:960285. https://doi.org/10.3389/fmicb.2022.960285

Nawaz M, Khan TS, Mudassar RHA, Kausar M, Usman A, Fatima T, Ahmad R, Ahmad J (2020) Plant disease detection using internet of thing (IoT). Int J Adv Comput Sci Appl 11:5

Omara J, Talavera E, Otim D, Turcza D, Ofumbi E, Owomugisha G (2023) A field-based recommender system for crop disease detection using machine learning [original research]. Front Artif Intell 6:1010804. https://doi.org/10.3389/frai.2023.1010804

Orchi H, Sadik M, Khaldoun M (2021) On using artificial intelligence and the internet of things for crop disease detection: a contemporary survey. Agriculture 12:9. https://doi.org/10.3390/agriculture12010009

Salman Z, Muhammad A, Piran MJ, Han D (2023) Crop-saving with AI: latest trends in deep learning techniques for plant pathology [review]. Front Plant Sci 14:4709. https://doi.org/10.3389/fpls.2023.1224709

Shoaib M, Shah B, Ei-Sappagh S, Ali A, Ullah A, Alenezi F, Gechev T, Hussain T, Ali F (2023) An advanced deep learning models-based plant disease detection: a review of recent research [review]. Front Plant Sci 14:1158933. https://doi.org/10.3389/fpls.2023.1158933

Singh T, Kumar K, Bedi SS (2021) A review on artificial intelligence techniques for disease recognition in plants. IOP Conf Ser Mater Sci Eng 1022(1):012032. https://doi.org/10.1088/1757-899X/1022/1/012032

Talaviya T, Shah D, Patel N, Yagnik H, Shah M (2020) Implementation of artificial intelligence in agriculture for optimisation of irrigation and application of pesticides and herbicides. Artif Intell Agric 4:58–73. https://doi.org/10.1016/j.aiia.2020.04.002

Tariq A, Mumtaz F, Majeed M, Zeng X (2023) Spatio-temporal assessment of land use land cover based on trajectories and cellular automata Markov modelling and its impact on land surface temperature of Lahore district Pakistan. Environ Monit Assess 195:1

Tassadduq SS, Akhtar S, Waheed M, Bangash N, Nayab DE, Majeed M, Abbasi S, Muhammad M, Alataway A, Dewidar AZ, Elansary HO, Yessoufou K (2022) Ecological distribution patterns of wild grasses and abiotic factors. Sustainability (Switzerland) 14:18

Thakur T, Mittal A (2020) Real time IoT application for classification of crop diseases using machine learning in cloud environment. Int J Innov Sci Mod Eng 6:1–4. https://doi.org/10.35940/ijisme.D1186.016420

Vasudevan N, Karthick T (2023) A hybrid approach for plant disease detection using E-GAN and CapsNet. Comput Syst Sci Eng 46(1):337–356. http://www.techscience.com/csse/v46n1/51322

Waheed M, Arshad F, Majeed M, Fatima S, Mukhtar N, Aziz R, Mangrio WM, Almohamad H, Dughairi AA, Al-Mutiry M, Abdo HG (2022) Community structure and distribution pattern of Woody vegetation in response to soil properties in semi-arid Lowland District Kasur Punjab, Pakistan. Land 11:12

Waheed M, Arshad F, Majeed M, Haq SM, Aziz R, Bussmann RW, Ali K, Subhan F, Jones DA, Zaitouny A (2023) Potential distribution of a noxious weed (*Solanum viarum* Du-nal), current status, and future invasion risk based on MaxEnt modeling. In: Geology, ecology, and landscapes. Taylor & Francis, p 1. https://doi.org/10.1080/24749508.2023.2179752

Yağ İ, Altan A (2022) Artificial intelligence-based robust hybrid algorithm design and implementation for real-time detection of plant diseases in agricultural environments. Biology 11(12):1732. https://www.mdpi.com/2079-7737/11/12/1732

Chapter 16
Sustainable AI-Driven Applications for Plant Care and Treatment

Muhammad Naveed, Nafeesa Zahid, Ibtihaj Fatima, Ayesha Saleem, Muhammad Majeed, Amina Abid, Khushbakht Javed, Rehmana Wazir, and Amina Qasim

Abstract Technology has emerged as a formidable ally in the endeavor, with artificial intelligence (AI) leading the charge in revolutionizing farming practices via the notion of precision agriculture. Since the dawn of the civilization, there has always been a basic human need in the field of agriculture as plants were a primary source of food. The plant disease poses a serious threat to food security because it legitimately affects harvest yield, which reduces the growth of a crop yield and leads to significant loss and consequent financial losses. As a result, there is a demand for quick and effective plant disease detection and evaluation strategies. AI intervention in agriculture is assisting farmers in regaining their farming efficiency and reducing adverse environmental influences. By substituting traditional methods with more effective ones, it is bringing about a revolution in agriculture. Plant disease is one of the most important and critical challenges that affects the agriculture and its trading system. Plant diseases can adversely affect the growth of plants and, in severe cases, might lead to the plant death. AI-driven technologies have the capability not only to detect the disease but also to assess its toxicity and precisely classify the specific type

M. Naveed · A. Saleem · A. Abid · K. Javed · R. Wazir
Department of Biotechnology, Faculty of Science and Technology, University of Central Punjab, Lahore, Punjab, Pakistan
e-mail: dr.naveed@ucp.edu.pk

N. Zahid
Department of Botany, University of Kotli, Kotli, Azad Jammu and Kashmir, Pakistan
e-mail: nafeesa.botany@uokajk.edu.pk

I. Fatima
Department of Botany, University of Education, Lahore, Punjab, Pakistan

M. Majeed (✉)
Department of Botany, University of Gujrat, Gujrat, Pakistan
e-mail: m.majeed@uog.edu.pk; https://www.researchgate.net/profile/Muhammad-Majeed-18

A. Qasim
Department of Botany, Minhaj University Lahore, Lahore, Pakistan
e-mail: 2022F-mulmpl-bot-019@mul.edu.pk

A. Khamparia et al. (eds.), *Microbial Data Intelligence and Computational Techniques for Sustainable Computing*, Microorganisms for Sustainability 47,
https://doi.org/10.1007/978-981-99-9621-6_16

of disease detected in a given plant sample. It investigates AI's revolutionary role in improving farming practices through precision agriculture. Precision agriculture is based on data-driven decision-making. The importance of AI extends beyond data collection to data processing and interpretation. AI algorithms analyze the acquired data to uncover patterns, correlations, and anomalies that the human eye may not see.

Keywords Plant diseases · Artificial intelligence · Environmental influences

16.1 Introduction

Regardless of the developmental stages of countries, agriculture system has always been serving as the backbone of economy (Arzani and Ashraf 2017). The significance of agriculture is evident in several domains as approximately 70% of the population rely on the cultivation of crops for their needs. This depicts the importance of agriculture as being the most vital source that could stand the chance in the era of rapidly growing population (Sekaran et al. 2021). These diseases can occur at various stages of plant development such as seedling, seed development, and its growth (Majeed et al. 2020a). When plants get affected with diseases, they undergo a range of mechanical, morphological, and biochemical changes (Van Esse et al. 2020). Moreover, plant stress can be categorized into two primary types: biotic stress, which involves interactions with living organisms in a way that affect plant growth, such as viruses, bacteria, and fungi. The other type is abiotic stress, including non-living or environmental factors (Haq et al. 2022). As the world's population is increasing, which limits the agricultural land resources, it becomes difficult to ensure the efficient production of agricultural products. For the production of agricultural products, detection of disease at an early stage, fertilization, spraying and detection of unwanted weed is important (Mosa et al. 2017). According to the research, plant disease could cause 20–40% of losses to the crops yield %, thereby reducing market availability for potential buyers. Delayed diagnosis of pathogenic, viral, or pest-related plant diseases necessitates the application of higher doses of pesticides, ultimately compromising crop quality (Rashid et al. 2019). Early and accurate identification of plant diseases not only enhances the overall quality of agricultural products but also minimizes the need for chemical sprays such as fungicides and herbicides. Since agriculture is essential, there is need for techniques that enhance the agriculture methods in terms of monitoring, planting, harvesting, and detecting plant diseases (Mohamed et al. 2021).

However, visual inspection based on the traditional method is used for detecting plant diseases in small-scale agriculture fields (Majeed et al. 2022a). However, it is difficult to use these methods in large-scale agriculture as this method is continuous and tiring. This challenge becomes even more difficult when diagnosing plants with similar leaves for both species and diseases, often resulting in visual errors (Martinelli et al. 2015). To address these issues, diverse image processing and artificial intelligence (AI) approaches, including deep learning and machine learning

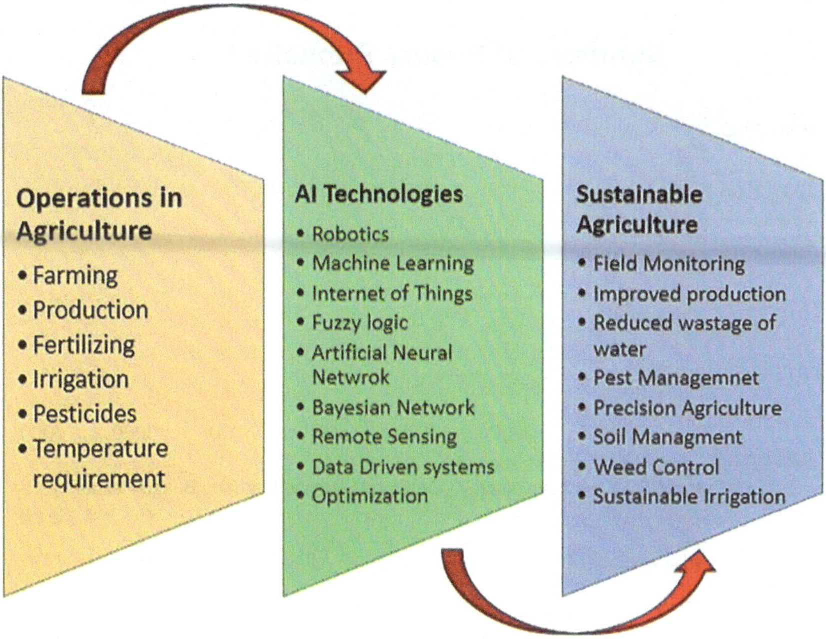

Fig. 16.1 Role of artificial intelligence (AI) in agriculture system

(ML), are used for real-time plant disease classification (Fig. 16.1) (Rakhmatulin et al. 2021).

The academic literature in this field had rapid growth from 2000 to 2015. Subsequently, there was a great impact of industry on different agriculture practices; hence there is a gradually increasing trend since 2017 and substantial 255.7% increase in the number of Scopus-indexed articles and reviews published between 2019 and 2020 (Ullah et al. 2022) (Fig. 16.2). This growth strongly indicates a pronounced scholarly interest in exploring the applications of AI within the context of sustainable agriculture practices.

The detection methods of plant disease can be categorized into two main groups based on their complexity: machine learning and deep learning (Jackulin and Murugavalli 2022). Basic machine learning techniques such as Support Vector Machine (SVM), K-Nearest Neighbor (KNN), random forest (RF), and Naïve Bayes (NB) rely on specifically designed features, necessitating the recognition of patterns. These specific features are Histogram of Oriented Gradient (HOG), Hue Saturation Value (HSV), Red-Green-Blue (RGB) color features, and Linear Binary Pattern (LBP) (Zhang et al. 2020). In machine learning, more data is required to achieve satisfactory results. Firstly, the data is collected and characterized according to the classes of disease. Subsequently, for the image to be processed, dataset is created for further extraction. Machine learning has the algorithms which are capable of recognizing any change in features through comparison and thus classifying the

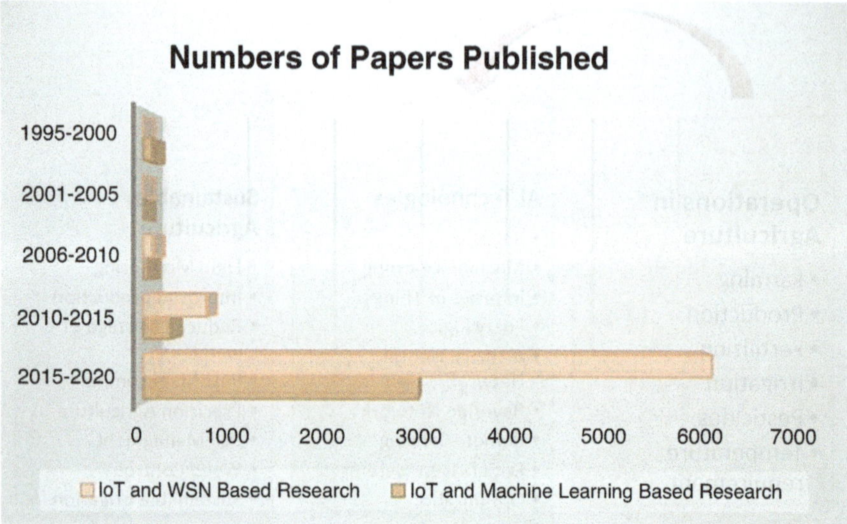

Fig. 16.2 Yearwise number of publications based on the AI agricultural practices

output data as healthy or diseased (Meng et al. 2020). Conversely, deep learning techniques such as Convolutional Neural Networks (CNNs) have found extensive application in research focused on plant disease detection. While the Support Vector Machine (SVM) was the predominant machine learning approach for an extended period, there was a notable shift around 2015 when Convolutional Neural Networks (CNNs) replaced SVM as the most widely adopted machine learning technique for disease detection (Tugrul et al. 2022). There are various types of CNN such as Inception-V3, AlexNet, ResNet50, and VGG16 (Xiao et al. 2018). However, deep machine learning needs large set of data, which is often considered a challenge. Nowadays, CNN has been used as the art of the model for the detection of plant diseases, especially since this task requires dealing with image data applications (Waheed et al. 2022). However, these technological approaches could be used for the detection of plant diseases. Thus, these methods can be viewed as innovative means of detecting diseases, as they rely on image-processing techniques rather than destructive serology and molecular methods (Alzubaidi et al. 2021).

16.1.1 Role of AI in Optimizing Farming Practices Through Precision Agriculture

Precision agriculture collects, analyzes, and interprets data using AI-driven solutions, allowing farmers to make educated decisions that optimize resource utilization, boost production, and encourage sustainable agricultural practices (Linaza et al. 2021). With its capacity to handle massive volumes of data and extract useful insights, AI serves as the backbone of this strategy (Khoja et al. 2022). One of AI's most important contributions is satellite imaging, drones, and Internet of Things (IoT) devices, which are used to collect real-time data on crop health, soil moisture, temperature, and other factors. These technologies enable farmers to monitor their fields in unprecedented detail by providing high-resolution pictures and remote sensing capabilities. AI systems give a thorough grasp of field circumstances by continually recording and analyzing data, resulting in proactive and accurate responses (Singh et al. 2021).

These insights are essential in anticipating and avoiding possible problems. For example, using historical and real-time data, AI can forecast disease outbreaks, insect infestations, and yield changes (Coulibaly et al. 2022). Farmers can use this forecasting capacity to respond early, employing tailored remedies and minimizing crop yield losses. Furthermore, based on data analysis, AI can identify appropriate planting density, irrigation rates, and fertilizer application, ensuring that resources are dispersed efficiently across the field (Tassadduq et al. 2022). Another incredible use of AI in precision agriculture is yield prediction. AI algorithms can reliably anticipate agricultural yields by analyzing a variety of parameters such as weather patterns, soil quality, and planting strategies. This forecast enables farmers to make smart decisions about harvesting schedules, storage, and selling, resulting in increased profitability (Sudha 2023; Mathenge et al. 2022). The various AI applications for plant care and treatment used in different agriculture and cultivating fields are shown in Table 16.1.

AI-powered solutions have a significant influence on resource management. For example, automated irrigation systems modify water applications depending on real-time soil moisture levels and weather forecasts. This accuracy reduces water waste and increases water-use efficiency, which is crucial in water-stressed areas. Similarly, the importance of AI in weed and pest management cannot be emphasized (Rathore 2017). Artificial intelligence-powered image identification systems identify and categorize weeds, pests, and illnesses in real time, allowing for focused treatments and lowering dependency on broad-spectrum pesticides. Precision agriculture encourages sustainable agricultural practices that limit environmental damage by minimizing the excess use of water, fertilizers, and pesticides (Majeed et al. 2020b). Furthermore, AI's capacity to optimize resource utilization is consistent with the ideas of the circular economy, in which inputs are utilized more efficiently, waste is eliminated, and the total environmental impact is decreased (Agrawal et al. 2021).

However, it is crucial to recognize that the use of AI in agriculture is fraught with difficulties. For small-scale farmers, the initial investment in technology, training, and infrastructure might be intimidating (Minkoff-Zern 2019). Furthermore, data privacy and security issues must be addressed to guarantee that sensitive agricultural data is not affected. Collaboration among technology developers, agricultural specialists, politicians, and farmers is critical to overcoming these obstacles and establishing an ecosystem that promotes the wider use of AI-powered precision agriculture (Gardezi et al. 2023). The use of AI in optimizing agricultural practices through precision agriculture is a transformational force with the potential to influence the future of agriculture. The combination of technology and agriculture has the

Table 16.1 Various applications of AI in agriculture for plant care and treatment

Applications	Description	References
Soil analysis	Based on climate, geography, and generally stable soil properties (such as soil texture, depth, and mineralogy), land management has long-term potential. This application assists farmers in better understanding the potential of their property as well as the possibility for climate variation change and mitigation methods	Karlen et al. (2001)
Farm manager	The farm manager app assists farmers in determining which tactics to use before planting begins. Without using your phone, this programmer views, organizes, and modifies any information about your field, such as yield, planting, and spraying conditions	Rehman et al. (2022)
Pest management	Village tree offers smart pest management solutions by gathering pest incidence information from farms. Furthermore, it uses a crowdsourcing technique, distributing photographs and location data to other farmers who may be affected	Sun et al. (2021)
Agrippa	By using the eFarmer Application, farmers may produce electronic maps of their fields, record a history of crops grown in the field (e.g., planting, fertilizing, harvesting, warehouses, petrol stations), and track the position of things in the field (e.g., soil sample for agrochemical laboratories)	Karetsos et al. (2014)
Semios	Network coverage, orchard pests, frost, diseases, and irrigation are all covered. As part of the monitoring services, event alerts are sent out in real time	Rehman et al. (2022)
Fertilizer management	EcoFert aids in fertilizer management, allowing it to be utilized to its greatest capacity. It determines the ideal fertilizer combination to cover the required nutrient suspension and takes into account the needs of various yields. Furthermore, it takes into account the cost of fertilizer, depending on current market prices	Rehman et al. (2022)

potential to feed the world's rising population while also assuring appropriate management of our planet's resources (Majeed et al. 2021a). As AI advances, its impact on precision agriculture will definitely expand, providing novel solutions to the difficulties of a changing world.

16.1.2 Agriculture Satellite Imagery: A Technological Revolution for Sustainable Farming

The potential for monitoring and management is at the heart of satellite imagery's importance in agriculture. Farmers can now monitor vast areas of land with unprecedented clarity and accuracy, allowing them to detect early signs of crop stress, disease, insect infestations, and nutritional deficiencies (Arshad et al. 2022). This proactive technique allows for early intervention, which reduces production losses and eliminates the need for unnecessary chemical inputs (Lamb et al. 2008). Satellite imaging enables farmers to make educated decisions regarding irrigation, fertilization, and pest management tactics by giving vital information on crop health and development. This improves resource efficiency and production, which is critical as the world's population continues to grow (Sinha et al. 2023). Satellite imagery's predictive capacity strengthens its importance in agriculture. This capacity to predict harvest quantities has a dramatic impact on supply chain management, allowing for improved coordination among farmers, distributors, and consumers (Raj et al. 2022). Furthermore, satellite imaging data-driven insights enhance precision agriculture practices. Farmers may use resources like water, fertilizer, and pesticides more efficiently by separating fields into zones based on changes in soil type and vegetation. This will not only reduce waste but also have a lower environmental effect, which is consistent with the spirit of sustainable farming (Jadhav et al. 2023).

Satellite photography also enables farmers to better address water management concerns. Monitoring soil moisture levels and crop water stress aids in the optimization of irrigation practices, ensuring that water resources are utilized efficiently (Khan et al. 2022a). This is especially important in water-stressed areas, since it allows for agricultural cultivation while minimizing water waste. Furthermore, using satellite data with weather forecasting improves farmers' capacity to adapt to extreme weather occurrences (Alzubaidi et al. 2021). Farmers can take preemptive actions to safeguard their crops and minimize possible losses if they get timely information about imminent droughts, strong rains, or frost. Satellite photography assists in monitoring changes in land cover, deforestation, and urban expansion, providing insights into the changing terrain. Such data is crucial to legislators and urban planners because it helps them make decisions regarding land use, environmental protection, and sustainable development. Furthermore, satellite photography aids agricultural research and development. These data may be used by researchers to investigate production patterns and measure the influence of various agricultural practices (Tiwari et al. 2023).

16.1.3 Use of Sensors and Data Analytics

Sensors are at the cutting edge of this revolutionary journey, acting as the digital world's eyes and ears. These devices collect information from the physical environment, such as temperature, humidity, soil moisture, air quality, and energy use (Liang and Shah 2023). Sensors put in fields, for example, give real-time information about soil conditions and crop health, allowing farmers to modify irrigation and fertilization schedules. However, the true value of these sensors comes in their capacity to generate vast amounts of data. Advanced analytical approaches are necessary to extract relevant insights from this abundance of data. This is where data analytics can help (Tariq et al. 2023). Data analytics is the process of processing, interpreting, and visualizing acquired data using modern algorithms and statistical approaches. In the context of resource allocation, data analytics helps decision-makers to spot patterns, trends, and anomalies that might otherwise

Table 16.2 IoT sensor types used in agriculture

Sr. no.	Sensor type	Sensor	Function	Application in agriculture	References
1	Optical	Photodiode	Use of light to measure soil properties	Determines the soil's clay, organic matter, and moisture content	Wai et al. (2021)
2	Mechanical	Tensiometer	Take action by using probes to detect soil compaction	Detects the root force employed in water absorption	Peixoto et al. (2019)
3	Electro-mechanical	Ion-selective electrodes and ion-selective fields affect transistor sensor	Electrodes are used to detect certain ions in soil	Detecting nitrogen, potassium, and phosphorus (NPK) levels in soils	Kim et al. (2013)
4	Di-electric soil moisture	Electrodes for frequency or time domain reflectometry	To assess moisture levels	To measure water content from soil	Kashyap and Kumar (2021)
5	Airflow	Measurements can be taken at a fixed location using a designated method	To measure soil permeability	Classifies soil kinds, moisture levels, and soil structure/compaction	Pawar and Deosarkar (2023)
6	Location	Global positioning system	Provides latitude, longitude, and altitude information as well as altitude	The GPS system allows for exact location	Nayfeh et al. (2023)

go undetected using traditional approaches. In healthcare, for example, data analytics may be used to analyze patient information to forecast disease outbreaks and allocate medical resources appropriately (Kuppusamy et al. 2021). These cutting-edge approaches of AI make the farming sector more adaptable. Soil moisture and fertility can be monitored by biosensors. Past data of weather fluctuation with non-linear dependencies can be predicted by neural networks. Therefore, economically important commodities like rice, wheat, and maize can be sowed at appropriate times by using the knowledge of AI. Farms produce thousands of data points on temperature, soil, water level, and weather on a daily basis. These data are used by AI and ML models for greater yield (Table 16.2). Some crops need an adequate amount of water, and water at a greater scale became tough, so the AI irrigation system was introduced.

16.2 Drones in Agriculture: Planting the Seeds of a High-Tech Future

The use of technology into agriculture has resulted in dramatic developments, with the drone being one of the most inventive technologies to emerge. Unmanned aerial vehicles, or drones, have found wide and effective applications in a variety of industries, including agriculture. Drones have revolutionized farming practices by capturing high-resolution pictures, collecting data, and providing insights from above, ushering in a new era of accuracy, efficiency, and sustainability (Khan and Shahriyar 2023).

16.2.1 Precision Farming

Drones have become an integral part of precision agriculture, allowing farmers to manage their crops. Drones, which are outfitted with modern cameras and sensors, may take comprehensive photos and data, providing insights into numerous aspects of crop health and growth (Bashir et al. 2023). These discoveries span from identifying nutrient shortages, illnesses, and insect infestations to determining irrigation requirements and monitoring plant stress. Farmers may make reliable decisions regarding resource allocation, waste reduction, and yield potential using this information (Maddikunta et al. 2021).

16.2.2 Crop Monitoring and Management

Drones provide effective crop monitoring over wide regions. It may acquire data that the human eye cannot see by using imaging technology such as multispectral and infrared cameras. Farmers can monitor tiny changes in plant health and soil conditions, allowing for preventive treatments. The capacity to detect problems early and administer focused therapies leads to increased production and less reliance on broad-spectrum chemicals (Budiharto et al. 2019).

16.2.3 Mapping and Surveying

Drones excel in producing precise and up-to-date maps of agricultural lands. They can make accurate topographical maps, determine changes in soil type and elevation, and even construct 3D representations of the ground using aerial surveys (Majeed et al. 2022b). These maps help in the construction of efficient drainage systems, the optimization of irrigation practices, and the uniform use of resources. Drone mapping capabilities also include monitoring changes in land cover, which aids in land use planning and management (Daponte et al. 2019).

16.2.4 Yield Prediction and Assessment

Drones help in yield prediction and assessment by collecting data on plant density, canopy cover, and flowering patterns. When this data is paired with modern analytics, growers may more correctly anticipate possible crops. They also aid in post-harvest inspections, providing information on yield variances across different portions of a crop. This information is useful for assessing the effectiveness of various planting and management practices (Ben Ayed and Hanana 2021).

16.2.5 Environmental Stewardship

Drones can help promote sustainable farming practices. They can reduce the environmental effect of fertilizers and pesticides by permitting accurate resource application. Furthermore, their capacity to monitor soil moisture levels and plant health assists in effective water consumption and decreases agriculture's ecological imprint (Dutta and Mitra 2021).

16.2.6 Internet of Things (IoT) in Agriculture: Real-Time Data Collection for Better Farming

The Internet of Things (IoT) has ushered in a new era of connectivity and data-driven decision-making across several industries, and agriculture has seen significant benefits from this technological revolution. The research article published in *Agronomy* investigated the potential of Internet of Things (IoT) technologies to play a pivotal role in the evolution of smart agriculture in the twenty-first century. Illustrative instances of IoT applications within agriculture include farm management, controlling greenhouse environments, monitoring animals and herds, tracking emissions, managing irrigation systems, operating autonomous machinery, and deploying drones (Fig. 16.3). These gadgets allow precision agriculture, decrease resource waste, and contribute to sustainable agricultural practices by permitting real-time data collection and analysis (Majeed et al. 2021b).

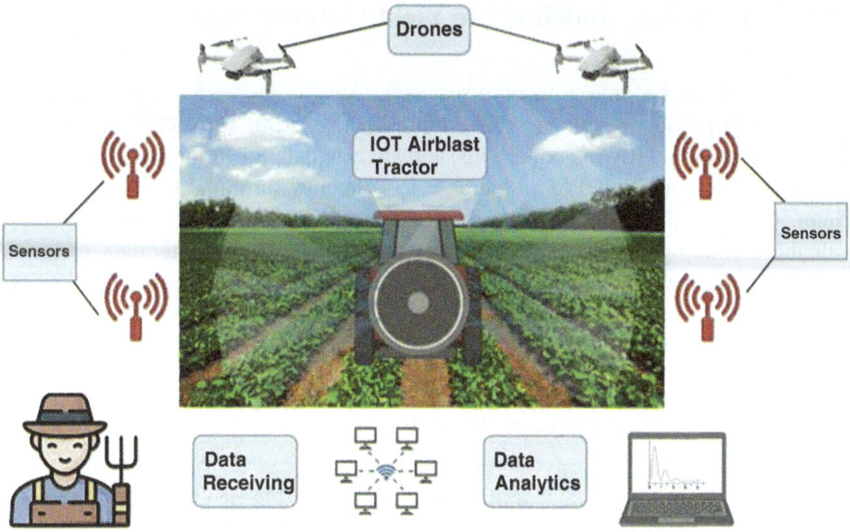

Fig. 16.3 The IoT-based smart agriculture monitoring system

16.2.7 Weather Monitoring and Forecasting

The weather has a significant impact on agricultural outputs. Temperature, humidity, wind speed, and precipitation data are collected by IoT weather stations. This information not only is important for on-farm decision-making but also helps with accurate weather predictions. Farmers can predict natural weather disasters and adjust their operations accordingly, safeguarding their crops and minimizing potential losses (Khan et al. 2022b).

16.2.8 Livestock Management

IoT devices are also used in animal husbandry. Smart collars with sensors can monitor animal health and behavior. Tracking movement patterns, spotting symptoms of disease, and even forecasting when animals are going to give birth are all part of this. These gadgets improve overall herd management and contribute to animal welfare by offering insights on animal well-being (Hassan et al. 2022a).

16.3 Predictive Modeling for Yield Optimization

Recently, AI has become a widely used technology in the agricultural sector. It is employed to produce healthier crops, minimize losses due to insects, monitor soil quality for optimal growing conditions, and analyze data for farmers. As a result, AI activities have contributed to an increase in efficiency throughout the food supply chain. Crop yield prediction and optimization through AI tools have become a valuable technology for modern agriculture. AI tools collect weather data, historical patterns, and growth models to provide more precise knowledge to farmers and agricultural stakeholders.

16.3.1 DSSAT AI Growth Models

Agricultural decision-makers required excessive amounts of knowledge to understand the possible outcomes of their resolution. This helps them organize plans and policies to achieve their goals. DSSAT (decision support system for agrotechnology transfer) model was developed to find crop production, sources, and risks linked to crop production. This microcomputer software package has a stimulation model of soil, weather database, crops, and soil strategy evolution programs integrated with the main shell program. It helps decision-makers to make efficient decisions by minimizing human resources used for complex analyses. It stimulates the growth and development of plants.

The CERES-Wheat model has been designed for growth stimulation. This model considers some key factors like plantation date, temperature, availability of water, and nitrogen level to predict the yield of wheat and optimize cultivation ways. CERES models simulated the yield of grain with different levels of root mean square error. This model has also been used for maize and rice for growth stimulation.

16.3.2 Crop Enhancement Through Predictive Modeling

In precise agriculture, yield prediction remains a major concern. It is important for decision-makers at the regional and national stages. A precise prediction model can provide a helping hand for a farmer to make decisions about what and when to grow. Various models have been introduced and validated as well. The crop yield depends on a number of factors like climate, weather, soil conditions, fertilizer use, and variety of seeds, so it requires more than one dataset for yield prediction. This shows that yield prediction comprises many complex steps. Nowadays, actual yield can be estimated through different models, but still there is a need for better performance prediction models. Different features are used to build predictive models. In agriculture, crops with high market value are mostly preferred; the right choice of crop is based on many factors. Data mining technologies play an important role in identifying the best crop choice or hybrid seed choice. A crop with a high growth probability can be determined by algorithms. Therefore, an AI forecasting approach for crops addresses issues such as crop selection, necessitating the implementation of an efficient decision-making method for optimal crop selection. Optimized decision-making methodology can be used to increase crop yield. In this approach, datasets are collected and then preprocessed to normalize the Z-score, and adaptive shearlet method is applied to extract the features. The discrete hybrid Deep Belief Network utilized the VGG NET algorithm to find out the best crop for growth. This approach can be utilized in three different datasets. Many studies have revealed that models for crop production prediction have been developed at the country level. A study was conducted in Saudi Arabia in which AI was used for modeling and predicting yield to increase food security (Arshad et al. 2022). They use artificial neural networks (ANNs) and multilayer perceptron (MLP) models to precisely predict the yield of crops, temperature, insecticides, and level of rainfall, depending on environmental information. Temperature fluctuation, rainfall level, and insecticide effects on yield can be successfully evaluated by using AI. Future values of different crop yields like potatoes, rice, wheat, and sorghum can be predicted by using the MLP model. AI approaches continually check soil nutrient levels and compare them with previous data. It uses different datasets to analyze the environmental effects of dosage and a suitable number of fertilizers for high-yield production. This will promote the eco-friendly agriculture system. With the passage of time, there is a rapid change in climate which has had a negative impact on agriculture, like difficulties in deciding which crop should be sowed. AI makes it simple to understand the effects of weather, seasonal sunshine, wind, and rain on the plant cycle. Farmers can get information from weather predictions as they examine the plant seeding time (Arif et al. 2021). Exact fertilization can be developed by clever seeding technology. AI tools may help farmers with time management and increase quality. Nowadays, drones collect data and monitor farm conditions. AI improves yield through soil and crop monitoring by collecting the data through drones, the internet, and field satellite image. These data are analyzed by AI tools to provide the most appropriate answer (Waheed et al. 2023). ML helps the farmer to assess the data of crops, check weather alteration, and efficiently manage the farm. AI also provides a sustainable way of water usage to prevent water shortage. AI helps find the optimal place for a plantation depending on geographical factors and soil chemistry. AI uses an ML algorithm to check the seed quality by scanning it and comparing it with a healthy one. With the help of this technology, farmers can make well-informed choices and lower the likelihood of crop failure, which ultimately increases agricultural output and profitability (Fig. 16.4).

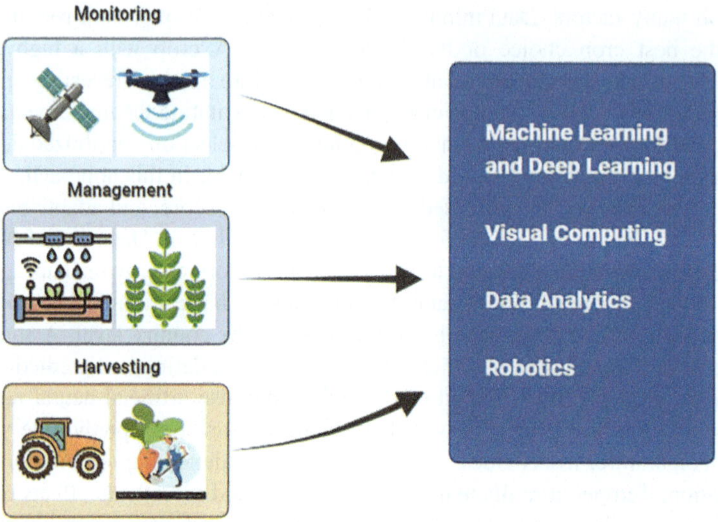

Fig. 16.4 Management and monitoring of plant care through AI-based machine and deep learning

16.4 Disease Detection and Treatment

Any developing country can be regarded as having its foundation in the agricultural sector. Therefore, farmers are required to have access to the most substantial technologies and practices in order to get the highest yield possible from their crops (Majeed et al. 2023). The diagnosis of various plant diseases is one of the many farming aspects where information technology in agriculture has made tremendous progress. Plant disease identification is one of the major challenges in agriculture and has a considerable impact on crop production. Artificial intelligence can be a major resource in combating crop diseases because of its capacity to identify problems, diagnose diseases, and determine their causes as well as their potential treatments (Mohanty et al. 2016). Computers are now able to learn without being specifically programmed because of machine learning, which is conceptually similar to how people learn. The use of algorithms that use machine learning to create models out of known data is called learning or training (Zhou 2021).

16.4.1 Machine Learning for Crop Disease Management

For the detection and diagnosis of plant diseases, numerous studies have been conducted using traditional machine learning techniques like random forests, artificial neural networks, fuzzy logic, K-means method, Support Vector Machines (SVM), and Convolutional Neural Networks (CNNs). For instance, in a study of tomato leaf disease classification by Tan et al. a total of 105 color features were recovered using color histograms and color moment methods, while 52 texture features were extracted using a gray level co-occurrence matrix (GLCM) and local binary pattern (LBP) technique. For ten different forms of tomato leaf disease, the efficacy of KNN, SVM, and random forest (RF) machine learning algorithms, as well as AlexNet, VGG16, ResNet34, EfficientNet-b0, and MobileNetV2 architectural deep learning methods, was investigated (Tan et al. 2021). Utilizing contemporary, innovative imaging and information technologies for disease detection is essential to precision agriculture. These smart, non-intrusive techniques use almost real-time observations to guard from crop damage brought by plant diseases.

16.4.2 Deep Learning Models to Extract Features

Another artificial intelligence approach is to utilize deep learning models, where researchers initially used a combination of deep learning techniques and conventional classifiers to classify plant diseases and pests. They extracted image features using pertained CNN models, which they then fed into machine learning classifiers like SVM to categorize (Yalcin and Razavi 2016). The efficiency of this strategy was demonstrated by experimental data. Others developed CNN-based Meta topologies with various feature extractors that could distinguish between healthy and diseased plants. Another study by Hassan et al. used features generated from a deep CNN model and input into an SVM classifier to identify and categorize several types of rice illnesses with 97.5% accuracy (Hasan et al. 2019).

16.4.3 Two-Stage Detection Network (Faster R-CNN)

The Faster R-CNN two-stage detection network's fundamental procedure involves using the backbone network to retrieve the input image's feature map. The anchor box confidence is then calculated using RPN's (Region Proposal Network) suggestions. The network is given the feature map of the proposed area, and the initial detection results are adjusted to produce the results for the placement and categorization of the lesions. To increase the identification of plant diseases and pests, common techniques improve on the backbone structure, feature map, anchor ratio, ROIpooling, and loss function (Liu and Wang 2021). For the identification of tomato diseases, beetroot leaf spot disease, rice blast, bacterial blight, sheath blight, and grape leaf diseases, several researches have used Faster R-CNN, resulting in high detection accuracy and speed improvements (Fuentes et al. 2019).

16.4.4 AI and Remote Sensing Applications for Plant Health Monitoring and Treatment

Plant diseases can be caused by bacteria, fungi, nematodes, viruses, pests, weeds, insects, and other pathogens. Farmers can identify the symptoms and signs of a plant's condition based on routine inspections. The following are some examples of tools and techniques that can be harnessed by farmers to revolutionize crop disease detection and management. An intelligent expert system with a fuzzy logic-based decision-making algorithm has been developed to diagnose crop diseases (Hassan et al. 2022b) (Table 16.3). Twenty-one prevalent diseases of wheat and cotton can be detected by the method. Fuzzy logic has been used as the backend decision-making engine for the proposed architecture, and a front-end Android application has been created using the jFuzzyLite library. For the evaluation, various aggregation and defuzzification techniques have been used (Toseef and Khan 2018). The most effective mix of these strategies is advised for application at a practical scale based on the experimental findings.

Table 16.3 IoT technologies, its applications, and advantages in smart agriculture

Sr. no.	IoT technology	Application in agriculture	Advantages in agriculture system	References
1	Wire senseless network (WSN): sensor nodes having radio communication capabilities	Sensors used in monitoring physical parameters	Management and data collection gathered from sensors	Jondhale et al. (2022)
2	Cloud computing: internet-based computing	Provides processing resources and data shared to computer on demand	Management and easy data collection gathered from cloud computing services such as cloud storage and fields maps	Sadeeq et al. (2021)
3	Protocols of communication: it is the backbone of the IoT systems, which provides connectivity	These protocols provide exchange of data within the network in a number of exchange data formats	Management and data collection of data collected from cloud computing services, sensors, cloud storage, etc.	de Kfouri et al. (2019)
4	Data analytics: the process of analyzing and examining large datasets	Have access to multiple forms of data types	Market trends, correlations, preferences, and other important information	Hariri et al. (2019)
5	Embedded system: a system consists of both software and hardware	The system executes particular functions, including management, controlling, and monitoring of different activities	Profitability and sustainability increase by the reduction of production cost	Pereira et al. (2019)

16.4.4.1 Plantix Application

The ability to diagnose plant diseases is an important aspect of the Plantix app in addition to certain other features. Berlin-based horticultural IT business PEAT developed the Plantix app. It is used to locate deficiencies and flaws in soil. The software analyzes plant images to identify a disease. A variety of these images are saved in a sophisticated cell and are coordinated with the image in the worker's diagnosis (Fujita et al. 2016).

16.4.4.2 Stress Detection

Plant stress produced by conditions such as drought, nutrient inadequacy, or pollution may be assessed using remote sensing and AI technology. Timely intervention based on stress indicators promotes optimal resource allocation, such as water and fertilizers, ultimately enhancing plant health (Khang et al. 2023).

16.4.4.3 Yield Prediction

AI algorithms anticipate agricultural yields using historical data and remote sensing inputs. This knowledge aids in efficient planning, allowing farmers to properly handle harvest logistics and marketing plans (Wang and Gamon 2019).

16.4.4.4 Weed Control

Remote sensing, combined with AI, can differentiate between crops and weeds, enabling more targeted herbicide delivery. This method reduces the environmental effect while also lowering the expenditures associated with needless treatments (Iqbal et al. 2020).

16.4.4.5 Nutrient Management

Precise fertilization and efficient nutrient management are essential in agriculture. Recently, nutrient management decision support technologies have been developed for modern agricultural systems using machine learning (ML). It is feasible to combine and interpret several bits of knowledge that have never been investigated for this purpose before the era of nutrient management (Ennaji et al. 2023). The texture, soil type, organic matter, exchangeable nutrients, pH, total inorganic or organic carbon, and the total soil capacity to hold exchangeable cations (CEC) are all considered in nutrient management algorithms (Qin et al. 2018). The amount of nitrogen lost during wet conditions significantly influences a soil's ability to retain water beyond its saturation point, typically corresponding to the depth of the water table. Deep learning was employed in a study (Condori et al. 2017) to identify maize with insufficient nitrogen levels. Four CNN models were pre-trained for this purpose using applied transfer learning, with growth stages (V4, V6, and R1) taken into consideration. Following digitization of the leaf samples, 384 pictures representing the 16 nitrogen treatments taken into consideration by the study were produced. Regarding each growth stage, the results revealed a considerable variation. However, when compared to conventional text-based methods, CNNs formed using RGB images produced excellent results (on average 61.8%) as opposed to these methods (Bashir et al. 2023).

16.4.4.6 Smart Irrigation

Considering the crop security and satisfaction of farmers, accurate predictions of irrigation needs and crop yields are essential. The planned predictions render a substantial contribution to raising agricultural yields and reducing production costs. The traditional methods are unable to meet the expanding demand as they use pesticides in greater quantity, which affects the oil content. Challenges in agriculture, such as crop pests, limited storage capacity, pesticide regulation, irrigation management, and water resource control, are a few fields that cause agricultural difficulties which can be resolved by artificial intelligence (Sinwar et al. 2020). Therefore, it is essential to use digital technologies at several points along the agricultural supply chain, including the automation of farm equipment, use of instruments and information from satellites from a distance, machine learning and artificial intelligence for more effective crop monitoring, water management, and traceability of agricultural food products (Majeed et al. 2022c). The sensors are linked to economical Arduino-based devices to store the collected data and run analysis algorithms to forecast the crop's water needs at a specific moment. Smart irrigation systems ensure precise delivery of the required amount of water to each plant, thereby ensuring their overall health. Regular monitoring is essential for sprinklers and drip modules to prevent any potential malfunctions. The non-uniformity of the land and crop types is the fundamental challenge in the construction of a smart irrigation system. Using machine learning approaches (PLSR and ANFIS) and sensors, Navarro-Helln et al. proposed a smart irrigation decision support system (SIDSS) for calculating the weekly irrigation needs. The technology was tried on citrus plants in Spain's southeast. The system's performance was evaluated using expert human assessment (Hinnell et al. 2010). Their approach revealed a decrease of 22% in irrigation needs relative to weather values from the prior year (i.e., irrigation needs for 2015 have been projected using data from 2014).

Various researchers have demonstrated how automation and AI could improve yield growth. There is a dire need to install smart irrigation systems that can irrigate larger areas with minimal water use in order to cope up with incurring water scarcity. Gathering information regarding the level of moisture in the soil, the water content of plants, the atmosphere's humidity, the temperature, etc., is necessary for smart irrigation. A Neural Network-based irrigation system (Neuro-Drip) has been presented by Hinnell et al. Sensors for measuring humidity, temperature, and soil moisture can all be used to collect data (Hinnell et al. 2010).

16.4.4.7 Climate Resilience

Remote sensing and artificial intelligence (AI) help to improve climate resilience by detecting changes in plant patterns over time. These tools aid in the early detection of alterations induced by climate change, hence assisting adaption methods (Khan et al. 2022a). In the near future it is believed that there will be a rise in the use of artificial intelligence models around the world given the recent decades' tremendous advancement of the field. Applications and integration have become more and more impressive in regard to agricultural developments including smart irrigation, soil and nutrient management, and yield enhancement. When applied correctly for decision-making or support, ML algorithms have great potential in any field of work. However, due to complicated definition concepts and computational procedures, the application of AI in this subject has been restricted despite its potential.

16.5 Conclusion

Before the advent of artificial intelligence (AI), agricultural systems lacked effective irrigation and proper monitoring of plant diseases, especially considering challenges related to varying plant heights and extreme weather conditions. These problems have been resolved with the help of AI technology. Now, soil moisture levels can be detected via remote sensors, and performance can be enhanced through automated irrigation systems aided by GPS. Agriculture-related applications of machine learning and deep learning are becoming increasingly prevalent. Precision farming employing artificial intelligence in agriculture increases crop output and quality while utilizing scarce resources. Furthermore, remote sensing employs cutting-edge techniques that help ranchers keep an eye on their crops without having to physically monitor the land. Many businesses today anticipate the expansion of agriculture with AI support. Remote sensing and artificial intelligence reclassify the traditional farming model by redefining common agricultural trends. Several commercially available goods are becoming increasingly well acknowledged as distinct over time due to the implementation of artificial intelligence revolutionary applications that can help identify and manage crop diseases to increase crop yield and quality. The use of artificial intelligence in agriculture is rapidly developing with numerous complex tactics and is predicted to bring about an extensive change in the near future.

References

Agrawal R, Wankhede VA, Kumar A, Luthra S, Majumdar A, Kazancoglu Y (2021) An exploratory state-of-the-art review of artificial intelligence applications in circular economy using structural topic modeling. Oper Manag Res 15:609–626

Alzubaidi L, Zhang J, Humaidi AJ, Al-Dujaili A, Duan Y, Al-Shamma O, Farhan L (2021) Review of deep learning: concepts, CNN architectures, challenges, applications, future directions. J Big Data 8:1–74

Arif U, Bhatti KH, Ajaib M, Wagay NA, Majeed M, Zeb J, Hameed A, Kiani J (2021) Ethnobotanical indigenous knowledge of Tehsil Charhoi, District Kotli, Azad Jammu and Kashmir, Pakistan. Ethnobot Res Appl 22:1–24. https://doi.org/10.32859/ERA.22.50.1-24

Arshad F, Waheed M, Harun N, Fatima K, Khan BA, Fatima K, Abbas Z, Jabeen S, Majeed M (2022) Indigenous farmer perception about fodder and foraging species of semi-arid lowlands of Pakistan: a case study of District Kasur, Pakistan. Taiwania 67:4

Arzani A, Ashraf M (2017) Cultivated ancient wheats (Triticum spp.): a potential source of health-beneficial food products. Compr Rev Food Sci Food Saf 16(3):477–488

Bashir SM, Altaf M, Hussain T, Umair M, Majeed M, Mangrio WM, Khan AM, Gulshan AB, Hamed MH, Ashraf S, Amjad MS, Bussmann RW, Abbasi AM, Casini R, Alataway A, Dewidar AZ, Al-Yafrsi M, Amin MH, Elansary HO (2023) Vernacular taxonomy, cultural and ethnopharmacological applications of avian and mammalian species in the vicinity of Ayubia National Park, Himalayan Region. Biology 12:4

Ben Ayed R, Hanana M (2021) Artificial intelligence to improve the food and agriculture sector. J Food Qual 2021:1–7

Budiharto W, Chowanda A, Gunawan AAS, Irwansyah E, Suroso JS (2019) A review and progress of research on autonomous drone in agriculture, delivering items and geographical information systems (GIS). In: Paper presented at the 2019 2nd world symposium on communication engineering (WSCE)

Condori RHM, Romualdo LM, Bruno OM, de Cerqueira Luz PH (2017) Comparison between traditional texture methods and deep learning descriptors for detection of nitrogen deficiency in maize crops. In: Paper presented at the 2017 workshop of computer vision (WVC)

Coulibaly S, Kamsu-Foguem B, Kamissoko D, Traore D (2022) Deep learning for precision agriculture: a bibliometric analysis. Intell Syst Appl 16:200102

Daponte P, De Vito L, Glielmo L, Iannelli L, Liuzza D, Picariello F, Silano G (2019) A review on the use of drones for precision agriculture. In: Paper presented at the IOP conference series: earth and environmental science

de Kfouri GO, Gonçalves DG, Dutra BV, de Alencastro JF, de Caldas Filho FL, e Martins LM, de Sousa Jr RT (2019) Design of a distributed HIDS for IoT backbone components. In: Paper presented at the FedCSIS (Communication Papers)

Dutta PK, Mitra S (2021) Application of agricultural drones and IoT to understand food supply chain during post COVID-19, pp 67–87

Ennaji O, Vergütz L, El Allali A (2023) Machine learning in nutrient management: a review. Artif Intell Agric 9:1–11. https://doi.org/10.1016/j.aiia.2023.06.001

Fuentes A, Yoon S, Kim SC, Park DS (2019) A robust deep-learning-based detector for real-time tomato plant diseases and pests recognition. Sens Agric 1(17):153

Fujita E, Kawasaki Y, Uga H, Kagiwada S, Iyatomi H (2016) Basic investigation on a robust and practical plant diagnostic system. In: Paper presented at the 2016 15th IEEE international conference on machine learning and applications (ICMLA)

Gardezi M, Joshi B, Rizzo DM, Ryan M, Prutzer E, Brugler S, Dadkhah A (2023) Artificial intelligence in farming: challenges and opportunities for building trust. Agron J 2023:21353

Haq SM, Yaqoob U, Majeed M, Amjad MS, Hassan M, Ahmad R, Morales-de la Nuez A (2022) Quantitative ethnoveterinary study on plant resource utilization by indigenous communities in high-altitude regions. Front Vet Sci 9:94404

Hariri RH, Fredericks EM, Bowers KM (2019) Uncertainty in big data analytics: survey, opportunities, and challenges. J Big Data 6(1):1–16

Hasan MJ, Mahbub S, Alom MS, Nasim MA (2019) Rice disease identification and classification by integrating support vector machine with deep convolutional neural network. In: Paper presented at the 2019 1st international conference on advances in science, engineering and robotics technology (ICASERT)

Hassan M, Haq SM, Ahmad R, Majeed M, Sahito HA, Shirani M, Mubeen I, Aziz MA, Pieroni A, Bussmann RW, Alataway A, Dewidar AZ, Al-Yafrsi M, Elansary HO, Yessoufou K (2022a) Traditional use of wild and domestic fauna among different ethnic groups in the Western Himalayas? Cross cultural analysis. Animals 12:17

Hassan M, Haq SM, Majeed M, Umair M, Sahito HA, Shirani M, Waheed M, Aziz R, Ahmad R, Bussmann RW, Alataway A, Dewidar AZ, El-Abedin TKZ, Al-Yafrsi M, Elansary HO, Yessoufou K (2022b) Traditional food and medicine: ethno-traditional usage of fish fauna across the valley of Kashmir: a Western Himalayan region. Diversity 14:6

Hinnell A, Lazarovitch N, Furman A, Poulton M, Warrick A (2010) Neuro-drip: estimation of subsurface wetting patterns for drip irrigation using neural networks. Irrig Sci 28:535–544

Iqbal N, Khaliq A, Cheema ZA (2020) Weed control through allelopathic crop water extracts and S-metolachlor in cotton. Inform Process Agric 7(1):165–172

Jackulin C, Murugavalli S (2022) A comprehensive review on detection of plant disease using machine learning and deep learning approaches. Meas Sens 24:100441

Jadhav P, Kachave V, Mane A, Joshi K (2023) Crop detection using satellite image processing. I-Managers J Image Process 10(2):50

Jondhale SR, Maheswar R, Lloret J, Jondhale SR, Maheswar R, Lloret J (2022) Fundamentals of wireless sensor networks. Received signal strength based target localization and tracking using wireless sensor networks. Springer, pp 1–19

Karetsos S, Costopoulou C, Sideridis A (2014) Developing a smartphone app for m-government in agriculture. J Agric Inform 5(1):129

Karlen D, Andrews S, Doran J (2001) Soil quality: current concepts and applications. Elsevier

Kashyap B, Kumar RJ (2021) Sensing methodologies in agriculture for soil moisture and nutrient monitoring. IEEE Access 9:14095–14121

Khan A, Shahriyar AK (2023) Optimizing onion crop management: a smart agriculture framework with IoT sensors and cloud technology. Appl Res Artif Intell Cloud Comput 6(1):49–67

Khan AM, Li Q, Saqib Z, Khan N, Habib T, Khalid N, Majeed M, Tariq A (2022a) MaxEnt modelling and impact of climate change on habitat suitability variations of economically important Chilgoza pine (*Pinus gerardiana* Wall.) in South Asia. Forests 13:5

Khan MHU, Wang S, Wang J, Ahmar S, Saeed S, Khan SU, Feng X (2022b) Applications of artificial intelligence in climate-resilient smart-crop breeding. Int J Mol Sci 23(19):11156

Khang A, Rath KC, Panda S, Sree PK, Panda SK (2023) Revolutionizing agriculture: exploring advanced technologies for plant protection in the agriculture sector. In: Handbook of research on AI-equipped IoT applications in high-tech agriculture. IGI Global, pp 1–22

Khoja AA, Haq SM, Majeed M, Hassan M, Waheed M, Yaqoob U, Bussmann RW, Alataway A, Dewidar AZ, Al-Yafrsi M, Elansary HO, Yessoufou K, Zaman W (2022) Diversity, ecological and traditional knowledge of pteridophytes in the Western Himalayas. Diversity 14:8

Kim H, Sudduth K, Hummel JW, Drummond S (2013) Validation testing of a soil macronutrient sensing system. Trans ASABE 56(1):23–31

Kuppusamy P, Shanmugananthan S, Tomar P (2021) Emerging technological model to sustainable agriculture. In: Artificial intelligence and IoT-based technologies for sustainable farming and smart agriculture. IGI Global, pp 101–122

Lamb DW, Frazier P, Adams P (2008) Improving pathways to adoption: putting the right P's in precision agriculture. Comput Electron Agric 61(1):4–9

Liang C, Shah T (2023) IoT in agriculture: the future of precision monitoring and data-driven farming. Eigenpub Rev Sci Technol 7(1):85–104

Linaza MT, Posada J, Bund J, Eisert P, Quartulli M, Döllner J, Moysiadis T (2021) Data-driven artificial intelligence applications for sustainable precision agriculture. Agronomy 11(6):1227

Liu J, Wang X (2021) Plant diseases and pests detection based on deep learning: a review. Plant Methods 17:1–18

Maddikunta PKR, Hakak S, Alazab M, Bhattacharya S, Gadekallu TR, Khan WZ, Pham Q-V (2021) Unmanned aerial vehicles in smart agriculture: applications, requirements, and challenges. IEEE Sens J 21(16):17608–17619

Majeed M, Bhatti KH, Amjad MS, Abbasi AM, Rashid A, Nawaz F , Ahmad KS (2020a) Ethno-veterinary practices of Poaceae taxa in Punjab, Pakistan

Majeed M, Bhatti KH, Amjad MS, Abbasi M, Id RWB, Nawaz F, Rashid A, Mehmood A, Id MM, Khan WM, Id SA (2020b) Ethno-veterinary uses of Poaceae in Punjab, Pakistan. PLoS One 15: e0241705. https://doi.org/10.1371/journal.pone.0241705

Majeed M, Bhatti KH, Pieroni A, Sukand R, Bussmann RW, Khan AM, Chaudhari SK, Aziz MA, Amjad MS (2021a) Gathered wild food plants among diverse religious groups in Jhelum District, Punjab, Pakistan. Foods 10:3

Majeed M, Tariq A, Anwar MM, Khan AM, Arshad F, Mumtaz F, Farhan M, Zhang L, Zafar A, Aziz M, Abbasi S, Rahman G, Hussain S, Waheed M, Fatima K, Shaukat S (2021b) Monitoring of land use? And cover change and potential causal factors of climate change in Jhelum District, Punjab, Pakistan, through GIS and multi-temporal satellite data. Land 10:10

Majeed M, Khan AM, Habib T, Anwar MM, Sahito HA, Khan N, Ali K (2022a) Vegetation analysis and environmental indicators of an arid tropical forest ecosystem of Pakistan. Ecol Indic 142:109291. https://doi.org/10.1016/j.ecolind.2022.109291

Majeed M, Lu L, Haq SM, Waheed M, Sahito HA, Fatima S, Aziz R, Bussmann RW, Tariq A, Ullah I, Aslam M (2022b) Spatiotemporal distribution patterns of climbers along an abiotic gradient in Jhelum District, Punjab, Pakistan. Forests 13:8

Majeed M, Tariq A, Haq SM, Waheed M, Anwar MM, Li Q, Aslam M, Abbasi S, Mousa BG, Jamil A (2022c) A detailed ecological exploration of the distribution patterns of wild Poaceae from the Jhelum District (Punjab), Pakistan. Sustainability (Switzerland) 14:7

Majeed M, Lu L, Anwar MM, Tariq A, Qin S, El-Hefnawy ME, El-Sharnouby M, Li Q, Alasmari A (2023) Prediction of flash flood susceptibility using integrating analytic hierarchy process (AHP) and frequency ratio (FR) algorithms. Front Environ Sci 10:1037547. https://doi.org/10.3389/fenvs.2022.1037547

Martinelli F, Scalenghe R, Davino S, Panno S, Scuderi G, Ruisi P, Goulart LR (2015) Advanced methods of plant disease detection. A review. Agron Sustain Dev 35:1–25

Mathenge M, Sonneveld BG, Broerse JE (2022) Application of GIS in agriculture in promoting evidence-informed decision making for improving agriculture sustainability: a systematic review. Sustainability 14(16):9974

Meng T, Jing X, Yan Z, Pedrycz W (2020) A survey on machine learning for data fusion. Inf Fusion 57:115–129

Minkoff-Zern L-A (2019) The new American farmer: immigration, race, and the struggle for sustainability. MIT Press

Mohamed ES, Belal A, Abd-Elmabod SK, El-Shirbeny MA, Gad A, Zahran MB (2021) Smart farming for improving agricultural management. Egypt J Remote Sens Space Sci 24(3):971–981

Mohanty SP, Hughes DP, Salathé M (2016) Using deep learning for image-based plant disease detection. Front Plant Sci 7:1419

Mosa KA, Ismail A, Helmy M, Mosa KA, Ismail A, Helmy M (2017) Introduction to plant stresses. Springer, pp 1–19

Nayfeh M, Li Y, Al Shamaileh K, Devabhaktuni V, Kaabouch N (2023) Machine learning modeling of GPS features with applications to UAV location spoofing detection and classification. Comput Secur 126:103085

Pawar A, Deosarkar SJ (2023) IoT-based smart agriculture: an exhaustive study. Springer, pp 1–14

Peixoto DS, Silva BM, de Oliveira GC, Moreira SG, da Silva F, Curi N (2019) A soil compaction diagnosis method for occasional tillage recommendation under continuous no tillage system in Brazil. Soil Tillage Res 194:104307
Pereira RI, Jucá SC, Carvalho PC (2019) IoT embedded systems network and sensors signal conditioning applied to decentralized photovoltaic plants. Measurement 142:195–212
Qin Z, Myers DB, Ransom CJ, Kitchen NR, Liang SZ, Camberato JJ, Franzen DW (2018) Application of machine learning methodologies for predicting corn economic optimal nitrogen rate. Agron J 110(6):2596–2607
Raj EFI, Appadurai M, Athiappan K (2022) Precision farming in modern agriculture. In: Smart agriculture automation using advanced technologies: data analytics and machine learning, cloud architecture, automation and IoT. Springer, pp 61–87
Rakhmatulin I, Kamilaris A, Andreasen C (2021) Deep neural networks to detect weeds from crops in agricultural environments in real-time: a review. Remote Sens 13(21):4486
Rashid S, Getnet K, Lemma S (2019) Maize value chain potential in Ethiopia: constraints and opportunities for enhancing the system. Gates Open Res 3(354):354
Rathore B (2017) Sustainable fashion marketing: AI-powered solutions for effective promotions. Int J New Media Stud Int Peer Rev Sch Indexed J 4(2):70–80
Rehman A, Saba T, Kashif M, Fati SM, Bahaj SA, Chaudhry H (2022) A revisit of internet of things technologies for monitoring and control strategies in smart agriculture. Agronomy 12(1):127
Sadeeq MM, Abdulkareem NM, Zeebaree SR, Ahmed DM, Sami AS, Zebari RR (2021) IoT and Cloud computing issues, challenges and opportunities: a review. Qubahan Acad J 1(2):1–7
Sekaran U, Lai L, Ussiri DA, Kumar S, Clay S (2021) Role of integrated crop-livestock systems in improving agriculture production and addressing food security—a review. J Agric Food Agric 5:100190
Singh RK, Berkvens R, Weyn M (2021) AgriFusion: an architecture for IoT and emerging technologies based on a precision agriculture survey. IEEE Access 9:136253–136283
Sinha D, Maurya AK, Abdi G, Majeed M, Agarwal R, Mukherjee R, Ganguly S, Aziz R, Bhatia M, Majgaonkar A, Seal S, Das M, Banerjee S, Chowdhury S, Adeyemi SB, Chen JT (2023) Integrated genomic selection for accelerating breeding programs of climate-smart cereals. Genes 14(7):1484
Sinwar D, Dhaka VS, Sharma MK, Rani G (2020) AI-based yield prediction and smart irrigation. Internet Things Anal Agric 2:155–180
Sudha T (2023) Artificial intelligence in human resource management. Future trends, breakthroughs and innovation in HRM
Sun Y, Ding W, Shu L, Li K, Zhang Y, Zhou Z, Han G (2021) On enabling mobile crowd sensing for data collection in smart agriculture: a vision. IEEE Syst J 16(1):132–143
Tan L, Lu J, Jiang H (2021) Tomato leaf diseases classification based on leaf images: a comparison between classical machine learning and deep learning methods. AgriEngineering 3(3):542–558
Tariq A, Mumtaz F, Majeed M, Zeng X (2023) Spatio-temporal assessment of land use land cover based on trajectories and cellular automata Markov modelling and its impact on land surface temperature of Lahore district Pakistan. Environ Monit Assess 195:1
Tassadduq SS, Akhtar S, Waheed M, Bangash N, Nayab DE, Majeed M, Abbasi S, Muhammad M, Alataway A, Dewidar AZ, Elansary HO, Yessoufou K (2022) Ecological distribution patterns of wild grasses and abiotic factors. Sustainability (Switzerland) 14:18
Tiwari AK, Mishra H, Nishad DC, Pandey A (2023) Sustainable water management in agriculture: irrigation techniques and water conservation. Dr. Ajay B. Jadhao 53
Toseef M, Khan MJ (2018) An intelligent mobile application for diagnosis of crop diseases in Pakistan using fuzzy inference system. Comput Electron Agric 153:1–11. https://doi.org/10.1016/j.compag.2018.07.034
Tugrul B, Elfatimi E, Eryigit R (2022) Convolutional neural networks in detection of plant leaf diseases: a review. Agriculture 12(8):1192

Ullah I, Aslam B, Shah SHIA, Tariq A, Qin S, Majeed M, Havenith HB (2022) An integrated approach of machine learning, remote sensing, and GIS data for the landslide susceptibility mapping. Land 11:8

Van Esse HP, Reuber TL, van der Does D (2020) Genetic modification to improve disease resistance in crops. New Phytol 225(1):70–86

Waheed M, Arshad F, Majeed M, Fatima S, Mukhtar N, Aziz R, Mangrio WM, Almohamad H, Dughairi AA, Al-Mutiry M, Abdo HG (2022) Community structure and distribution pattern of woody vegetation in response to soil properties in semi-arid lowland District Kasur Punjab, Pakistan. Land 11:12

Waheed M, Arshad F, Majeed M, Haq SM, Aziz R, Bussmann RW, Ali K, Subhan F, Jones DA, Zaitouny A (2023) Potential distribution of a noxious weed (*Solanum viarum* Du-nal), current status, and future invasion risk based on MaxEnt modeling. In: Geology, ecology, and landscapes. Taylor & Francis, p 1. https://doi.org/10.1080/24749508.2023.2179752

Wai MHX, Huong A, Ngu X (2021) Soil moisture level prediction using optical technique and artificial neural network. Int J Electr Comput Eng 11(2):1752–1760

Wang R, Gamon JA (2019) Remote sensing of terrestrial plant biodiversity. Remote Sens Environ 231:111218

Xiao L, Bahri Y, Sohl-Dickstein J, Schoenholz S, Pennington J (2018) Dynamical isometry and a mean field theory of CNNS: how to train 10,000-layer vanilla convolutional neural networks. In: Paper presented at the international conference on machine learning

Yalcin H, Razavi S (2016) Plant classification using convolutional neural networks. In: Paper presented at the 2016 fifth international conference on Agro-Geoinformatics (Agro-Geoinformatics)

Zhang J, Yin Z, Chen P, Nichele SJIF (2020) Emotion recognition using multi-modal data and machine learning techniques: a tutorial and review. Inf Fusion 59:103–126

Zhou Z-H (2021) Machine learning. Springer Nature

Chapter 17
Use Cases and Future Aspects of Intelligent Techniques in Microbial Data Analysis

Muhammad Naveed, Zaibun-nisa Memon, Muhammad Abdullah, Syeda Izma Makhdoom, Arooj Azeem, Sarmad Mehmood, Maida Salahuddin, Zeerwah Rajpoot, and Muhammad Majeed

Abstract Microbes, including bacteria, archaea, fungi, and viruses, are fundamental to our ecosystems, health, and industries. Microbial data analysis has become indispensable in understanding their roles and interactions. In this era of big data, advanced techniques, such as high-throughput sequencing, metagenomics, and bioinformatics, have accelerated microbial research. This chapter explores the significance of intelligent techniques, particularly machine learning and artificial intelligence, in revolutionizing microbial data analysis. The aim of this chapter is to showcase the pivotal role of intelligent techniques in microbial data analysis across diverse domains, from ecology and public health to biotechnology. We delve into case studies that highlight the practical applications of these techniques and the transformative impact they have had on microbial research. Several case studies are presented, illustrating the applications of intelligent techniques in microbial research. These include predicting disease risk through gut microbiome analysis, antibiotic resistance prediction, environmental microbiology for ecosystem management, bioprocess optimization in biotechnology, and AI-powered antibiotic susceptibility testing. Each case study demonstrates how intelligent techniques have enhanced data analysis, prediction, and decision-making in their respective domains.

M. Naveed · S. I. Makhdoom · A. Azeem · S. Mehmood · M. Salahuddin · Z. Rajpoot
Department of Biotechnology, Faculty of Science and Technology, University of Central Punjab, Lahore, Punjab, Pakistan
e-mail: dr.naveed@ucp.edu.pk

Z.-n. Memon
Department of Zoology, Shah Abdul Latif University, Khairpur Mirs, Sindh, Pakistan
e-mail: zaib.nisa@salu.edu.pk

M. Abdullah
Biodiversity Park, Cholistan Institute of Desert Studies (CIDS), The Islamia University of Bahawalpur, Bahawalpur, Pakistan
e-mail: abdullahfrw@iub.edu.pk

M. Majeed (✉)
Department of Botany, University of Gujrat, Gujrat, Pakistan
e-mail: m.majeed@uog.edu.pk; https://www.researchgate.net/profile/Muhammad-Majeed-18

A. Khamparia et al. (eds.), *Microbial Data Intelligence and Computational Techniques for Sustainable Computing*, Microorganisms for Sustainability 47,
https://doi.org/10.1007/978-981-99-9621-6_17

Microbial data analysis, driven by intelligent techniques, has ushered in a new era of understanding and harnessing the power of microorganisms. The future of microbial data analysis holds immense promise, with emerging trends including the integration of omics data, explainable AI, personalized microbiome analysis, and the development of ethical and regulatory frameworks. Collaborative research and data sharing are expected to further advance our understanding of the microbial world, offering solutions to some of the most critical challenges of our time.

Keywords Microbiome · Intelligent · Data analysis · Bioprocess · Optimization · Personalized

17.1 Introduction

Microbes, the small, frequently invisible organisms that live everywhere on our globe, have a significant impact on how the world is now. Numerous ecosystems, human health, and industrial processes depend on these microscopic organisms, which also include bacteria, archaea, fungus, viruses, and other microbes. Data analysis has become a significant tool in the understanding and use of microorganisms. The significance of microbial data analysis in the age of big data cannot be emphasized (Majeed et al. 2021a).

In order to research microbial communities, their roles, and interactions, scientists use a wide variety of tools and procedures, which collectively make up microbial data analysis. Thanks to technological advancements like high-throughput sequencing, metagenomics, metatranscriptomics, and bioinformatics, this discipline has quickly developed. Drug development requires examination of microbial data, especially to comprehend how the microbiome affects drug metabolism and effectiveness (Tirtawijaya et al. 2018).

Ecological processes can be better understood by examining the diversity and distribution of microbes, which are essential parts of ecosystems. Analysis of microbial data assists in tracking microbial community changes brought by environmental disturbances and offers insightful information for conservation efforts (Khoja et al. 2022). Our bodies are home to microbes, and they have a significant impact on our health. Understanding the human microbiome, which is linked to several health issues such as obesity, autoimmune illnesses, and mental health disorders, requires analysis of microbial data. Microbial data analysis is also necessary for probiotic creation and personalized treatment (Belkaid and Hand 2014).

Analyzing microbial data is essential for monitoring and controlling infectious disease epidemics. It helps in tracking the transmission of pathogens, locating the origin of infections, and creating efficient therapies. Microbes are sensitive environmental change indicators. Analysis of microbial data can be used to determine how pollution, land use, and climate change affect microbial communities, assisting in the development of environmental policies (Haberbeck et al. 2017).

Science continues to learn new things about microbes every day. Through the use of data analysis, scientists are now able to discover new species, examine their genetic makeup, and learn more about their special skills. Given that many medications are made from naturally occurring chemicals generated by bacteria, examination of microbial data is crucial in the drug development process. In this research, mining microbial genomes for new therapeutic compounds is a promising strategy (Arif et al. 2021).

Microbial data analysis is a cornerstone of contemporary science and technology, with applications in everything from industry and environmental protection to ecology and medicine. The enormous volume of data produced by microbial investigations offers a wealth of knowledge just waiting to be discovered. Our capacity to harness the power of these microscopic creatures for the benefit of people and the environment grows as our knowledge of the microbial world expands. Therefore, the value of microbial data analysis cannot be overstated, and further developments in this area hold the promise of tackling some of the most important problems of our day (Waheed et al. 2023).

17.1.1 Role of Intelligent Techniques

The subject of microbial data analysis has undergone a revolution thanks to intelligent techniques, which now provide strong tools and procedures to solve the riddles of microbial communities. It is becoming more and more difficult for researchers to manually extract significant insights from the huge information generated by high-throughput sequencing, metagenomics, and other cutting-edge techniques used in microbial data analysis. This is where intelligent methods have intervened to improve the effectiveness and precision of microbiological data analysis, such as machine learning and artificial intelligence (AI; Ullah et al. 2022).

For instance, machine learning algorithms have been crucial in taxonomy classification, enabling researchers to quickly detect and categorize microorganisms. These methods are able to identify minor patterns and similarities across microbial sequences that may be difficult for people to notice manually by training algorithms on labeled datasets. This greatly expedites the identification of existing species and even makes it possible to find new ones (Qu et al. 2019).

Additionally, intelligent approaches excel at predictive modeling, enabling researchers to foresee microbial interactions and behaviors. For instance, in ecological studies, machine learning models can forecast how changes in environmental conditions would affect the dynamics of microbial communities, assisting with the preservation of biodiversity and the evaluation of the health of ecosystems (Lucas 2020). Predictive modeling based on microbiome data in the healthcare sector assists in predicting disease risks, treatment outcomes, and tailored medicine interventions based on a person's microbial composition (Majeed et al. 2020a).

When tackling the enormous complexity of microbiological data, artificial intelligence (AI) is also a factor. With the help of AI-driven data integration technologies, information from diverse sources, including environmental metadata, metatranscriptomics, and genomic data, may be unified to give a comprehensive picture of microbial communities. Understanding how bacteria react to environmental changes is vital in areas like bioremediation and agriculture and is made possible by this integrated approach (Sahayasheela et al. 2022).

17.2 Microbial Data Analysis

Microbes, the microscopic organisms that inhabit various environments, play a crucial role as sensitive indicators of ecosystem changes (Falkowski et al. 2008). Their incredible diversity and adaptability make them valuable sensors for detecting shifts in environmental conditions. Microbes can swiftly respond to alterations in chemical and physical surroundings, providing researchers with precise data. In this commentary, we explore the immense potential of microbial data analysis, drawing an intriguing parallel with the olfactory system. Just as populations of sensory neurons in our noses respond to specific odors, microbial communities can be considered as nature's sensors, capable of reporting environmental changes with remarkable fidelity (Haq et al. 2022).

Moreover, this commentary emphasizes the role of machine learning in unlocking the insights hidden within microbial data. Machine learning techniques hold the promise of deciphering microbial responses to environmental fluctuations, enhancing our understanding of ecosystems and aiding in the detection of significant environmental changes (Bzdok and Ioannidis 2019).

17.2.1 Using AI for Public Health Surveillance

The twenty-first century has witnessed an unprecedented surge in data collected from public health surveillance. This exponential growth can be attributed to advancements in information technology and the establishment of comprehensive data collection systems (Bates et al. 2014). In this context, the utilization of artificial intelligence (AI) emerges as a potent solution for addressing the challenges posed by this deluge of data. AI offers the potential to revolutionize disease monitoring, making it more reliable and efficient (Majeed et al. 2021b).

By harnessing AI, we can not only manage the vast volumes of infectious disease data effectively but also enable proactive surveillance and projection of disease outbreaks (Kamel Boulos et al. 2011). Robust data management platforms, in conjunction with AI methods, have the potential to provide valuable support to government agencies, healthcare providers, and medical professionals in their efforts to respond to diseases swiftly and with precision (Khan et al. 2022).

17.2.2 Microbial Networks: Nature's Intelligent Systems

Living organisms, from the tiniest microbes to complex multicellular organisms like humans, persist and thrive due to intricate interactions within dynamic, environment-responsive networks (Jeong et al. 2000). These networks operate across multiple scales and dimensions, facilitating communication and adaptation. While we often associate "intelligence" with the human brain, this commentary suggests that similar traits may exist in the intricate macromolecular networks of microbes (Fig. 17.1).

Despite lacking brains, microbes exhibit intelligent characteristics such as memory, anticipation, adaptation, and reflection. The commentary explores how microbial networks demonstrate these traits and how they may confer advantages in their respective environments. It proposes that if we broaden our definition of "intelligence" to exclude terms like "human" and "brain," all forms of life, including microbes, display some or all characteristics consistent with intelligence (Guttenberg et al. 2020). Additionally, it highlights how advances in genome-wide data analysis offer valuable insights into microbial intelligence, which may pave the way for synthetic biology applications and the creation of intelligent molecular networks.

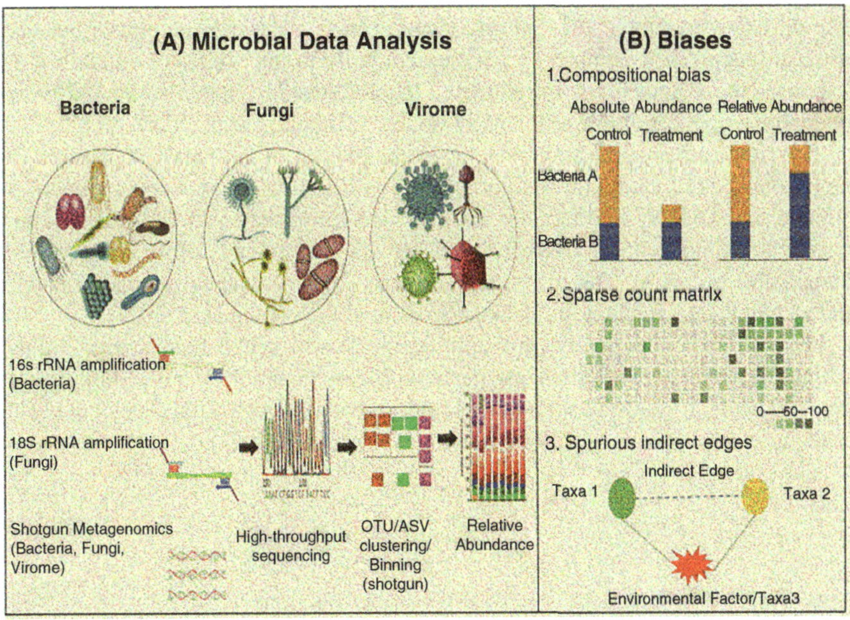

Fig. 17.1 Overview of (**a**) taxonomic profiling and (**b**) biases in microbial co-occurrence networks (Matchado et al. 2021)

17.2.3 Microbial Data Analysis for Disease Association

The human microbiota, comprising a diverse array of microorganisms, exhibits remarkable variability within individuals. These microorganisms residing in the human body play a pivotal role in maintaining health and potentially contributing to disease (Human Microbiome Project Consortium 2012). However, understanding the intricate relationships between microbes and diseases remains a complex challenge.

To address this challenge, the commentary introduces a novel approach for predicting disease-microbe associations. This method involves integrating disease and microbe networks and employing advanced network topology analysis. By doing so, researchers can prioritize candidate microbes and unveil potential disease-causing microorganisms. This not only enhances our comprehension of microbial interactions and mechanisms but also offers clinical solutions for understanding disease mechanisms, diagnosis, and therapeutic interventions (Zhou et al. 2018).

17.2.4 AI-Powered Antibiotic Susceptibility Testing

Antimicrobial resistance stands as a significant global health threat, primarily driven by the misuse of antibiotics (Laxminarayan et al. 2013). Traditional methods like disk diffusion antibiotic susceptibility testing (AST) are widely employed but face criticism due to their complexity and susceptibility to inter-operator variability. In response, the commentary introduces a groundbreaking solution—an AI-based smartphone application for offline AST analysis (Lutgring et al. 2020).

This innovative application leverages the power of artificial intelligence to capture images of antibiotic susceptibility tests using the smartphone's camera. It offers a user-friendly interface for analysis and incorporates an embedded expert system to validate data coherence and provide interpreted results. The result? Remarkable accuracy, with high agreement rates against hospital-standard systems and a significant reduction in inter-operator variability (Tang et al. 2021).

Perhaps most importantly, this application is well suited for resource-limited settings, potentially revolutionizing patients' access to antibiotic susceptibility testing worldwide. It showcases the tremendous potential of AI to address global health challenges and improve healthcare outcomes (Lam et al. 2022).

17.3 Use Cases of Intelligent Techniques

17.3.1 Microbiome Studies

Microbiome studies have emerged as a critical field of research with applications ranging from human health to environmental ecology. These studies involve analyzing the diverse communities of microorganisms that inhabit different ecosystems. Here, we emphasize the pivotal role of artificial intelligence (AI) and machine learning in advancing our understanding of these complex microbial communities (Majeed et al. 2022a).

Microorganisms in the human microbiome play a critical role in skin health and the development of various skin diseases. The complexity of microbiome data has historically challenged traditional analysis methods. However, recent advancements in artificial intelligence (AI) and machine learning are revolutionizing microbiome analysis in dermatology, enabling more accurate disease diagnosis and prognosis (Waheed et al. 2022).

17.3.1.1 Predicting Disease Risk Through Gut Microbiome Analysis

In a study by Pasolli and colleagues they harnessed the power of AI and machine learning to analyze the gut microbiome data of over 150,000 individuals from diverse geographical locations, age groups, and lifestyles. They developed predictive models that could identify individuals at higher risk for specific diseases, such as type 2 diabetes and inflammatory bowel disease, based on their gut microbiome composition. This research highlights how AI-driven microbiome analysis can contribute to personalized disease risk assessment and early intervention (Majeed et al. 2020b).

17.3.1.2 Microbial Biomarkers for Disease Prediction

In a study published in *Nature*, researchers used machine learning to analyze the gut microbiomes of patients with IBD. They applied supervised learning to identify microbial signatures associated with disease status. By training their model on a dataset comprising healthy individuals and IBD patients, the researchers discovered specific microbial taxa and functional genes that were strongly correlated with IBD. This study showcased how intelligent techniques can reveal microbial biomarkers for complex diseases (Tassadduq et al. 2022).

17.3.1.3 Targeted Microbiome Interventions

Zmora and colleagues used machine learning to examine how individual variations in gut microbiome composition influence responses to probiotics. By analyzing host and microbiome features, they developed predictive models to identify individuals who would benefit from specific probiotic interventions. This personalized approach to microbiome modulation illustrates the potential of AI in tailoring microbial therapies for improved health outcomes (Majeed et al. 2021c).

17.3.1.4 Discovering Novel Microbial Functions

Forslund and colleagues applied machine learning to analyze metagenomic data from individuals with type 2 diabetes before and after metformin treatment. Their study revealed shifts in the gut microbiota associated with both the disease and treatment. Machine learning algorithms helped identify specific microbial functions influenced by metformin, shedding light on the interplay between host metabolism, drug treatment, and the gut microbiome. This research showcases how AI can unravel novel microbial functions and their roles in health and disease (Hassan et al. 2022a).

17.3.1.5 Functional Metagenomics in Environmental Microbiomes

In a study published in *Nature Methods*, scientists applied machine learning to predict the functional potential of microbial communities in an aquatic ecosystem. They used metagenomic data to train their model, which could infer the presence of metabolic pathways and functional genes within the microbiome. This approach allowed researchers to gain insights into how these microbial communities were contributing to nutrient cycling and energy flow in the ecosystem (Khan et al. 2022).

17.3.1.6 Recognition of Fungi with Convolutional Neural Networks (CNN)

Convolutional neural networks (CNN) are a specialized form of machine learning that can analyze and classify images based on their unique features. In dermatology, CNN can be applied to diagnose skin conditions by annotating images with corresponding medical records and pathological results, generating standardized data for analysis. This approach has proven particularly effective in diagnosing fungal infections, such as onychomycosis, which is the most common nail disease caused by fungi (Hassan et al. 2022b). Traditional clinical diagnosis methods for onychomycosis involve microscopy and fungal culture, which are time consuming and susceptible to false negatives. CNN models, such as VGG16 and InceptionV3, have demonstrated superior accuracy, sensitivity, and specificity compared to traditional methods. CNN can also complement manual microscopic examinations, improving diagnostic accuracy and efficiency (Bashir et al. 2023).

17.3.1.7 Combined Application of Microbial Genome Sequencing and Machine Learning

Machine learning can analyze sequencing data to identify patterns in pathogen genomes, enabling the subtyping of pathogens for precise treatment. This approach has been applied in the diagnosis of syphilis, a disease caused by *Treponema pallidum*. Serological tests used in clinical diagnosis lack the ability to determine the specific strain of the pathogen. By collecting and sequencing *T. pallidum* samples from patients worldwide and applying machine learning techniques, researchers successfully classified spirochete branches, providing crucial information for personalized treatment decisions. Additionally, machine learning can analyze 16S sequencing results to reveal differences in microbial species and composition ratios between patients and healthy individuals. This approach has been used to diagnose diseases like human papillomavirus (HPV) infection and predict HPV types. It can also assess skin status and predict skin diseases based on microbial composition. Metagenomic sequencing, another technique, has been used to diagnose acne by analyzing changes in skin microbiota (Sinha et al. 2023).

17.3.1.8 AI and Skin Diseases in the Context of the Gut–Skin Axis

The gut microbiota plays a vital role in maintaining overall health, and disruptions in the gut–skin axis have been linked to various skin diseases, including atopic dermatitis, psoriasis, vitiligo, and acne vulgaris. Research has shown that changes in skin microbes are related to the progression of vitiligo, and machine learning techniques can help diagnose and predict disease progression based on skin microbiome change. Similarly, machine learning has been used to analyze the gut microbiomes of patients with atopic dermatitis, revealing characteristic microbial signatures associated with the disease. These findings suggest that gut microbiota analysis can not only predict skin diseases but also distinguish between disease subtypes, enhancing diagnostic accuracy and guiding treatment decisions (Majeed et al. 2023).

17.3.2 *Antibiotic Resistance Prediction*

Predicting antibiotic resistance patterns in microorganisms is crucial for effective treatment strategies. Case studies demonstrate the role of intelligent techniques in antibiotic resistance prediction.

17.3.2.1 Genomic Analysis of Antibiotic Resistance in Pathogens

In a clinical setting, a team of researchers used machine learning to predict antibiotic resistance in a collection of bacterial isolates. By analyzing the genetic profiles of these bacteria, the team trained a model to predict the resistance of each isolate to multiple antibiotics. The model demonstrated high accuracy in predicting antibiotic susceptibility, aiding clinicians in selecting the most effective treatments for infected patients (Sinha et al. 2023).

17.3.2.2 Real-Time Antibiotic Resistance Prediction

In a hospital setting, a real-time antibiotic resistance prediction system was implemented using artificial intelligence. This system integrated patient-specific data, such as antibiotic exposure history and genetic information of infecting bacteria, into a machine learning model. Clinicians received instant predictions of antibiotic resistance, enabling them to adjust treatment plans promptly. This case study showcased the potential of intelligent techniques in improving patient outcomes.

Microbial data analysis, powered by intelligent techniques, has revolutionized our understanding of microbiological ecosystems and their impact on various fields. In this section, we will delve into specific use cases where these intelligent techniques have been pivotal, backed by real-world examples and scholarly references (Jamil et al. 2022).

17.3.2.3 Tracking Resistance Trends

Intelligent techniques also aid in tracking antibiotic resistance trends. A study by Hu et al. (2020) used machine learning to analyze resistance data from multiple healthcare facilities. By identifying emerging resistance patterns, they contributed to more informed antibiotic stewardship practices, minimizing the risk of multidrug-resistant infections. The main focus of the study involved the application of advanced machine learning models to the prepared dataset. These models were designed to identify hidden patterns and trends in antibiotic resistance. Supervised learning techniques, such as logistic regression and decision trees, were used to predict the likelihood of antibiotic resistance for specific antibiotics and bacterial

Table 17.1 Applications of machine learning and AI in microbial and disease analysis

Title	Significance	Tools used	References
Predicting disease risk through gut microbiome analysis	Personalized disease risk assessment and early intervention	AI, machine learning, gut microbiome data	Pasolli et al. (2019)
Microbial biomarkers for disease prediction	Identification of microbial signatures for complex diseases	Machine learning, gut microbiome data	Lloyd-Price et al. (2019)
Targeted microbiome interventions	Tailoring microbial therapies for improved health outcomes	Machine learning, host and microbiome features	Zmora et al. (2018)
Discovering novel microbial functions	Uncovering novel microbial functions and their roles in health and disease	Machine learning, metagenomic data	Forslund et al. (2015)
Functional metagenomics in environmental microbiomes	Insights into microbial contributions to nutrient cycling and energy flow in ecosystems	Machine learning, metagenomic data	
Recognition of fungi with convolutional neural networks (CNN)	Accurate diagnosis of fungal infections in dermatology	Convolutional neural networks (CNN), dermatological images	
Combined application of microbial genome sequencing and machine learning	Precision diagnosis of pathogens and assessment of skin status	Machine learning, genomic and 16S sequencing data	
AI and skin diseases in the context of the gut–skin Axis	Prediction and distinction of skin diseases based on gut microbiota	Machine learning, gut microbiome data	
Genomic analysis of antibiotic resistance in pathogens	Accurate prediction of antibiotic resistance in bacterial isolates	Machine learning, genetic profiles	
Real-time antibiotic resistance prediction	Prompt adjustment of treatment plans for patients	Artificial intelligence (AI), patient data, genetic information	
Tracking resistance trends	Informed antibiotic stewardship practices to minimize drug resistance	Machine learning, resistance data	Hu et al. (2020)

strains. Unsupervised learning methods, such as clustering and principal component analysis, were employed to uncover hidden relationships among bacterial strains, resistance profiles, and patient attributes as shown in Table 17.1.

17.3.3 Environmental Microbiology

Environmental microbiology is a multidisciplinary field that delves into the intricate relationships between microorganisms and their natural habitats, from terrestrial soils to aquatic environments, and even extreme ecosystems like hydrothermal vents. The microorganisms in these environments play pivotal roles in nutrient cycling, degradation of organic matter, and the overall health of ecosystems. They impact global processes such as climate change, biogeochemical cycling, and pollution remediation (Majeed et al. 2022b).

Intelligent techniques, which encompass a variety of data-driven and computational approaches, have become indispensable tools in environmental microbiology. These techniques have revolutionized the way we analyze and predict microbial dynamics, providing unprecedented insights into the hidden world of microorganisms in natural ecosystems. Given below is the significance of intelligent techniques in understanding and managing our environment, illustrated through case studies and practical applications.

17.3.3.1 Machine Learning Empowers Microbial Community Analysis in Soil Ecosystems

In a research conducted by Zhang et al. (2021), machine learning emerged as a powerful tool for unraveling the complex relationships between soil microbial communities and their environment. Traditional methods for analyzing soil microorganisms are labor intensive and time consuming, making large-scale studies challenging. In this study, high-throughput sequencing technology generated extensive microbial data, which machine learning efficiently processed. The application of machine learning techniques, such as Random Forest and Gradient Boosting Machines, allowed researchers to identify the key environmental factors influencing soil microbial communities in Chinese tea plantations, quantify the impacts of land management practices, and discover specific microbial biomarkers for different soil conditions and management practices. This case study illustrates the critical importance of machine learning in enabling data-driven decisions for sustainable agriculture and land management, ultimately contributing to informed and environmentally responsible soil ecosystem conservation and optimization (Majeed et al. 2023).

17.3.3.2 Advancements in Predicting Nutrient Cycling in Aquatic Environments

Yao et al. (2021) conducted a research that emerged as a groundbreaking tool for predicting nutrient concentrations and cycling rates in polluted river ecosystems, addressing challenges posed by the complexity and variability of aquatic environments. By harnessing the power of intelligence techniques such as Random Forest and Gradient Boosting Machines, researchers accurately forecasted nutrient concentrations based on water quality and weather data and predicted nutrient cycling rates considering pollutant inputs. These accurate and real-time predictions hold paramount importance for managing and preserving water quality, preventing eutrophication, and enabling timely interventions to safeguard aquatic ecosystems. This study highlights how machine learning enhances our capacity to proactively manage and conserve vital aquatic resources, highlighting its crucial role in ensuring the sustainable management of our planet's waterways (Hassan et al. 2022a).

17.3.3.3 Machine Learning and AI Revolutionize Bioremediation of Contaminated Sites

In a recent study by Smith et al. (2022), the integration of artificial intelligence (AI) and machine learning emerged as a game-changing solution for optimizing the bioremediation of highly contaminated industrial sites. Traditional approaches to bioremediation often prove time consuming and resource intensive, but AI and machine learning techniques, including Artificial Neural Networks (ANNs) and Genetic Algorithms, enabled precise modeling of complex interactions between environmental factors and microbial responses. This allowed researchers to predict optimal conditions for pollutant degradation, design targeted bioremediation strategies, and even identify genetic modifications to enhance native microbial species' pollutant-degrading capabilities. The transformative impact of AI and machine learning in this study not only expedites cleanup processes but also minimizes the ecological impact of contaminants, exemplifying their vital role in advancing sustainable and environmentally responsible bioremediation efforts for industrial contaminated sites.

17.3.4 Biotechnology and Bioprocess Optimization

Biotechnology stands at the forefront of scientific innovation, utilizing the remarkable capabilities of microorganisms to address a myriad of challenges and to drive progress in diverse sectors, including biofuel production, pharmaceuticals, and bioprocessing. Within this dynamic field, the optimization of bioprocesses is of paramount importance to maximize efficiency, reduce costs, and minimize environmental impact. In this context, intelligent techniques such as artificial intelligence (AI) and machine learning have emerged as indispensable tools. This section explores the pivotal role of intelligent techniques in biotechnology and bioprocess optimization, elucidated through case studies that underscore their significance in the industrial and pharmaceutical sectors (Majeed et al. 2023).

17.3.4.1 Optimization of Bioethanol Fermentation

In a groundbreaking study by Chen et al. (2022), machine learning and artificial intelligence emerged as transformative tools for optimizing the bioethanol fermentation process, addressing challenges associated with intricate parameter interactions. By efficiently processing extensive datasets, machine learning models, including Artificial Neural Networks (ANNs) and Genetic Algorithms, accurately predicted optimal fermentation conditions for bioethanol production. This precision enabled researchers to fine-tune the process, resulting in higher bioethanol yields, reduced production costs, and minimized environmental impact, ultimately advancing the sustainability and economic viability of bioethanol production. This case study indicates the pivotal role of data-driven decision-making in revolutionizing bioenergy research and positioning bioethanol as a renewable and environmentally responsible alternative to fossil fuels, offering a path to a more sustainable energy future (Arif et al. 2021).

17.3.4.2 Accelerating Drug Discovery Through Metabolic Engineering

In a study by Sharma et al. (2022), AI emerged as a potent tool for accelerating drug discovery through metabolic engineering, addressing the complexities of enhancing secondary metabolite production in Streptomyces strains. By analyzing extensive omics data, including genomics, transcriptomics, and metabolomics, machine learning techniques such as Random Forest Regression and Genetic Algorithm Optimization accurately identified genetic modifications that led to significantly increased secondary metabolite yields. This precise strain design not only expedited drug discovery processes but also reduced development time and costs, potentially making life-saving medications more accessible and cost effective (Sinha et al. 2023). The study underscores the transformative impact of data-driven decision-making in revolutionizing the pharmaceutical industry, offering a promising path to more efficient drug development, and improved global healthcare outcomes through innovative secondary metabolite production in microbial strains as shown in Table 17.2.

17.4 Challenges in Microbial Data Analysis

17.4.1 Data Quality and Quantity

Ensuring data quality and quantity is a foundational challenge in microbial data analysis with AI technology. The microbial world's diversity and complexity demand comprehensive, high-quality datasets for meaningful analysis (Thompson et al. 2017). Challenges arise from various sources, including environmental

Table 17.2 Applications of machine learning and AI in environmental and microbial sciences

Title of case study	Importance	Tools used	References
Machine learning empowers microbial community analysis in soil ecosystems	Machine learning revolutionizes soil microbial community analysis, enabling data-driven decisions for sustainable agriculture and land management	High-throughput sequencing, machine learning (random Forest, gradient boosting machines)	Zhang et al. (2021)
Advancements in predicting nutrient cycling in aquatic environments	Machine learning plays a crucial role in real-time nutrient prediction for water quality management and ecological conservation	Machine learning (random Forest, gradient boosting machines)	Yao et al. (2021)
Machine learning and AI revolutionize bioremediation of contaminated sites	AI and machine learning optimize bioremediation, reducing cleanup time, costs, and ecological impacts in contaminated industrial sites	Artificial intelligence (artificial neural networks, genetic algorithms)	Smith et al. (2022)
Optimization of bioethanol fermentation	Machine learning enhances bioethanol production efficiency, reducing costs, and environmental impact, advancing sustainable energy solutions	Machine learning (artificial neural networks, genetic algorithms)	Chen et al. (2022)
Accelerating drug discovery through metabolic engineering	Machine learning expedites secondary metabolite production in microbial strains, potentially improving global healthcare outcomes	Machine learning (random Forest regression, genetic algorithms)	Sharma et al. (2022)

sampling, clinical specimens, and genomic sequencing. Contamination, sequencing errors, and biases can affect data integrity, making robust data preprocessing and quality control crucial (Karstens et al. 2019).

Moreover, AI models, especially deep learning approaches, require substantial quantities of data for effective training (Esteva et al. 2019). Collecting diverse microbial samples for model training is resource intensive and time consuming. Researchers are addressing these challenges through advancements in data collection methods, bioinformatics, and collaboration across research domains (Pasolli et al. 2019).

17.4.2 Interpretability

Interpreting AI models used in microbial data analysis is pivotal but challenging. Often considered "black boxes," these models lack transparency in explaining their decision-making processes. In microbial research, understanding how AI models reach conclusions about microbial interactions, disease associations, or other insights is essential (Holzinger et al. 2017).

Recent research has focused on developing interpretable AI models and methods for microbial data analysis. Explainable AI (XAI) techniques, such as feature importance analysis and rule-based models, aim to enhance transparency and trust in AI-generated findings (Ribeiro et al. 2016). These approaches enable researchers to glean insights from AI models while preserving interpretability.

17.4.3 Ethical and Privacy Concerns

AI's use in microbial data analysis raises ethical and privacy concerns, especially when handling human-associated microbial data. Protecting individual privacy is paramount, necessitating robust data governance and consent frameworks (Cabili et al. 2020). Ensuring that data is anonymized and de-identified is crucial to mitigate privacy risks (Brinegar et al. 2021). Ethical considerations extend to issues like data ownership, informed consent, and responsible data sharing practices (Nebert et al. 2020). Striking a balance between scientific advancement through AI-driven analysis and safeguarding individual privacy remains an ongoing challenge.

17.4.4 Validation and Generalization

Validating AI models in microbial data analysis is complex. Models trained on specific datasets may excel on that data but struggle to generalize to new environments or populations. Achieving robust model generalization is crucial for their utility across diverse microbial ecosystems.

Furthermore, rigorous validation against ground truth data, experimental results, or clinical outcomes is essential to ensure the reliability of AI-generated insights. This validation process is particularly critical in clinical settings where AI-driven decisions impact patient care. Researchers are actively working on solutions to these challenges, employing techniques like cross-validation and external validation datasets to assess model performance and generalization.

Addressing these challenges is paramount for realizing the full potential of AI in microbial data analysis. The latest research and interdisciplinary collaborations are advancing methods and best practices to overcome these obstacles, facilitating breakthroughs in microbiology, biotechnology, and healthcare.

17.5 Future Aspects and Trends

Artificial intelligence (AI) has revolutionized disease diagnosis, ushering in a new era of healthcare that is increasingly accurate, efficient, and personalized. With the rise in the usage of AI, several promising future aspects and trends are poised to shape the landscape of disease diagnosis. Microbial data analysis is a dynamic field, and several exciting future aspects and trends are shaping its evolution.

17.5.1 Integration of Omics Data

Microbial data analysis is increasingly embracing multi-omics data integration. This involves combining data from various "omics" levels, such as genomics, metagenomics, transcriptomics, proteomics, and metabolomics. Integration enables a more holistic view of microbial communities and their functions. For example, integrating metagenomic and meta transcriptomic data can reveal which microbial genes are actively expressed under different conditions (Franzosa et al. 2018). This approach offers a deeper understanding of microbial communities' roles in health, disease, and environmental processes.

17.5.2 Explainable AI in Microbial Data Analysis

As AI techniques become more pervasive in microbial data analysis, the need for interpretability and transparency grows. Explainable AI (XAI) methods are gaining importance to provide insights into how AI models make decisions. In microbiome research, XAI can help researchers understand the features and patterns that drive predictions or classifications, enhancing trust and facilitating the translation of AI-driven findings into actionable insights (Ribeiro et al. 2016).

17.5.3 Personalized Microbiome Analysis

The concept of personalized medicine is extending to the microbiome. Microbiome data can be used to tailor healthcare interventions to individual patients. For instance, by analyzing an individual's microbiome, healthcare providers can prescribe personalized dietary recommendations, probiotics, or antibiotics to treat conditions like irritable bowel syndrome (Jeffery et al. 2012). This trend is likely to grow as the link between the microbiome and various health factors becomes clearer.

17.5.4 Ethical and Regulatory Frameworks

The responsible use of microbiome data is a pressing concern. Ethical considerations, including informed consent and data privacy, are paramount. The development of ethical guidelines and regulatory frameworks for microbiome data collection, sharing, and analysis is essential (Haeuser and Dawson 2020). Researchers and policymakers must work together to establish ethical best practices that respect individual rights and ensure the secure handling of sensitive microbiome data (Waheed et al. 2023).

17.5.5 Collaborative Research and Data Sharing

Collaboration and data sharing initiatives are crucial for advancing microbial data analysis. The establishment of global microbiome data repositories and open science initiatives will foster data sharing, enabling researchers worldwide to access, analyze, and build upon each other's work. Collaboration can lead to more comprehensive datasets and insights into the microbiome's role in diverse ecosystems and health conditions (Knight et al. 2018).

Incorporating these future aspects and trends into microbial data analysis will contribute to a more comprehensive understanding of microbial communities and their implications for human health, environmental sustainability, and biotechnological advancements. Researchers and practitioners in this field should be vigilant in staying current with these developments to make the most of intelligent techniques in microbiome research.

17.6 Conclusion

Microbial data analysis, empowered by intelligent techniques such as machine learning and artificial intelligence, has ushered in a new era of understanding and harnessing the power of microorganisms. From environmental microbiology to biotechnology and healthcare, the applications of these techniques are vast and transformative. As we conclude this exploration, it is evident that the future of microbial data analysis holds immense promise and potential. The significance of microbial data analysis in various fields, including ecology, public health, and biotechnology, cannot be overstated. Intelligent techniques have enabled us to delve deeper into the microbial world, uncovering hidden patterns, predicting behaviors, and accelerating scientific discoveries. From predicting disease risks based on gut microbiome composition to revolutionizing antibiotic resistance prediction and optimizing bioprocesses, intelligent techniques have become indispensable tools for researchers and practitioners alike. However, several challenges remain in the field of microbial data analysis, including data quality and quantity, interpretability of machine learning models, ethical and privacy concerns, and the need for robust validation and generalization. Addressing these challenges will be crucial in ensuring the responsible and ethical use of intelligent techniques in microbial research. Looking ahead, the future of microbial data analysis is bright. Emerging trends such as the integration of omics data, the development of explainable AI for microbial research, personalized microbiome analysis, and the establishment of ethical and regulatory frameworks will shape the landscape of this field. Collaborative research and data sharing will also play a pivotal role in advancing our understanding of microorganisms and their impact on the world.

References

Arif U, Bhatti KH, Ajaib M, Wagay NA, Majeed M, Zeb J, Hameed A, Kiani J (2021) Ethnobotanical indigenous knowledge of Tehsil Charhoi, District Kotli, Azad Jammu and Kashmir, Pakistan. Ethnobot Res Appl 22:1–24. https://doi.org/10.32859/ERA.22.50.1-24

Bashir SM, Altaf M, Hussain T, Umair M, Majeed M, Mangrio WM, Khan AM, Gulshan AB, Hamed MH, Ashraf S, Amjad MS, Bussmann RW, Abbasi AM, Casini R, Alataway A, Dewidar AZ, Al-Yafrsi M, Amin MH, Elansary HO (2023) Vernacular taxonomy, cultural and ethnopharmacological applications of avian and mammalian species in the vicinity of Ayubia National Park, Himalayan Region. Biology 12:4

Bates DW, Saria S, Ohno-Machado L, Shah A, Escobar G (2014) Big data in health care: using analytics to identify and manage high-risk and high-cost patients. Health Aff 33(7):1123–1131. https://doi.org/10.1377/hlthaff.2014.0041

Belkaid Y, Hand TW (2014) Role of the microbiota in immunity and inflammation. Cell 157(1): 121–141

Brinegar K, DeDeo S, Lazer D (2021) Ethnicity and representation in academic machine learning discourse. arXiv:2104.05560

Bzdok D, Ioannidis JP (2019) Exploration, inference, and prediction in neuroscience and biomedicine. Trends Neurosci 42(4):251–262. https://doi.org/10.1016/j.tins.2019.02.003

Cabili MN, Dunagin MC, McClanahan PD, Biaesch A, Padovan-Merhar O, Regev A, Raj A (2020) Localization and abundance analysis of human lncRNAs at single-cell and single-molecule resolution. Genes Dev 34(7–8):440–451

Chen Y, Wang Q, Jiang H et al (2022) Machine learning models for enhanced bioethanol production through fermentation process optimization. Bioenergy Res:1–12

Esteva A, Robicquet A, Ramsundar B, Kuleshov V, DePristo M, Chou K, Dean J (2019) A guide to deep learning in healthcare. Nat Med 25(1):24–29

Falkowski PG, Fenchel T, Delong EF (2008) The microbial engines that drive Earth's biogeochemical cycles. Science 320(5879):1034–1039. https://doi.org/10.1126/science.1153213

Forslund K, Hildebrand F, Nielsen T et al (2015) Disentangling type 2 diabetes and metformin treatment signatures in the human gut microbiota. Nature 528(7581):262–266

Franzosa EA, Hsu T, Sirota-Madi A, Shafquat A, Abu-Ali G, Morgan XC, Huttenhower C (2018) Sequencing and beyond: integrating molecular 'omics' for microbial community profiling. Nat Rev Microbiol 16(11):591–606

Guttenberg N, Warfield K, Pickett BE, Ledogar RJ (2020) Brain size and network properties of human population genetic networks. Nat Commun 11(1):1–12. https://doi.org/10.1038/s41467-020-19135-w

Haberbeck LU, Wang X, Michiels C, Devlieghere F, Uyttendaele M, Geeraerd AH (2017) Cross-protection between controlled acid-adaptation and thermal inactivation for 48 *Escherichia coli* strains. Int J Food Microbiol 241:206–214

Haeuser E, Dawson N (2020) Microbiome data should be regulated as personal data. Nat Med 26(12):1806–1808

Haq SM, Yaqoob U, Majeed M, Amjad MS, Hassan M, Ahmad R, Morales-de la Nuez A (2022) Quantitative ethnoveterinary study on plant resource utilization by indigenous communities in high-altitude regions. Front Vet Sci 9:94404

Hassan M, Haq SM, Ahmad R, Majeed M, Sahito HA, Shirani M, Mubeen I, Aziz MA, Pieroni A, Bussmann RW, Alataway A, Dewidar AZ, Al-Yafrsi M, Elansary HO, Yessoufou K (2022a) Traditional use of wild and domestic fauna among different ethnic groups in the Western Himalayas? Cross cultural analysis. Animals 12:17

Hassan M, Haq SM, Majeed M, Umair M, Sahito HA, Shirani M, Waheed M, Aziz R, Ahmad R, Bussmann RW, Alataway A, Dewidar AZ, El-Abedin TKZ, Al-Yafrsi M, Elansary HO, Yessoufou K (2022b) Traditional food and medicine: ethno-traditional usage of fish fauna across the valley of Kashmir: a Western Himalayan region. Diversity 14:6

Holzinger A, Langs G, Denk H, Zatloukal K, Müller H (2017) Causability and explainability of artificial intelligence in medicine. Wiley Interdiscip Rev Data Mining Knowl Discov 7(6):e1212

Hu Y, Yang X, Qin J, Lu N, Cheng G, Wu N, Lin L (2020) Metagenome-wide analysis of antibiotic resistance genes in a large cohort of human gut microbiota. Nat Commun 11(1):1–12

Jamil MD, Waheed M, Akhtar S, Bangash N, Chaudhari SK, Majeed M, Hussain M, Ali K, Jones DA (2022) Invasive plants diversity, ecological status, and distribution pattern in relation to edaphic factors in different habitat types of district Mandi Bahauddin, Punjab, Pakistan. Sustainability (Switzerland) 14:20

Jeffery IB, O'Toole PW, Öhman L, Claesson MJ, Deane J, Quigley EM, Simrén M (2012) An irritable bowel syndrome subtype defined by species-specific alterations in faecal microbiota. Gut 61(7):997–1006

Jeong H, Tombor B, Albert R, Oltvai ZN, Barabási AL (2000) The large-scale organization of metabolic networks. Nature 407(6804):651–654. https://doi.org/10.1038/35036627

Kamel Boulos MN, Resch B, Crowley DN, Breslin JG, Sohn G, Burtner R, Lu Z (2011) Crowdsourcing, citizen sensing and sensor web technologies for public and environmental health surveillance and crisis management: trends, OGC standards and application examples. Int J Health Geogr 10(1):67. https://doi.org/10.1186/1476-072X-10-67

Karstens L, Asquith M, Davin S, Fair DA, Gregory WT, Wolfe AJ, McWeeney SK (2019) Controlling for contaminants in low-biomass 16S rRNA gene sequencing experiments. mSystems 4(4):e00290–e00219

Khan AM, Li Q, Saqib Z, Khan N, Habib T, Khalid N, Majeed M, Tariq A (2022) MaxEnt modelling and impact of climate change on habitat suitability variations of economically important Chilgoza pine (*Pinus gerardiana* wall.) in South Asia. Forests 13:5

Khoja AA, Haq SM, Majeed M, Hassan M, Waheed M, Yaqoob U, Bussmann RW, Alataway A, Dewidar AZ, Al-Yafrsi M, Elansary HO, Yessoufou K, Zaman W (2022) Diversity, ecological and traditional knowledge of pteridophytes in the Western Himalayas. Diversity 14:8

Knight R, Vrbanac A, Taylor BC, Aksenov A, Callewaert C, Debelius J, Flores R (2018) Best practices for analysing microbiomes. Nat Rev Microbiol 16(7):410–422

Lam TB, Hultcrantz M, Wallis C et al (2022) Robot-assisted radical prostatectomy versus open radical prostatectomy: a systematic review and meta-analysis. Eur Urol 17:2617–2631. https://doi.org/10.1016/j.eururo.2022.01.057

Laxminarayan R, Duse A, Wattal C, Zaidi AK, Wertheim HF, Sumpradit N, Cars O (2013) Antibiotic resistance—the need for global solutions. Lancet Infect Dis 13(12):1057–1098. https://doi.org/10.1016/S1473-3099(13)70318-9

Lucas TC (2020) A translucent box: interpretable machine learning in ecology. Ecol Monogr 90(4): e01422

Lutgring JD, Machado MJ, Benitez AJ (2020) Evaluation of the VITEK® 2 automated susceptibility testing system against carbapenemase-producing enterobacteriaceae with a modified carbapenem inactivation method. J Clin Microbiol 58(5):e02088–e02019. https://doi.org/10.1128/JCM.02088-19

Majeed M, Bhatti KH, Amjad MS, Abbasi AM, Rashid A, Nawaz F, Ahmad KS (2020a) Ethno-veterinary practices of Poaceae taxa in Punjab, Pakistan

Majeed M, Bhatti KH, Amjad MS, Abbasi M, Id RWB, Nawaz F, Rashid A, Mehmood A, Id MM, Khan WM, Id SA (2020b) Ethno-veterinary uses of Poaceae in Punjab, Pakistan. PLoS One 15: e0241705. https://doi.org/10.1371/journal.pone.0241705

Majeed M, Bhatti KH, Amjad MS (2021a) Impact of climatic variations on the flowering phenology of plant species in Jhelum district, Punjab, Pakistan. Appl Ecol Environ Res 19:5

Majeed M, Bhatti KH, Pieroni A, Sõukand R, Bussmann RW, Khan AM, Chaudhari SK, Aziz MA, Amjad MS (2021b) Gathered wild food plants among diverse religious groups in Jhelum District, Punjab, Pakistan. Foods 10:3

Majeed M, Tariq A, Anwar MM, Khan AM, Arshad F, Mumtaz F, Farhan M, Zhang L, Zafar A, Aziz M, Abbasi S, Rahman G, Hussain S, Waheed M, Fatima K, Shaukat S (2021c) Monitoring of land use? And cover change and potential causal factors of climate change in Jhelum District, Punjab, Pakistan, through GIS and multi-temporal satellite data. Land 10:10

Majeed M, Lu L, Haq SM, Waheed M, Sahito HA, Fatima S, Aziz R, Bussmann RW, Tariq A, Ullah I, Aslam M (2022a) Spatiotemporal distribution patterns of climbers along an abiotic gradient in Jhelum District, Punjab, Pakistan. Forests 13:8

Majeed M, Tariq A, Haq SM, Waheed M, Anwar MM, Li Q, Aslam M, Abbasi S, Mousa BG, Jamil A (2022b) A detailed ecological exploration of the distribution patterns of wild Poaceae from the Jhelum District (Punjab), Pakistan. Sustainability (Switzerland) 14:7

Majeed M, Lu L, Anwar MM, Tariq A, Qin S, El-Hefnawy ME, El-Sharnouby M, Li Q, Alasmari A (2023) Prediction of flash flood susceptibility using integrating analytic hierarchy process (AHP) and frequency ratio (FR) algorithms. Front Environ Sci 10:1037547. https://doi.org/10.3389/fenvs.2022.1037547

Matchado MS et al (2021) Network analysis methods for studying microbial communities: a mini review. Comput Struct Biotechnol J 19:2687–2698. https://doi.org/10.1016/j.csbj.2021.05.001

Nebert DW, Zhang G, Vesell ES, Dixon K (2020) The human cytochrome P450 (CYP) allele nomenclature database: a one-stop site for nomenclature, functional-allelic-variant reference, and genotype–phenotype relations. Hum Genomics 14(1)

Pasolli E, Asnicar F, Manara S et al (2019) Extensive unexplored human microbiome diversity revealed by over 150,000 genomes from metagenomes spanning age, geography, and lifestyle. Cell 176(3):649–662

Qu K, Guo F, Liu X, Lin Y, Zou Q (2019) Application of machine learning in microbiology. Front Microbiol 10:827

Ribeiro MT, Singh S, Guestrin C (2016) “Why should I trust you?” Explaining the predictions of any classifier. In: Proceedings of the 22nd ACM SIGKDD international conference on knowledge discovery and data mining, pp 1135–1144

Sahayasheela VJ, Lankadasari MB, Dan VM, Dastager SG, Pandian GN, Sugiyama H (2022) Artificial intelligence in microbial natural product drug discovery: current and emerging role. Nat Prod Rep 39:2215

Sharma A, Gupta A, Patel D et al (2022) Machine learning-driven metabolic engineering for enhanced secondary metabolite production in streptomyces. 3 Biotech 12:449

Sinha D, Maurya AK, Abdi G, Majeed M, Agarwal R, Mukherjee R, Ganguly S, Aziz R, Bhatia M, Majgaonkar A, Seal S, Das M, Banerjee S, Chowdhury S, Adeyemi SB, Chen JT (2023) Integrated genomic selection for accelerating breeding programs of climate-smart cereals. Genes 14:7

Smith J, Brown L, Garcia M et al (2022) Integrating artificial intelligence for enhanced bioremediation of industrial contaminated sites. Environ Sci Eng 5:100016

Tang H, Zhao X, Dube L, Boudreau RA, Fang R, Xu L (2021) An integrated smartphone-based platform for rapid antimicrobial susceptibility testing. Adv Sci 8(18):2101407. https://doi.org/10.1002/advs.202101407

Tassadduq SS, Akhtar S, Waheed M, Bangash N, Nayab DE, Majeed M, Abbasi S, Muhammad M, Alataway A, Dewidar AZ, Elansary HO, Yessoufou K (2022) Ecological distribution patterns of wild grasses and abiotic factors. Sustainability (Switzerland) 14:18

Thompson LR, Sanders JG, McDonald D, Amir A, Ladau J, Locey KJ, Knight R (2017) A communal catalogue reveals Earth's multiscale microbial diversity. Nature 551(7681):457–463

Tirtawijaya G, Meinita MDN, Marhaeni B, Haque MN, Moon IS, Hong Y-K (2018) Neurotrophic activity of the Carrageenophyte *Kappaphycus alvarezii* cultivated at different depths and for different growth periods in various areas of Indonesia. Evid Based Complement Alternat Med 2018:1098076

Ullah I, Aslam B, Shah SHIA, Tariq A, Qin S, Majeed M, Havenith HB (2022) An integrated approach of machine learning, remote sensing, and GIS data for the landslide susceptibility mapping. Land 11:8

Waheed M, Arshad F, Majeed M, Fatima S, Mukhtar N, Aziz R, Mangrio WM, Almohamad H, Dughairi AA, Al-Mutiry M, Abdo HG (2022) Community structure and distribution pattern of woody vegetation in response to soil properties in semi-arid Lowland District Kasur Punjab, Pakistan. Land 11:12

Waheed M, Arshad F, Majeed M, Haq SM, Aziz R, Bussmann RW, Ali K, Subhan F, Jones DA, Zaitouny A (2023) Potential distribution of a noxious weed (*Solanum viarum* Du-nal), current status, and future invasion risk based on MaxEnt modeling. In: Geology, ecology, and landscapes. Taylor & Francis, p 1. https://doi.org/10.1080/24749508.2023.2179752

Yao M, Liu Z, Hou L et al (2021) Machine learning models for predicting nutrient concentrations and cycling rates in polluted river ecosystems. Glob Biogeochem Cycles 35(9): e2021GB007146

Zhang B, Sun L, Zheng W et al (2021) Machine learning reveals the environmental and management impacts on soil microbial communities in Chinese tea plantations. Environ Microbiol 23(7):3435–3449

Zhou Y, Gao H, Mihindukulasuriya KA, La Rosa PS, Wylie KM, Vishnivetskaya T, Jansson JK (2018) Biogeography of the ecosystems of the healthy human body. Genome Biol 19(1):1–14. https://doi.org/10.1186/s13059-018-1556-2

Zmora N, Zilberman-Schapira G, Suez J et al (2018) Personalized gut mucosal colonization resistance to empiric probiotics is associated with unique host and microbiome features. Cell 174(6):1388–1405

Chapter 18
Early Crop Disease Identification Using Multi-fork Tree Networks and Microbial Data Intelligence

S. S. Ittannavar, B. P. Khot, Vibhor Kumar Vishnoi, Swati Shailesh Chandurkar, and Harshal Mahajan

Abstract In the early stages of crop disease, timely acquisition of information about crop diseases, determination of the causes and severity of infection, and targeted treatment are essential for preventing a decline in crop yield caused by disease spread. To address the issue of low accuracy in traditional deep learning networks for early crop disease identification, we propose an improved attention mechanism-based multi-fork tree network method. This method combines the attention mechanism with a residual network to recalibrate disease feature maps, resulting in SMLP_Res (Squeeze-Multi-layer Perceptron ResNet). Additionally, we extend the high-feature extraction-capable SMLP_ResNet (Squeeze-Multi-Layer Perceptron ResNet) network with a multi-fork tree structure, simplifying the task of early crop disease identification and effectively extracting early disease features. In our experiments, we use two datasets, Plant Village and AI Challenger 2018, to train and validate three network models: 18-layer ResNet, SE_ResNet, and SMLP_ResNet, as well as their equivalent multi-fork tree structure models, to assess the impact of SMLP_Res and the multi-fork tree structure on crop disease identification models. The experimental analysis shows that the three network models, 18-layer ResNet, SE_ResNet, and SMLP_ResNet, all achieve an accuracy rate of over 99% in disease identification on the Plant Village dataset, where disease features are more

S. S. Ittannavar (✉) · B. P. Khot
ECE Department, Hirasugar Institute of Technology, Belgaum, India

V. K. Vishnoi
College of Computing Sciences and Information Technology, Teerthanker Mahaveer University, Moradabad, India
e-mail: vibhor.computers@tmu.ac.in

S. S. Chandurkar
Pimpri Chinchwad College of Engineering, Pune, India
e-mail: swati.chandurkar@pccoepune.org

H. Mahajan
Indira College of Engineering and Management, Pune, India
e-mail: mahajan.harshal@indiraicem.ac.in

A. Khamparia et al. (eds.), *Microbial Data Intelligence and Computational Techniques for Sustainable Computing*, Microorganisms for Sustainability 47,
https://doi.org/10.1007/978-981-99-9621-6_18

pronounced. However, their accuracy rates on the early disease dataset AI Challenger 2018 do not exceed 87%. SMLP_ResNet, due to the inclusion of the SMLP_Res module, provides more comprehensive feature extraction for crop disease data, resulting in better detection performance. Among the three early disease identification models with multi-fork tree structures, all models show significant improvements in accuracy on the AI Challenger 2018 dataset. The multi-fork tree SMLP_ResNet outperforms the other two models, achieving the best performance with a cherry early disease identification accuracy rate of 99.13%. The proposed multi-fork tree SMLP_ResNet crop early disease identification model simplifies the recognition task, suppresses noise transmission, and achieves high accuracy.

Keywords Crop disease identification · Early detection · Multi-fork tree networks · Microbial data intelligence · Deep learning · Feature extraction · Disease classification

18.1 Introduction

Diseases represent a significant constraint on the quality and sustainable growth of crop output. During the initial phases of disease infestation, it is crucial to promptly acquire illness-related data and accurately diagnose the underlying factors contributing to the infection. Through precise evaluation of leaf disease severity, implementation of suitable chemical interventions, and reduction of pesticide usage, it is possible to mitigate environmental pollution and successfully prevent and manage illnesses, hence averting a decrease in crop productivity resulting from disease propagation. Crop diseases can arise due to a multitude of reasons, and they have the potential to exhibit a range of symptoms under varying environmental conditions and at different developmental phases of the crop. The conventional method of disease identification in agriculture involves the expertise of agricultural professionals who possess substantial practical knowledge. This approach is characterised by its time-consuming nature and a tendency to yield erroneous results.

The advent of information technology has facilitated the utilisation of image processing and machine learning methodologies in the field of crop disease diagnostics. These techniques have proven to be highly valuable in enabling rapid, precise, and non-invasive identification of diseases affecting crops. In the realm of conventional disease identification, scholars have employed infrared imaging technology. The utilisation of thermal infrared imaging by the author (Karanth et al. 2023) facilitated the acquisition of temperature data pertaining to both healthy and diseased areas of rapeseed plants, ultimately leading to the successful early diagnosis of rapeseed club root disease. The author (Gavidia et al. 2023) utilised hyper spectral imaging to identify and categorise rice blast disease. This was achieved by employing Principal Component Analysis (PCA) and Competitive Adaptive Reweighted Algorithm (CARS) to choose feature variables for disease grading clustering algorithms. These algorithms are commonly employed to group data with comparable characteristics. However, they encounter difficulties such as sensitivity to sample centroids, substantial fluctuations in accuracy (Kutyauripo et al. 2023), and intricate feature distance descriptions. Consequently, scholars have made enhancements to clustering algorithms in order to get disease identification. The SAK-means approach, as proposed by the author (Holzinger et al. 2023), utilises Hadoop parallel processing to autonomously ascertain the appropriate number of sample categories and identify photos related to citrus red spider mite disease. Nevertheless, this methodology is limited in its ability to detect many diseases due to the dataset's exclusive inclusion of single-crop, single-disease information. The researchers utilised a dataset consisting of four distinct diseases, namely angular leaf spot, anthracnose, bacterial wilt, and powdery mildew, along with a set of healthy leaves (Harner et al. 2023; Jagadeesan et al. 2019; Maier-Hein et al. 2022). They successfully conducted multi-class disease recognition by employing a Support Vector Machine (SVM) with a radial basis function kernel. Support Vector Machines (SVMs) are limited to addressing binary classification tasks, hence introducing challenges in the training procedure.

All the above-mentioned disease recognition methods are limited to single crops (Ren et al. 2022; Rowan and Galanakis 2020). When the data environment becomes complex, and the number of diseases to be identified increases, disease recognition accuracy still needs improvement. Convolutional neural networks (CNNs) have made significant strides in the field of computer vision, including image classification, object detection, and super-resolution reconstruction, gradually integrating deep learning techniques into crop disease recognition. Using CNNs has yielded favourable results in plant leaf (Rowan and Galanakis 2020) and disease recognition (Nychas et al. 2016). The introduction of the AlexNet network (Saeed et al. 2022; Varga and Csukas 2022) greatly improved the accuracy of crop disease recognition compared to traditional algorithms. Building upon this, methods such as VGG (Rowan et al. 2022), ResNet (Pakseresht et al. 2023), and GoogLeNet (Mrabet 2023) have improved network feature extraction capabilities by altering network depth and structure. Researchers have applied DCNN (Deep Convolutional Neural Networks) in banana disease (Joseph 2023), corn disease (Fernández-Ríos et al. 2022), and citrus disease (De Roever 1998) recognition, achieving good results. However, these deep convolutional network models achieve disease recognition only when visual features are prominent and are not suitable for early disease recognition in multiple crops.

To construct a highly accurate early crop disease recognition model, this study conducts experiments using the Plant Village dataset (Marvin et al. 2009), which contains crop disease data with prominent features, and the AI Challenger dataset, which exhibits significant differences in hierarchical features. We introduce attention mechanisms into the crop disease recognition model to suppress background noise information and select effective disease features. We also adopt tree structures as classification algorithms in the data structure domain (Das et al. 2022; Tirkey et al. 2023; Nigam et al. 2023; Abbasi et al. 2023) and integrate them with an improved attention mechanism to create a disease recognition model. This aims to enhance the ability to identify early crop diseases, achieve early disease detection and treatment, and reduce pesticide usage.

18.2 Crop Disease Data and Disease Recognition Methods

18.2.1 Experimental Data

The quality of disease diagnosis is heavily influenced by the characteristics of the sample data. The bigger the variety of disease cases, the smaller the distinctions across samples, and the more difficult disease detection becomes. We trained multi-crop disease detection neural network models on Plant Village and AI Challenger 2018, two datasets with varying degrees of disease recognition difficulty, and evaluated their accuracy and resilience (Islam et al. 2023; Zhou et al. 2021; Reumaux et al. 2023).

Apple, blueberry, cherry, corn, grape, peach, potato, raspberry, soya bean, pumpkin, strawberry, and tomato are just a few of the crops whose illnesses are represented in the 54,309 crop photos available in the Plant Village dataset. Moulds, fungus, and bacteria all have a role in causing these illnesses. In addition, there are 14 photos of healthy crop leaves included in the collection. There are 38 different types of data in this set (Gao et al. 2021; Kaveney et al. 2023).

There are almost 30,000 photos of agricultural illnesses in the AI Challenger 2018 dataset. These diseases affect a total of ten different crops: apples, cherries, grapes, citrus fruits, peaches, strawberries, tomatoes, chili peppers, corn, and potatoes. There are a total of 59 classes in the dataset, which are based on the kind of crop, the type of disease, and the severity of the disease (Xiong et al. 2020; Modi et al. 2023; Dutta et al. 2022).

The parameters of the two datasets are listed in Table 18.1.

Figure 18.1 displays a comparison of disease images from Plant Village and AI Challenger 2018. A noticeable similarity can be observed in the disease severity of late-stage diseases between AI Challenger 2018 and Plant Village. In the case of the ten crops in AI Challenger 2018, each crop exhibits different disease conditions, and diseases are classified into early and late stages based on their characteristics. Late-stage diseases are easily recognisable due to their distinct features, but the similarity between the features of healthy leaves and early-stage disease leaves is relatively high. Additionally, there is a wide variety of diseases, posing a significant challenge to disease recognition. Therefore, selecting datasets with distinct disease feature representations validates the effectiveness of the early disease recognition model in recognising diseases efficiently (Degu et al. 2023; Mensah et al. 2023; Noshiri et al. 2023; Kendler et al. 2022).

Table 18.1 Crop disease experimental datasets

Dataset	Plant Village	AI Challenger 2018
Number of samples	54,306	35,861
Number of classes	38	59
Number of cores	14	10
Number of diseases	24	27

Fig. 18.1 Comparison of images in the experimental datasets

18.2.2 Disease Recognition Methods

When constructing network models, it's common to opt for deep models to enhance learning capabilities. However, as networks become deeper, their structure becomes more complex and redundant, leading to issues like overfitting and gradient vanishing (Zhao et al. 2023). In the process of feature extraction, early layers of the network might weaken feature information, resulting in feature loss and network degradation. Residual Networks (ResNets) introduced the concept of residual structures, combining shortcut connections and identity mappings to fuse features from previous layers, thus extracting richer feature information. The Squeeze and Excitation Networks (SENet) model enhances model performance by focusing on relationships between feature channels, granting Convolutional Neural Networks (CNNs) the ability to recognise critical features. The core idea of the SE attention mechanism lies in the compression (Squeeze) and excitation (Excitation) phases, where it automatically determines the importance of different channels to enhance feature expression. Combining the characteristics of these two models, an improved attention mechanism called SMLP_Res is proposed for early crop disease recognition. This module enhances the SE module using multi-layer perceptron (MLP) and residual structures to differentiate the importance of feature map channels, reduce feature loss during transmission, and incorporate the concept of multiple trees by breaking down complex networks into simpler ones to improve early disease recognition accuracy (Agarwal et al. 2020).

18.2.2.1 Improved Attention Module (SMLP)

The SE attention mechanism extracts detailed features from the entire feature map through two stages: compression and excitation. Because early disease recognition places a high demand on the network's ability to extract details, the SMLP module replaces the excitation part of SE with MLP. MLP leverages the mathematical property of approximating arbitrary function mappings, learning the importance of each channel in the feature map, assigning corresponding weights, and enhancing the expression of significant features while weakening noise information representation, thus separating mixed features (SMLP). The structure of the SMLP module is depicted in Fig. 18.2.

GAP (Global Average Pooling) in SMLP encodes global information while preserving the feature information of the input feature map U. Fsq(uc) compresses the feature map U into a $1 \times 1 \times C$ format and encodes each channel uc to integrate all feature information. The expression is as follows:

$$G_{rq}(d) = \frac{1}{j \times X_i} \sum\nolimits_{i=1}^{j} \sum\nolimits_{j=1}^{X} v_c(i,j). \tag{18.1}$$

The encoded feature map is then input into the MLP to learn nonlinear relationships between channels. The structure of MLP is illustrated in Fig. 18.3, where the $1 \times 1 \times D$ feature map is mapped to y^{l-1} and serves as the input for layer l. This layer has n nodes, corresponding weights X^l, bias vectors c^l, and al as the output feature map for layer l. σ represents the activation function, and the expression for the l-th layer output in MLP is as follows:

$$\alpha^i = \sigma\left(X^i y^{i-1} + c^i\right) \tag{18.2}$$

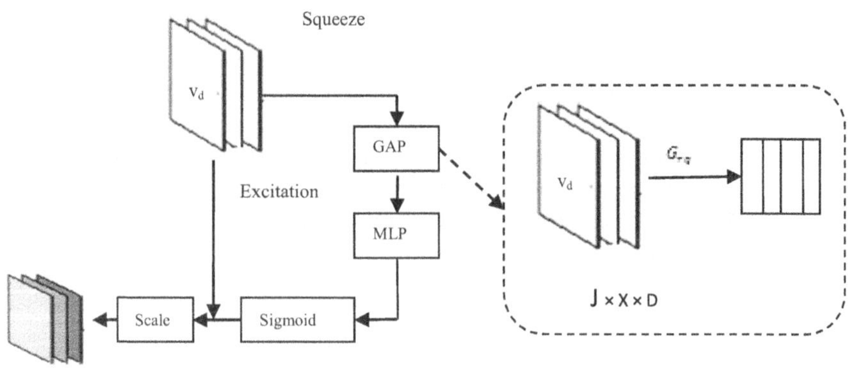

Fig. 18.2 SMLP module

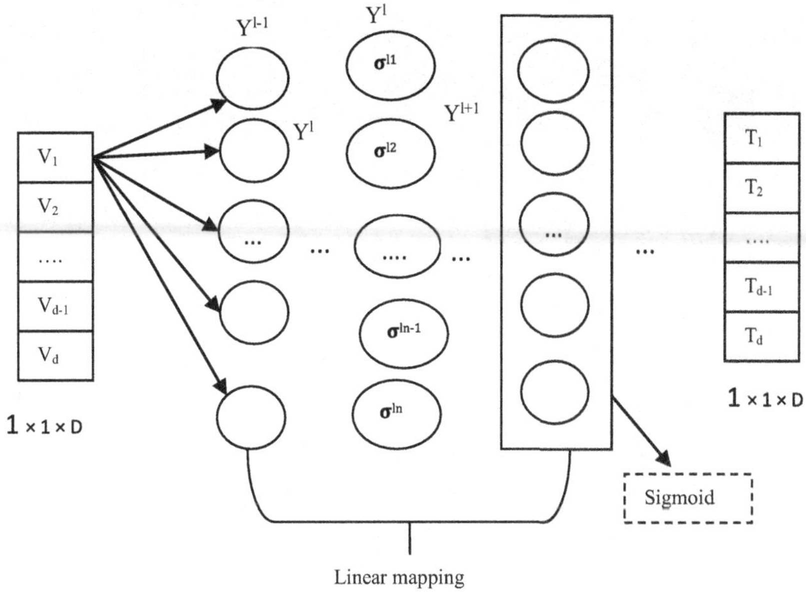

Fig. 18.3 Schematic diagram of MLP structure

$$\alpha^i = X^i y^{i-1} + c^i. \tag{18.3}$$

To represent the nonlinear dependencies between different channels in the feature map, the Sigmoid function is employed for nonlinear mapping, preserving the diversity of channel relationships, as shown in Eq. (18.4). This function maps the nonlinear dependencies between channels to a weight vector s in a multi-dimensional space, where $t,t \in (0,1)$. G_{Scale} applies s to the input feature map V, constructing a feature map with weight information.

$$\text{Sigmoid}\left(\alpha^i\right) = \frac{1}{1 + \mathrm{e}^{-\left(\alpha^i\right)}} \tag{18.4}$$

$$G_{\text{scale}}(t, x) = t \cdot x. \tag{18.5}$$

SMLP adjusts feature information at the channel level, achieving feature enhancement and recalibration of feature maps.

18.2.2.2 Improved Residual Network Structure (SMLP_ResNet)

Deep networks can extract richer feature information, but as the network depth increases, issues like gradient vanishing may occur, leading to a decrease in the recognition accuracy of network models. The residual structure, which incorporates

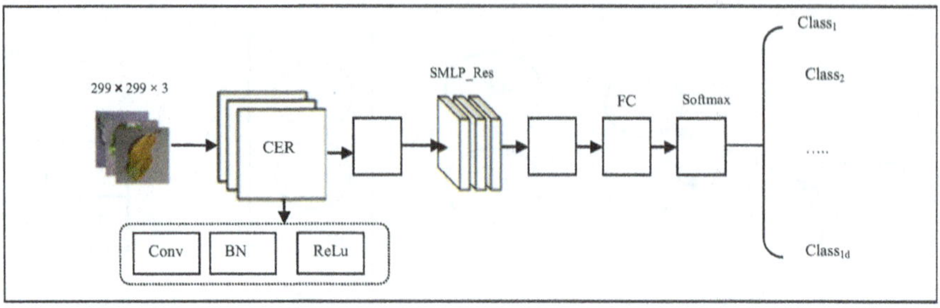

(a) SMLP_ResNet model

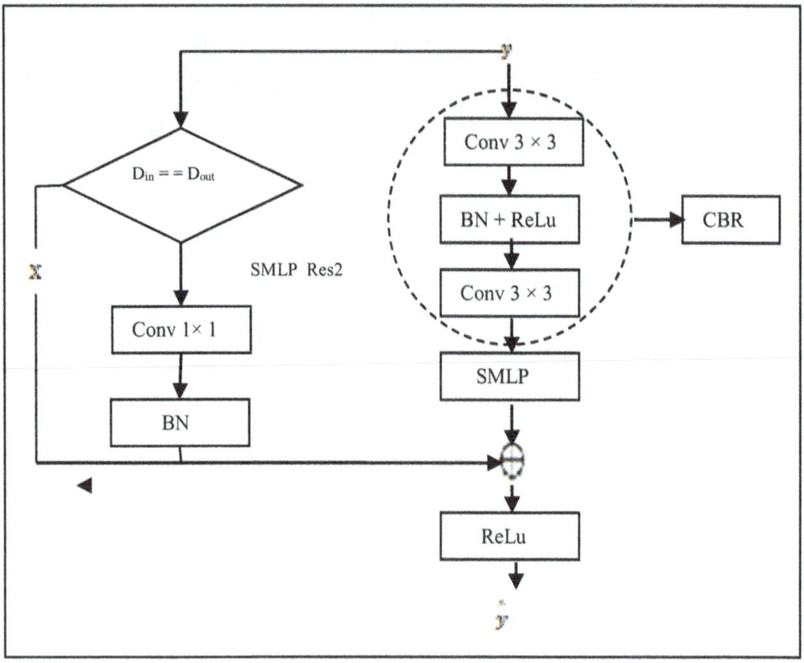

(b) SMLP_Res module

Fig. 18.4 SMLP_ResNet model structure diagram. (**a**) SMLP_ResNet model. (**b**) SMLP_Res module

shortcut connections and identity mappings, continuously merges feature information from previous layers, effectively alleviating the network degradation caused by depth and enhancing the feature extraction capability of the network model.

The combination of the residual module and SMLP is defined as the SMLP_Res residual module. Based on this module, Batch Normalization (Manavalan 2020; Rani and Singh 2022; Bondad et al. 2023) and ReLU (Rectified Linear Unit) activation functions, among others, are integrated to construct the SMLP_ResNet classification network model, as shown in Fig. 18.4a.

Table 18.2 SMLP_ResNet disease model parameter table

Parameter	SMLP_ResNet18	SMLP_ResNet50	SMLP_ResNet101
n	2	3	3
m	2	4	4
l	2	6	23
w	2	3	3

In Fig. 18.4a, input images, after data augmentation and resizing to a uniform size, are fed into the SMLP_ResNet classification network model. SMLP_ResNet consists of five modules: CBR, SMLP_Res, Fully Connected (FC), pooling layers, and Softmax. CBR and SMLP_Res perform different levels of disease feature extraction. During feature extraction, two pooling methods, maximum pooling and average pooling, are used. Maximum pooling preserves texture features of feature maps, while average pooling retains the overall feature information of the images. FC is used to learn feature information from different categories, and the Softmax function accomplishes the mapping to achieve the final category output.

Figure 18.4b provides the SMLP_Res module, which has two structures: SMLP_Res1 and SMLP_Res2. SMLP_Res1 represents cases where the input and output feature map channels are the same, i.e. $d_{\text{in}} = d_{\text{out}}$, and it directly fuses the input and output feature maps at the channel level. SMLP_Res2 represents cases where $d_{\text{in}} \neq d_{\text{out}}$. It adjusts the number of channels in the feature map using 1×1 convolutions to make $d_{\text{in}} = d_{\text{out}}$ before merging the feature maps. The parameters n, m, l, and w are used to specify the quantity of SMLP_Res1 and SMLP_Res2 structures when building the model. Different network depths can be constructed by setting these four parameters. For example, to build an 18-layer SMLP_ResNet network model, you can set $n = 2$, $m = 2$, $l = 2$, and $w = 2$. The specific parameter settings are listed in Table 18.2.

Table 18.2 lists three commonly used network model depths: 18 layers, 50 layers, and 101 layers, along with their parameter settings. When applying convolutional kernels to feature maps, taking a single channel as an example, Eq. (18.6) represents the convolution operation on the feature map y of channel y_c.

$$b_{i,j} = \sum_{m=0}^{f-1} \sum_{n=0}^{f-1} w_{m,n} y_{d_{i+m,j+n}} + w_c. \tag{18.6}$$

Here, $b_{i,j}$ represents the pixel at the i-th row and j-th column of the feature map after convolution, f is the size of the convolutional kernel, $w_{m,n}$ represents the weight parameters of the convolutional kernel, and $y_{d_{i+m,j+n}}$ represents the pixel at the $(i + m)$-th row and $(j + n)$-th column of the input feature channel y_c. w_c is the bias for fine-tuning the convolution result. After passing through the Convolution (Conv) layer, the feature map undergoes normalisation in the Batch Normalization (BN) layer, mitigating issues like slow convergence and low generalisation caused by sample distribution. For a batch of sample data $\beta = \{y_1, \ldots, y_m\}$,

$$\mu_\beta \leftarrow \frac{1}{m}\sum\nolimits_{i=1}^{m} y_i \tag{18.7}$$

$$\sigma_\beta^2 \leftarrow \frac{1}{m}\sum\nolimits^{m} (y_i - \mu_\beta)^2 \tag{18.8}$$

$$\widehat{y}_i \leftarrow \frac{y_i - \mu_\beta}{\sqrt{\sigma_\beta^2 + \varepsilon}} \tag{18.9}$$

$$x_i \leftarrow \gamma \widehat{y}_i + \beta \equiv BN_{\gamma,\beta}(y_i). \tag{18.10}$$

Equations (18.7) and (18.8) calculate the mean and variance of the input batch data, denoted as $\mu\beta$ and $\sigma^2\beta$, respectively. Equation (18.9) is used for data normalisation, where ε is a correction parameter to prevent division by zero. However, since xi is constrained below the normal distribution, it affects the expression of data information. To address this issue, scaling factor γ and shift factor β are introduced to enhance network expressiveness, as shown in Eq. (18.10). The feature map processed by the BN layer is activated by the ReLU function, as shown in Eq. (18.11).

$$\text{ReLU} = \max(0, y), y \in S. \tag{18.11}$$

The feature map processed by CBR and BN is then input into the SMLP module to calibrate the feature map. The choice of SMLP_Res1 or SMLP_Res2 determines how the feature map x is fused at the channel level. After activation by the ReLU function, the output feature map $x\wedge$ combines information from multiple sources, achieving feature extraction in the SMLP_ResNet network.

18.2.2.3 Disease Recognition Model Based on Multi-branch SMLP_ResNet

Most crop disease recognition models use separate network models. When the number of crops and disease types increases, the classification count grows exponentially, increasing the network's burden and decreasing disease recognition accuracy. Crop disease recognition tasks can be decomposed into two classification tasks: crop recognition and disease recognition. Combining a multi-branch tree model, the crop recognition model serves as the root node for crop classification, while the disease recognition model serves as child nodes for recognising diseases in the same crop. The multi-branch tree model employs multiple classification network models for early disease recognition, improving disease recognition accuracy.

Classification models implement crop and disease recognition tasks, and a loss function is used in the classification task to measure the distance between predicted results and true categories. By adjusting model parameters, the loss function value is minimised. CNN models output results in vector structure rather than probability distribution. Therefore, the Softmax function is used to convert the vector into a probability distribution, and the loss is calculated using the cross-entropy loss function (Eq. 18.12).

$$\text{loss} = -\sum_i x_i \ln b_i, \tag{18.12}$$

where b_i is the i-th prediction value output by Softmax, and x_i represents the corresponding true classification. The loss value is backpropagated to adjust the network model parameters. Based on the SMLP_ResNet classification model, the structure of the early disease recognition model with multi-branch SMLP_ResNet.

The SMLP_ResNet architecture is designed as a hierarchical network consisting of multiple layers, with each layer representing a tree structure. The first network is tasked with the identification of different crop varieties; however the subsequent network is specifically built to acquire knowledge on the detection of diverse diseases that impact a singular crop. The crop disease data undergoes a hierarchical processing approach, wherein it is divided into two unique categories: crop type data and crop disease data. Following this, the data pertaining to the disease is divided into two sets: a training set and a test set. This division is carried out using a random partitioning technique, with a ratio of 4:1. The training dataset was employed to train the multi-tree SMLP_ResNet model. The test dataset was utilised to assess the efficacy of the trained multi-tree SMLP_ResNet model in detecting the influence of the model on agricultural crops. The first phase in the procedure entails feeding augmented images of tomato diseases into the multi-tree SMLP_ResNet model to facilitate the early detection of illnesses. The first layer of the network is responsible for recognising the input crop as a tomato and then relaying this information to the subsequent layer of the network. The secondary network assumes the responsibility of detecting many health issues, including both the initial and advanced phases of powdery mildew disease. In order to assess the precision of the multi-tree SMLP_ResNet model's predictions, the cross-entropy function is utilised to quantify the disparity between the true label assigned to a picture and the label projected by the model. The network parameters are subsequently modified utilising the gradient descent technique. When the observation of losing value occurs, the model is stored when reaching a predetermined threshold. To boost the sickness detection network's potential for generalisation, a series of image processing techniques are applied, including augmentation and standardisation. These techniques result in a consistent image size of 224 × 224 × 3. The augmented dataset is employed as input to train the multi-tree SMLP_ResNet network model for the aim of model development. The SMLP_ResNet network model is improved through the integration of several trees, hence enhancing the network's capacity to extract features. This functionality allows individual sub-networks to effectively manage distinct learning tasks, leading to enhanced accuracy in total network recognition. Furthermore, the proposed model improves the precision of early illness categorisation and minimises the influence of feature similarity and other variables on disease identification accuracy through the reduction of data kinds and quantity.

18.3 Experiment

The crop disease recognition model in this experiment is trained using an NVIDIA Tesla P100 model GPU. The video memory capacity is 16 GB. The construction of a training platform for deep learning algorithms is founded upon the utilisation of the Ubuntu 16.04 64-bit operating system and the PyTorch frameworks. The current version of Python is 3.7.6. The PyTorch version utilised is 1.3.0, while the CUDA API version employed is 10.0. Additionally, the CuDNN version employed is 7.5.1.

18.3.1 Experimental Dataset and Parameter Settings

The SMLP_ResNet, ResNet, and SE_ResNet models, each consisting of 18 layers, are trained using Plant Village and AI Challenger 2018 datasets. The purpose of this training is to evaluate the influence of the SMLP_Res module on the capacity to recognise diseases. The AI Challenger 2018 dataset is utilised to train two network models: the multi-tree SMLP_ResNet and the multi-tree ResNet with identical structures. Additionally, the SE_ResNet network is introduced. The performance of the multi-tree SMLP_ResNet network model in disease recognition is evaluated by comparing it with three disease identification models consisting of 18 layers, including the multi-tree disease identification model. In order to enhance the training efficiency of the model, the utilisation of the gradient descent technique, specifically Stochastic Gradient Descent (SGD), is employed to expedite the convergence of the loss function and determine the optimal network weight. Through a series of iterative experiments, the initial learning rate is established at 0.05, while the momentum factor is set to 0.9. Considering the limited video memory capacity of the graphics card, which amounts to 16 GB, network architectures consisting of 18, 50, and 101 layers are utilised. The batch size is configured as 256, 128, and 64, correspondingly. Each training sample is iterated once, with a total of 101 iteration rounds, also known as epochs. The neural network undergoes training through the implementation of adaptive decrease of the learning rate in order to achieve the most optimal model. The evaluation of the crop disease network model involves assessing its average accuracy and precision, utilising evaluation metrics such as rate, recall rate, and weighted average score F1.

18.3.2 Disease Identification Experiment Based on SMLP_ResNet

The training results of the SMLP_ResNet network were compared with the ResNet and SE_ResNet network models to test the performance of the SMLP_ResNet network model. Three different network models were trained on the Plant Village

Table 18.3 Comparison of disease identification methods

Evaluation index	Plant Village dataset			AI Challenger 2018 dataset		
	ResNet	SE_ResNet	SMLP_ResNet	ResNet	SE_ResNet	SMLP_ResNet
Accuracy	99.05	99.19	99.32	83.83	85.53	86.93
Precision	98.57	99	99.1	79.8	82.64	84.15
Recall	98.52	98.7	98.78	78.16	81.77	83.42
F1 score	98.53	98.87	98.92	78.52	81.92	83.6

and AI Challenger 2018 datasets, and the experimental results are listed in Table 18.3.

The analysis of the data shown in Table 18.3 reveals that the SMLP_ResNet model had superior performance compared to the other models across both datasets, under identical experimental settings. The SMLP_ResNet model exhibited an increase in accuracy ranging from 0.13 to 0.27% when compared to the ResNet and SE_ResNet models in the Plant Village dataset. The SMLP_ResNet model exhibited a notable level of precision, with a rate of 99.32%. Furthermore, it demonstrated a higher level of accuracy with a precision rate of 99.10%, recall rate of 98.78%, and an F1 score of 98.91% in comparison to the other two models. Significantly, a marginal disparity of 1% in accuracy was observed between the aforementioned models. The SMLP_ResNet model exhibited significant improvements in precision, recall, and F1 score when compared to the ResNet model, leading to an estimated increase of 5%. The experimental results suggest that the SE_ResNet network architecture, which integrates the SE module, exhibits enhanced performance in comparison to the ResNet model. In addition, the SMLP_ResNet model, which combines the SMLP module with residual structures, demonstrated higher recognition results. This observation implies that the SMLP_Res architecture enhances the capacity for feature extraction to a certain extent.

ResNet, SE_ResNet, and SMLP_ResNet achieved accuracy rates exceeding 99% in the domain of crop disease recognition, as observed in the utilisation of the Plant Village dataset. Nevertheless, it is important to acknowledge that under the AI Challenger 2018 dataset, the accuracy rates of the three network models did not exceed 87%. This implies that disease recognition has a comparatively diminished level of accuracy. The observed discrepancy can potentially be ascribed to the composition of the Plant Village dataset, which has a total of 38 distinct types of crop diseases. Every individual sample in this dataset displays distinct illness characteristics that are easily distinguishable through visual observation. The dataset utilised in the AI Challenger 2018 competition comprises a total of 59 unique classifications, hence facilitating a comparative analysis. Moreover, the delineations between diseases in their initial stages and leaves in a state of good health exhibit a considerably diminished level of prominence. Therefore, it is evident that the accuracy of all three models in disease recognition on the AI Challenger 2018 early disease dataset is relatively lower.

18.3.3 Early Disease Recognition Experiment Based on Multi-branch SMLP_ResNet

Accurate crop disease recognition in the Plant Village dataset was obtained by employing the 18-layer ResNet, SE_ResNet, and SMLP_ResNet network models. Nevertheless, the level of accuracy in disease recognition within the early disease sample, specifically AI Challenger 2018, was suboptimal. The present study involved the construction of a crop early disease recognition network model utilising a multi-branch structure. The experimental outcomes were analysed to investigate the influence of the multi-branch structure on the accuracy of the model.

In order to examine the impact of the multi-branch structure model on early crop disease detection, this study employed three network models, namely ResNet, SE_ResNet, and SMLP_ResNet, as the foundational frameworks for constructing similar structured multi-branch models. The three crop disease recognition models underwent training using the AI Challenger 2018 dataset, following identical settings. The analysis focused on examining the impact of multi-branch structure network models on model performance by evaluating the findings of early disease recognition in crops. The figure presented in Fig. 18.5 demonstrates the level of precision achieved in the early detection of diseases by the three multi-branch models. The ResNet network model's multi-branch topology showed reduced accuracy in disease recognition across all crops. In contrast, the SMLP_ResNet network model with several branches consistently demonstrated superior accuracy compared to the other two networks. The multi-branch SMLP_ResNet network model showed notable enhancement in accurately identifying crop diseases in apple, cherry, pepper, and pear crops. The model achieved a maximum accuracy of 99.13%. Upon analysing the outcomes pertaining to the identification of diseases in grape and strawberry plants, it was observed that the ResNet network model exhibited the least accuracy in recognition, amounting to 76.4%. This discrepancy was

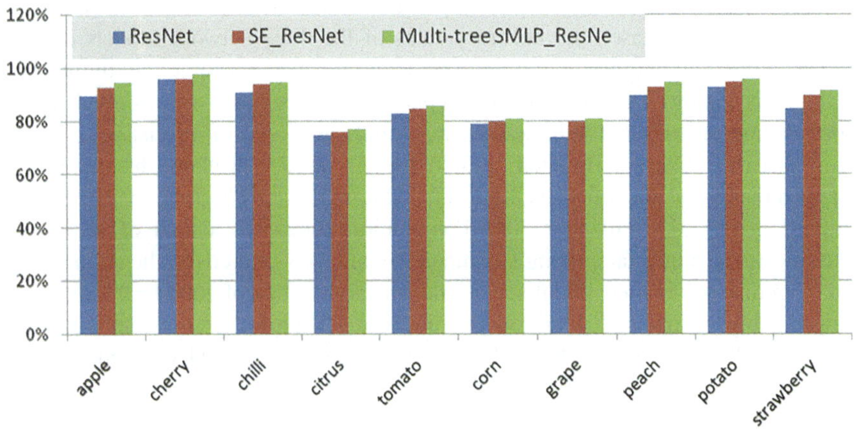

Fig. 18.5 Crop disease recognition accuracy

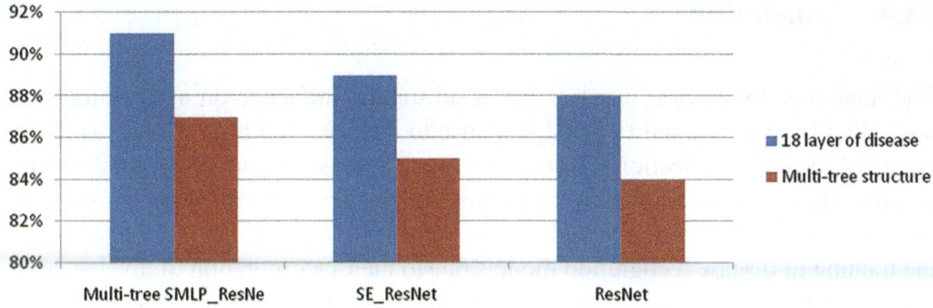

Fig. 18.6 Accuracy of different models in the AI Challenger 2018 dataset

particularly evident when compared to the SE_ResNet network model. In relation to the classification of citrus diseases, it was observed that all three network models showed relatively low identification accuracy, with none above 78%. Among these models, the multi-branch SMLP_ResNet network model achieved the highest recognition accuracy, reaching 77.66%.

Analysing the experimental data reveals that the ResNet network model has relatively weak feature extraction capabilities, resulting in a significant difference in disease recognition accuracy compared to the other two network models. In contrast, the SE_ResNet network model, which incorporates an attention mechanism, extracts crop disease data features more thoroughly, leading to a noticeable improvement in disease recognition accuracy. The improved attention mechanism in the multi-branch SMLP_ResNet network model effectively suppresses noise transmission and efficiently extracts early disease feature information, resulting in higher accuracy. Figure 18.6 provides a comparison of the average accuracy of the three 18-layer disease recognition models and the multi-branch disease recognition models.

All multi-branch illness recognition models have an accuracy of 87% or above, with the highest average accuracy achieved by the multi-branch SMLP_ResNet disease recognition model (average accuracy: 90.1%). The AI Challenger 2018 dataset shows that the multi-branch structure network models significantly outperform the three 18-layer disease recognition models in terms of early crop disease recognition accuracy. Experimental results show that multi-branch ResNet, SE_ResNet, and SMLP_ResNet disease recognition models on the AI Challenger 2018 dataset significantly improve accuracy and meet expectations for multi-crop early disease recognition compared to the three 18-layer disease recognition models. In terms of accuracy in early disease recognition across many crops, the experimental results show that the multi-branch SMLP_ResNet network model performs better than the multi-branch ResNet network model and the multi-branch SE_ResNet network model.

18.4 Conclusion

The identification of crop diseases has a substantial influence on agricultural productivity. The prompt and targeted response to address crop health issues is facilitated by the early detection of crop diseases. The issue pertaining to the early identification of diseases in many crops presents an even greater level of complexity. Datasets such as Plant Village and AI Challenger 2018 are frequently employed in the training of disease recognition models due to their incorporation of diverse crops and their corresponding prevalent diseases. In contrast to the Plant Village dataset, the AI Challenger 2018 dataset encompasses a greater number of categories and a substantial quantity of early illness samples. Consequently, this phenomenon imposes greater requirements on the network's ability to extract features, thereby augmenting the intricacy of crop disease recognition assignments. An enhanced attention mechanism multi-branch crop early disease recognition model was developed by using the SE attention mechanism and a multi-branch structure. Initially, the SMLP_Res module was established with the aim of augmenting the disease model's feature extraction capabilities. The application of the attention mechanism to multi-channel disease data involved the utilisation of MLP analysis to quantify various disease picture channel properties. This approach aimed to enhance the expression of significant features while simultaneously inhibiting the transmission of noise information. Furthermore, the integration of Residual (Res) structures was employed to establish connections across various layers, hence enhancing the transfer of cross-layer feature information and mitigating the loss of features during the transmission process. In order to address the complexity of crop disease recognition, a two-level multi-branch recognition model was developed. This model effectively simplifies the problem by dividing it into separate tasks of crop recognition and disease recognition. The utilisation of this approach resulted in a decrease in the learning burden of individual models, leading to an enhancement in the accuracy of early disease recognition through multi-crop techniques. Based on the empirical findings obtained from the 18-layer models and the multi-branch models, it can be deduced that the SMLP_Res module demonstrates a notable capability in extracting illness features. The disease recognition accuracy achieved on the AI Challenger 2018 dataset attained a peak value of 86.93%, exhibiting a 3% enhancement in comparison to the alternative two models. On the other hand, the disease recognition model within the multi-branch model had the highest level of accuracy, reaching 90.07%. The study demonstrated enhancements in the management of many crop diseases, providing evidence of the efficacy and resilience of SMLP_ResNet. The implementation of a multi-branch model in constructing a disease recognition model has been found to significantly enhance the accuracy of crop disease recognition.

References

Abbasi R, Martinez P, Ahmad R (2023) Crop diagnostic system: a robust disease detection and management system for leafy green crops grown in an aquaponics facility. Artif Intell Agric 10: 1–12. https://doi.org/10.1016/j.aiia.2023.09.001

Agarwal M, Gupta SK, Biswas KK (2020) Development of efficient CNN model for tomato crop disease identification. Sustain Comput Inform Syst 28:100407. https://doi.org/10.1016/j.suscom.2020.100407

Bondad J, Harrison MT, Whish J, Sprague S, Barry K (2023) Integrated crop-disease models: new frontiers in systems thinking. Farm Syst 1(1):100004. https://doi.org/10.1016/j.farsys.2023.100004

Das KP, Sharma D, Satapathy BK (2022) Electrospun fibrous constructs towards clean and sustainable agricultural prospects: SWOT analysis and TOWS based strategy assessment. J Clean Prod 368:133137. https://doi.org/10.1016/j.jclepro.2022.133137

De Roever C (1998) Microbiological safety evaluations and recommendations on fresh produce. Food Control 9(6):321–347. https://doi.org/10.1016/S0956-7135(98)00022-X

Degu T, Alemu T, Desalegn A, Amsalu B, Assefa A (2023) Association of cropping practices, cropping areas, and foliar diseases of common bean (*Phaseolus vulgaris* L.) in Ethiopia. J Agric Food Res 14:100765. https://doi.org/10.1016/j.jafr.2023.100765

Dutta K, Talukdar D, Bora SS (2022) Segmentation of unhealthy leaves in cruciferous crops for early disease detection using vegetative indices and Otsu thresholding of aerial images. Measurement 189:110478. https://doi.org/10.1016/j.measurement.2021.110478

Fernández-Ríos A, Laso J, Hoehn D, Amo-Setién FJ, Abajas-Bustillo R, Ortego C, Fullana-i-Palmer P, Bala A, Batlle-Bayer L, Balcells M, Puig R, Aldaco R, Margallo M (2022) A critical review of superfoods from a holistic nutritional and environmental approach. J Clean Prod 379 (Part 1):134491. https://doi.org/10.1016/j.jclepro.2022.134491

Gao R, Wang R, Feng L, Li Q, Wu H (2021) Dual-branch, efficient, channel attention-based crop disease identification. Comput Electron Agric 190:106410. https://doi.org/10.1016/j.compag.2021.106410

Gavidia JC, Chinelatto GF, Basso M, da Ponte Souza JP, Soltanmohammadi R, Vidal AC, Goldstein RH, Mohammadizadeh S (2023) Utilizing integrated artificial intelligence for characterizing mineralogy and facies in a pre-salt carbonate reservoir, Santos Basin, Brazil, using cores, wireline logs, and multi-mineral petrophysical evaluation. Geoenergy Sci Eng 231: 212303. https://doi.org/10.1016/j.geoen.2023.212303

Harner I, Anast J, Brehm-Stecher B (2023) Food safety applications of genomic technologies. In: Reference module in food science. Elsevier. https://doi.org/10.1016/B978-0-12-822521-9.00202-1

Holzinger A, Keiblinger K, Holub P, Zatloukal K, Müller H (2023) AI for life: trends in artificial intelligence for biotechnology. New Biotechnol 74:16–24. https://doi.org/10.1016/j.nbt.2023.02.001

Islam MM, Adil MA, Talukder MA, Ahamed MK, Uddin MA, Hasan MK, Sharmin S, Rahman MM, Debnath SK (2023) DeepCrop: Deep learning-based crop disease prediction with web application. J Agric Food Res 14:100764. https://doi.org/10.1016/j.jafr.2023.100764

Jagadeesan B, Gerner-Smidt P, Allard MW, Leuillet S, Winkler A, Xiao Y, Chaffron S, Van Der Vossen J, Tang S, Katase M, McClure P, Kimura B, Chai LC, Chapman J, Grant K (2019) The use of next generation sequencing for improving food safety: translation into practice. Food Microbiol 79:96–115. https://doi.org/10.1016/j.fm.2018.11.005

Joseph A (2023) Chapter 6: Salinity tolerance of inhabitants in thalassic and athalassic saline and hypersaline waters and salt flats. In: Joseph A (ed) Water worlds in the solar system. Elsevier, pp 255–310. https://doi.org/10.1016/B978-0-323-95717-5.00017-7

Karanth S, Benefo EO, Patra D, Pradhan AK (2023) Importance of artificial intelligence in evaluating climate change and food safety risk. J Agric Food Res 11:100485. https://doi.org/10.1016/j.jafr.2022.100485

Kaveney B, Barrett-Lennard E, Minh KC, Duy MD, Thi KPN, Kristiansen P, Orgill S, Stewart-Koster B, Condon J (2023) Inland dry season saline intrusion in the Vietnamese Mekong River Delta is driving the identification and implementation of alternative crops to rice. Agric Syst 207:103632. https://doi.org/10.1016/j.agsy.2023.103632

Kendler S, Aharoni R, Young S, Sela H, Kis-Papo T, Fahima T, Fishbain B (2022) Detection of crop diseases using enhanced variability imagery data and convolutional neural networks. Comput Electron Agric 193:106732. https://doi.org/10.1016/j.compag.2022.106732

Kutyauripo I, Rushambwa M, Chiwazi L (2023) Artificial intelligence applications in the agrifood sectors. J Agric Food Res 11:100502. https://doi.org/10.1016/j.jafr.2023.100502

Maier-Hein L, Eisenmann M, Sarikaya D, März K, Collins T, Malpani A, Fallert J, Feussner H, Giannarou S, Mascagni P, Nakawala H, Park A, Pugh C, Stoyanov D, Vedula SS, Cleary K, Fichtinger G, Forestier G, Gibaud B, Grantcharov T, Hashizume M, Heckmann-Nötzel D, Kenngott HG, Kikinis R, Mündermann L, Navab N, Onogur S, Roß T, Sznitman R, Taylor RH, Tizabi MD, Wagner M, Hager GD, Neumuth T, Padoy N, Collins J, Gockel I, Goedeke J, Hashimoto DA, Joyeux L, Lam K, Leff DR, Madani A, Marcus HJ, Meireles O, Seitel A, Teber D, Ückert F, Müller-Stich BP, Jannin P, Speidel S (2022) Surgical data science—from concepts toward clinical translation. Med Image Anal 76:102306. https://doi.org/10.1016/j.media.2021.102306

Manavalan R (2020) Automatic identification of diseases in grains crops through computational approaches: a review. Comput Electron Agric 178:105802. https://doi.org/10.1016/j.compag.2020.105802

Marvin HJP, Kleter GA, Frewer LJ, Cope S, Wentholt MTA, Rowe G (2009) A working procedure for identifying emerging food safety issues at an early stage: implications for European and international risk management practices. Food Control 20(4):345–356. https://doi.org/10.1016/j.foodcont.2008.07.024

Mensah PK, Akoto-Adjepong V, Adu K, Ayidzoe MA, Bediako EA, Nyarko-Boateng O, Boateng S, Donkor EF, Bawah FU, Awarayi NS, Nimbe P, Nti IK, Abdulai M, Adjei RR, Opoku M, Abdulai S, Amu-Mensah F (2023) CCMT: dataset for crop pest and disease detection. Data Brief 49:109306. https://doi.org/10.1016/j.dib.2023.109306

Modi RU, Kancheti M, Subeesh A, Raj C, Singh AK, Chandel NS, Dhimate AS, Singh MK, Singh S (2023) An automated weed identification framework for sugarcane crop: a deep learning approach. Crop Prot 173:106360. https://doi.org/10.1016/j.cropro.2023.106360

Mrabet R (2023) Chapter 2: Sustainable agriculture for food and nutritional security. In: Farooq M, Gogoi N, Pisante M (eds) Sustainable agriculture and the environment. Academic Press, pp 25–90. https://doi.org/10.1016/B978-0-323-90500-8.00013-0

Nigam S, Jain R, Marwaha S, Arora A, Haque MA, Dheeraj A, Singh VK (2023) Deep transfer learning model for disease identification in wheat crop. Ecol Informat 75:102068. https://doi.org/10.1016/j.ecoinf.2023.102068

Noshiri N, Beck MA, Bidinosti CP, Henry CJ (2023) A comprehensive review of 3D convolutional neural network-based classification techniques of diseased and defective crops using non-UAV-based hyperspectral images. Smart Agric Technol 5:100316. https://doi.org/10.1016/j.atech.2023.100316

Nychas G-JE, Panagou EZ, Mohareb F (2016) Novel approaches for food safety management and communication. Curr Opin Food Sci 12:13–20. https://doi.org/10.1016/j.cofs.2016.06.005

Pakseresht A, Yavari A, Kaliji SA, Hakelius K (2023) The intersection of blockchain technology and circular economy in the agri-food sector. Sustain Prod Consump 35:260–274. https://doi.org/10.1016/j.spc.2022.11.002

Rani AP, Singh NS (2022) Protecting the environment from pollution through early detection of infections on crops using the deep belief network in paddy. Total Environ Res Themes 3:100020. https://doi.org/10.1016/j.totert.2022.100020

Ren Q-S, Fang K, Yang X-T, Han J-W (2022) Ensuring the quality of meat in cold chain logistics: a comprehensive review. Trends Food Sci Technol 119:133–151. https://doi.org/10.1016/j.tifs.2021.12.006

Reumaux R, Chopin P, Bergkvist G, Watson CA, Öborn I (2023) Land Parcel Identification System (LPIS) data allows identification of crop sequence patterns and diversity in organic and conventional farming systems. Eur J Agron 149:126916. https://doi.org/10.1016/j.eja.2023.126916

Rowan NJ, Galanakis CM (2020) Unlocking challenges and opportunities presented by COVID-19 pandemic for cross-cutting disruption in agri-food and green deal innovations: Quo Vadis? Sci Total Environ 748:141362. https://doi.org/10.1016/j.scitotenv.2020.141362

Rowan NJ, Murray N, Qiao Y, O'Neill E, Clifford E, Barceló D, Power DM (2022) Digital transformation of peatland eco-innovations ('Paludiculture'): enabling a paradigm shift towards the real-time sustainable production of 'green-friendly' products and services. Sci Total Environ 838(Part 3):156328. https://doi.org/10.1016/j.scitotenv.2022.156328

Saeed R, Feng H, Wang X, Zhang X, Fu Z (2022) Fish quality evaluation by sensor and machine learning: a mechanistic review. Food Control 137:108902. https://doi.org/10.1016/j.foodcont.2022.108902

Tirkey D, Singh KK, Tripathi S (2023) Performance analysis of AI-based solutions for crop disease identification, detection, and classification. Smart Agric Technol 5:100238. https://doi.org/10.1016/j.atech.2023.100238

Varga M, Csukas B (2022) Foundations of programmable process structures for the unified modeling and simulation of agricultural and aquacultural systems. In: Information processing in agriculture. Elsevier. https://doi.org/10.1016/j.inpa.2022.10.001

Xiong Y, Liang L, Wang L, She J, Wu M (2020) Identification of cash crop diseases using automatic image segmentation algorithm and deep learning with expanded dataset. Comput Electron Agric 177:105712. https://doi.org/10.1016/j.compag.2020.105712

Zhao H, Huang X, Yang Z, Li F, Ge X (2023) Synergistic optimization of crops by combining early maturation with other agronomic traits. Trends Plant Sci 28(10):1178–1191. https://doi.org/10.1016/j.tplants.2023.04.011

Zhou J, Li J, Wang C, Wu H, Zhao C, Teng G (2021) Crop disease identification and interpretation method based on multimodal deep learning. Comput Electron Agric 189:106408. https://doi.org/10.1016/j.compag.2021.106408

Chapter 19
Guarding Maize: Vigilance Against Pathogens Early Identification, Detection, and Prevention

Khalil Ahmed, Mithilesh Kumar Dubey, and Sudha Dubey

Abstract Maize, also known as corn, is a crucial crop worldwide. It is noticeable that due to diseases maize production is affected. Maize diseases can have a significant impact on yield. Common diseases include those caused by viruses, bacteria, fungi, and other pathogens. Proper artificial intelligence-based crop management and disease control are essential for maximizing maize production. Corn leaf stage methods are used to assess the growth and development of maize plants. These stages are often described using a system called the V-stage or "the Vegetative stage." Corn is a widely cultivated cereal crop that has a rich history dating back 7000 years, originating from wild grasses in Central America and having significant cultural, economic, and nutritional importance worldwide. Around the globe, India is the fifth largest maize (corn) producer according to the USDA 2022–2023 report, which produces 3,20,00,000 million metric tons of corn annually. The United States holds the first position with 35,38,36,000 million metric tons of production. According to the USDA report maize production decreased by 1216.87 million tons last year. The prophecy indicated that in 2023 maize production also decreased by 1216.87 million tons, which is 4.52% of the globe. This continuous decrease in maize crops is due to diseases caused by bacteria, fungi, and viruses. Corn seed undergoes various stages from germination to the dotage stage, absorbing water and other essential nutrients for growth. Due to climatic conditions corn plants suffer from such diseases, identification of disease and detection at the prior stage are very important so that quality and yield of maize crops are not affected. For this, we introduce the anatomy of maize plants, the key features of a crop, leaf stage and

K. Ahmed · M. K. Dubey (✉)
School of Computer Application, Lovely Professional University, Phagwara, Punjab, India
e-mail: mithilesh.21436@lpu.co.in

S. Dubey
Department of Sociology, Lovely Professional University, Phagwara, Punjab, India

A. Khamparia et al. (eds.), *Microbial Data Intelligence and Computational Techniques for Sustainable Computing*, Microorganisms for Sustainability 47,
https://doi.org/10.1007/978-981-99-9621-6_19

methods, the maturity cycle of maize plants, and a brief overview of the diseases caused by fungi, bacteria, and viruses in maize.

Keywords Maize disease · Leaf stage · Early identification · *Zea mays*

19.1 Introduction

Maize, scientifically known as *Zea mays*, is a widely cultivated cereal crop that has significant cultural, economic, and nutritional importance around the world (Hallauer and Carena 2009). The history of the maize crop spans thousands of years. Originating 7000 years ago from wild grasses in Central America (McHale et al. 2012a), the areas in which ancient civilizations such as the Maya, Inca, and Aztecs engaged in the farming of maize, it has a long history of cultivation by indigenous people in America. It was a vital crop for the ancient civilization of America. It is an annual plant with a distinctive morphology, including tall, jointed stems, long leaves arranged alternatively, distinctive cobs with rows of kernels, and a tassels-like inflorescence known as the male flower. The female flower is enclosed in a husk, later known as corn or maize kernels (Zhang et al. 2012) as shown in Fig. 19.1.

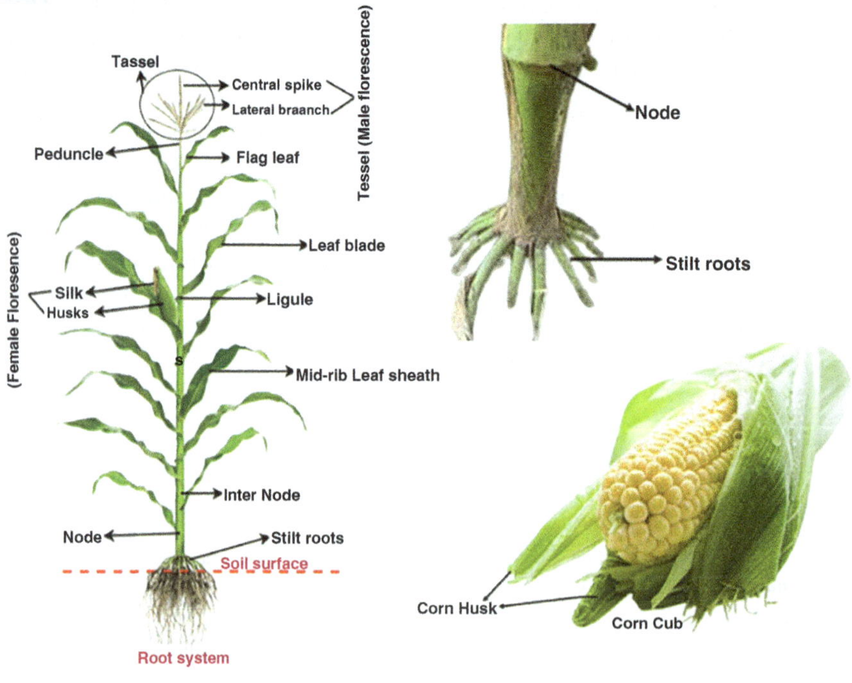

Fig. 19.1 Anatomy and morphology of maize plant

Maize is grown in extensive climates and geographical regions, one of the most often used cultivated crops around the globe. The major maize-producing countries include the United States, China, Brazil, and India (Ranum et al. 2014). The crop's adaptability and high-yield potential have contributed to its widespread cultivation, providing essential carbohydrates, dietary fibers (B-complex vitamins, especially niacin and thiamine), and some protein. Despite human use, it is as a raw product, and ethanol production, and in the manufacturing of industrial products. Maize is the stable food for billions of people worldwide. This diversity has prompted the creation of numerous sorts of maize, including tremendous genetic diversity helps to the creation of different varieties of maize plus dent, flint, popcorn, and sweet corn, each with specific attributes and uses. Genetic diversity also contributes to the plant's adaptability and resilience to different environmental conditions (McHale et al. 2012b). Successful maize cultivation involves careful consideration of factors such as soil types, temperature, availability of water, and disease management. Proper planting, fertilization, irrigation, and disease control are critical for achieving optimal yields.

19.1.1 Key Features of Maize

Versatility Maize is versatile and can be used in various forms, such as fresh, dried, or processed (FAOSTAT 2023). It can be consumed as whole kernels, ground into flour, or used to produce various food products like corn syrup, corn oil, and popcorn.

Cultural Significance Maize has a deep cultural and historical significance for many indigenous communities in the Americas (Ranum et al. 2014). It played a central role in the diet, rituals, and economics of these civilizations.

Botanical Structure A maize plant consists of different parts. The ear is the female inflorescence where the kernels develop, while the tassel is the male inflorescence that produces pollen as in Fig. 19.1. Maize plants have long, narrow leaves and a fibrous root system (Doebley et al. 2006). The fibrous root system maximizes the plant's ability to access soil nutrients, contributing to its overall growth and productivity. These leaves are usually around 30–100 cm in length, depending on factors such as genetics, environmental conditions, and the level of growth.

Cultivation and Climatic Conditions Maize is grown in a wide range of climates, from tropical to temperate regions. It requires well-drained soil and sufficient sunlight. It is considered a C4 plant (Sadras and Calderini 2009), which means it is efficient at photosynthesis, making it well suited for environments with higher temperatures.

Economic Importance Maize is a staple crop in many countries, serving as a major food source for living. It also has industrial applications in sectors such as food processing, animal feed, and biofuel production.

19.1.2 A Closer Look at World Maize Harvesting

The world corn production in the years 2022–2023 will be 1161.86 million metric tons around the globe according to data from the USDA. According to USDA data, in December 2022, the maize crop was less by 6.52 million metric tons than the previous month. In the last year maize production decreased by 1216.87 million tons (Dotasara and Choudhary 2023). The forecast for the end of 2023 anticipates a global maize production decrease of 55.00 million tons, equivalent to a 4.52% reduction. Statistical data of maize production is shown in Fig. 19.2.

India is the fifth largest impresario of maize in the world. Maize is grown in various states across the country, contributing significantly to both food and fodder requirements. Maize is cultivated in the Kharif season in India, which starts with the onset of the monsoon (June/July) and extends until October, and during winter (rabi) in October or November. The crops are then harvested in spring. In India, Kharif maize accounts for approximately 83% of total maize cultivation, whereas rabi maize makes up the remaining 17%. More than 70% of Kharif maize is grown without irrigation, relying solely on rainwater, which exposes it to a range of pests and environmental challenges as shown in Fig. 19.3. The exact timing of maize harvesting can vary based on the region and local climate conditions. The production of maize in India has been increasing steadily over the years due to its diverse uses. Almost 70% of the rural population in India is directly dependent on agriculture, which contributes 50% of in-country employment and 18% of the country's GDP (Monterrubio-Solís et al. 2023). Around the globe four billion people are indirectly dependent on maize for food as per (FAO) 2021 report. The maize crop is facing significant challenges in terms of reduced production due to the outbreak of various diseases worldwide. These diseases have the potential to cause extensive damage to maize crops, leading to yield losses and economic setbacks. Maize is a high-yield crop, and farmers must protect the plant while growing and avoid production damage.

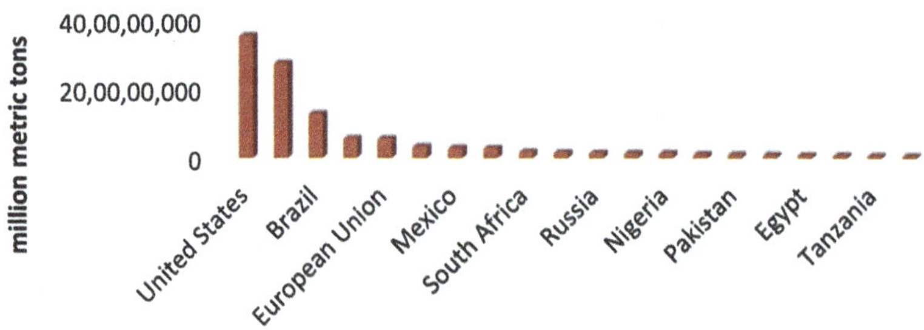

Fig. 19.2 Maize growth and yield data 2022–2023 (USDA)

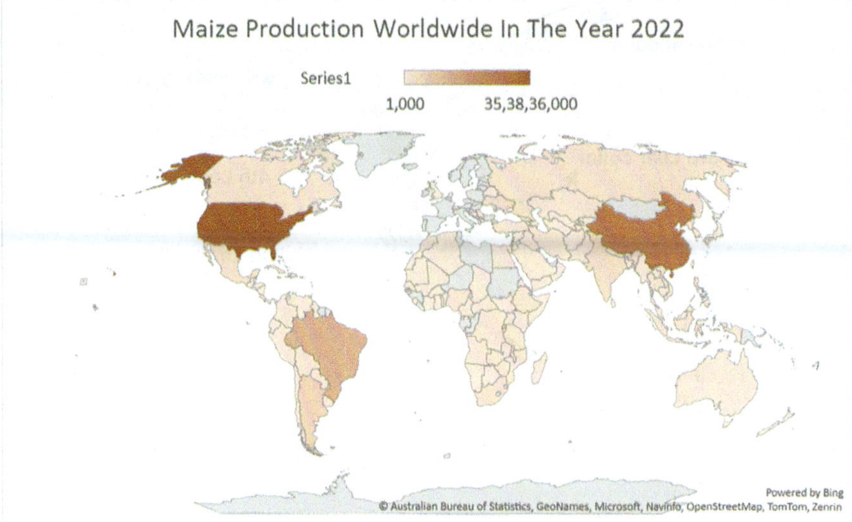

Fig. 19.3 A map of maize production worldwide in the years 2022–2023 (USDA)

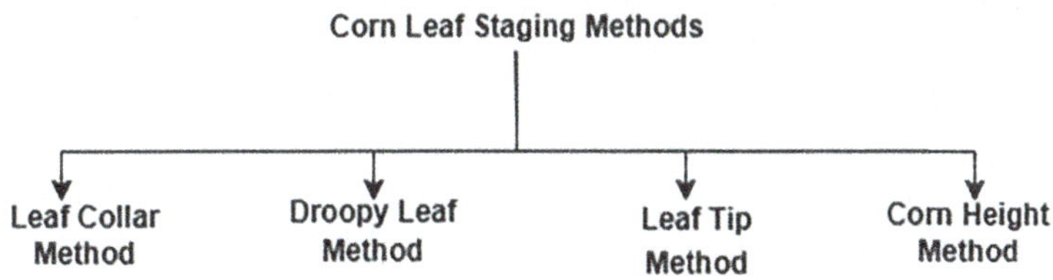

Fig. 19.4 Methods of maize leaf staging

19.1.3 Determining Corn Leaf Stages

Corn leaf stages refer to the various growth stages that a corn plant goes through during its development (Forde and Clarkson 1999). Farmers and agronomists frequently utilize these stages to monitor and control crop growth and development. The most commonly used system to describe corn leaf stages is the one introduced by the USDA, which uses a numerical scale from VE (emergence) to R6 (physiological maturity) as shown in Fig. 19.4 (Smith and Zeeman 2006).

19.1.3.1 Leaf Collar Method

This approach involves determining the developmental stage of corn leaves by tallying the count of the plant leaves that exhibit noticeable leaf collars. The process commences with the smallest genuine leaf characterized by its short length and rounded tip and concludes with the highest leaf that possesses a discernible leaf

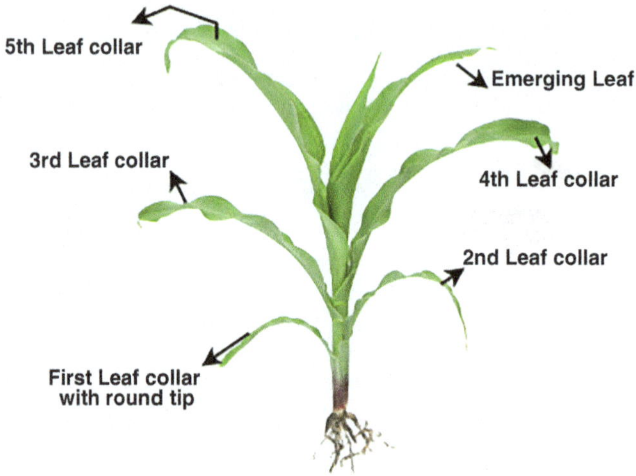

Fig. 19.5 Leaf collar method v5

collar. The leaf collar, resembling a light-colored band, is situated toward the base of an exposed leaf blade point of contact between the leaf blade and the plant's stem Leaves positioned within the cluster, which have not yet fully unfurled and lack an observable leaf collar, are not taken into account using this leaf staging technique (Kumar et al. 2022). This is typically denoted as "V" stages, for example, V2, signifying two leaves with visible leaf collars. In the United States, the leaf collar method is widely favored and extensively employed as shown in Fig. 19.5.

19.1.3.2 Droopy Leaf Method

This form of leaf staging, like the leaf collar approach, starts with a short initial leaf. Leaf counting then changes, concluding with the leaf that is at least 40–50% exposed from the whorl, rather than the highest leaf with a visible collar. The tip of this "indicator" leaf generally "droops" or hangs down in knee-high corn or older, thus I refer to this as the "droopy" leaf approach (Gao et al. 2023).

19.1.3.3 Leaf Tip Method

Tally, the points of the leaflet edges, starts from the plant's base and moves upward. Be sure to include any recently sprouted leaves that emerge among the group. Be cautious when applying the leaf tip technique, as various hybrid plants exhibit unique ways of growing (ur Rehman et al. 2021). For example, certain hybrids are deliberately cultivated to have shorter spaces between leaves, leading to a higher count of the leaflet edges prior to their curvature and transformation into fully mature leaves.

19.1.3.4 Corn Height Method

In this method the vertical distance from the base level to the highest point of the downward-pointing tip on the uppermost leaf of the arch is calculated. It's crucial to understand that we are not measuring the highest point of the entire plant's structure (Naseem et al. 2022a). Please remember that utilizing this method to assess the developmental phase of corn may not be highly accurate, as the crop's height can vary due to factors in the environment and how it's managed.

19.1.4 The Growth Cycle of the Maize Plant

The maize plant undergoes several distinct stages during its life cycle as different varieties and environmental conditions can slightly alter the timing of these stages. On average, it generally requires 100–120 days for the majority of maize cultivars to mature from planting to harvest. However, the environment and other circumstances, such as the product's intended purpose, have a significant impact on the precise time of harvest.

Seed Germination The life cycle begins when a maize seed germinates as shown in Fig. 19.6. The seed absorbs water from the soil moisture, the seed becomes soft, and the diameter also increases. Due to the softness, the germination period of maize seed takes 4–7 days, and the radicle emerges first followed by the shoot (Maitra et al. 2020).

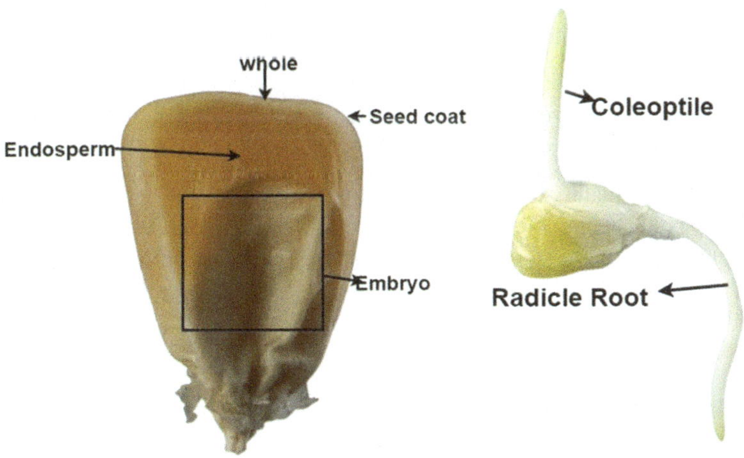

Fig. 19.6 Seed germination and radicle root emerging process

Vegetative Growth The young seedling develops leaves, and the stem elongates. The plant focuses on establishing a strong root system and growing above-ground structures to capture sunlight for photosynthesis (Naseem et al. 2022b).

Tillering As the plant grows, it produces side shoots known as tillers. Each tiller has its own set of leaves and can potentially form ears of corn.

Stem Elongation The stem continues to grow, and the plant gains height. More leaves are produced at the top of the plant.

Pre-flowering During this phase, the plant transitions from vegetative growth to reproductive growth. The tassel, which contains the male flowers (anthers), emerges from the top of the plant.

Silking Female reproductive structures, called silk, emerge from the ear of corn. Each silk corresponds to a potential kernel. Pollination occurs when pollen from the tassel lands on the silks.

Pollination The wind carries pollen from the tassel to the silks. Each pollen grain that lands on a silk forms a pollen tube that travels down the silk to fertilize an individual ovule, leading to kernel development.

Kernel Development Fertilized ovules develop into kernels. Each kernel undergoes stages of growth, starting with a milky fluid and progressing to a dough-like consistency.

Grain Filling Kernels continue to grow and accumulate starch, sugars, and other nutrients (World Agricultural Production 2023). The plant's focus shifts to providing nutrients to the developing kernels.

Maturity The maize plant reaches maturity when the kernels have fully developed and dried. The leaves may start to turn brown, and the plant's energy is concentrated in the mature kernels.

Harvest Once the maize reaches the desired moisture content and maturity level, it's ready for harvest as shown in Fig. 19.7.

Senescence The plant's remaining vegetative components begin to wither and die after harvest. This completes the life cycle of the maize plant. Please keep in mind that the timing of these stages can vary based on factors such as temperature, day length, and local growing conditions. Additionally, different varieties of maize may exhibit slight variations in their growth patterns.

19.2 Overview of Maize Crop Diseases

The most common disease in maize crops can be caused by fungi, bacteria, viruses, and mollicutes, which can result in the most prevalent diseases in maize (Santos et al. 2022). Microorganisms can spread diseases such as rust, smut, and blight in maize.

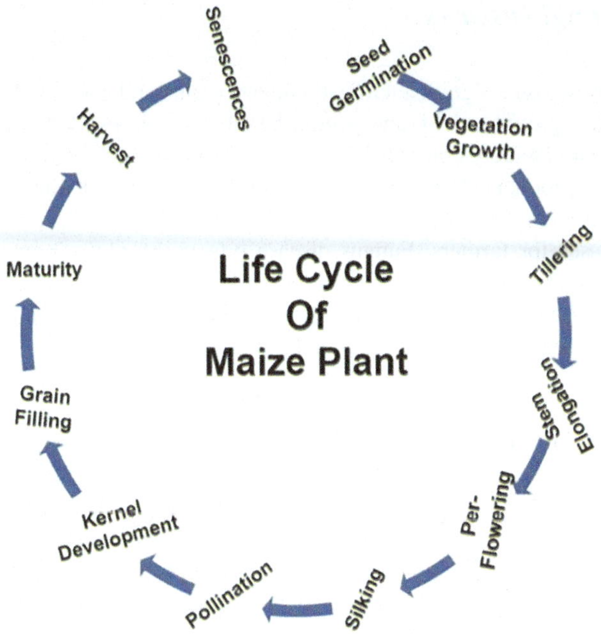

Fig. 19.7 The maize corn life cycle

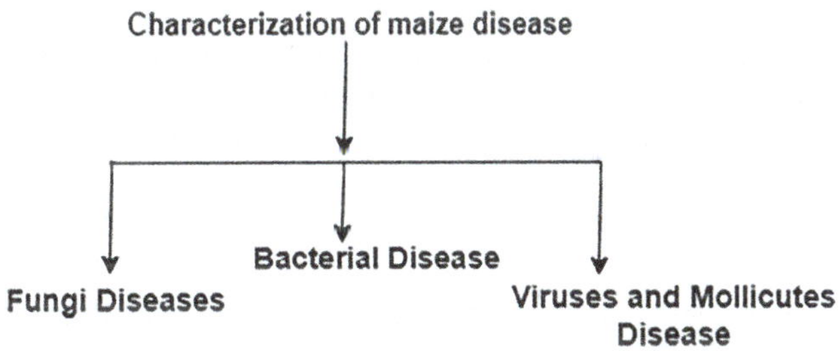

Fig. 19.8 Maize disease categorization

Bacteria are unicellular microorganisms that can cause diseases such as bacterial wilt and leaf spot. Viruses are infectious agents that cause diseases, e.g., maize dwarf mosaic virus and maize streak virus. Mollicutes are a type of bacteria that lack a cell wall and can cause diseases such as corn stunt disease (Zhao et al. 2020). These pathogens can negatively impact maize production and result in yield losses. Broadly maize diseases are classified into three categories as shown in Fig. 19.8.

19.2.1 *Fungi Diseases*

Fungal diseases pose a significant threat to maize plants and can have a considerable impact on their growth, yield, and overall health. Maize, also known as corn, is a staple crop that is widely cultivated for various purposes such as food, animal feed, and industrial products. However, its susceptibility to fungal infections makes it vulnerable to a range of diseases that can lead to reduced crop productivity and economic losses for farmers. Fungus disease infects the plant from root to tassel (FAOSTAT 2023). These fungi often thrive in humid and moist conditions, which create an ideal environment for their growth and spread. Some common fungal diseases that affect maize plants include the following.

19.2.1.1 Brown Spot

Brown spot disease, also known as brown spot leaf blight. Caused by the fungus *Bipolaris maydis*, known as *Helminthosporium maydis* (Subedi et al. 2023). The most distinctive symptom of this disease is the appearance of brown, oval-to-irregularly shaped spots on the leaves of maize plants. A golden halo frequently surrounds these dots. Over time, the brown spots can coalesce, leading to the formation of larger lesions. The fungus thrives in high-humidity environments, most active at 20–30 °C (68–86 °F) (Afzaal et al. 2022). Frequent rainfall or irrigation leads to moisture on the leaves providing a suitable environment for the fungus to grow and reproduce. The fungus can survive on crop residues and in the soil as a source of infection for new crops in subsequent growing seasons. It can also be transmitted through contaminated seeds; severe infections can cause the leaves to wither and die, resulting in blighting. To manage brown spot disease in maize, several strategies can be employed: resistant varieties, planting maize varieties that are resistant to the disease, can be an effective way to prevent infection. Crop rotation is another method to avoid planting maize in the same field repetitively to reduce the buildup of fungal spores in the soil. In severe cases, fungicides can be used to control the disease. However, this should be done under the guidance of agricultural experts. Proper irrigation—avoid overhead irrigation—is another strategy as it can create a humid environment that favors the fungus. Drip irrigation or furrow irrigation is preferable. Crop residue management is to remove and destroy crop residues after harvest to reduce the source of fungal spores for the next season as shown in Fig. 19.9.

19.2.1.2 Downy Mildew

Downy mildew is a fungal disease that affects various plants, including maize (corn). It is a common fungal disease caused by various species of the oomycete pathogen, typically caused by the pathogen *Peronosclerospora* spp. (Kominko et al. 2021). A

Fig. 19.9 *Helminthosporium maydis* infect corn leaf

general overview of the various types of downy mildews that can affect corn plants. This group of fungi causes downy mildew in corn. Among them, *Peronosclerospora maydis* is the most notable species. It primarily affects maize and can cause yellowing and wilting of leaves, along with reduced yield. Species within this group, such as *Sclerophthora macrospora*, can cause downy mildew in corn. Symptoms include stunting, yellowing of leaves, and the formation of a white to grayish fungal growth on the undersides of leaves. Some members of this genus, like *Pseudoperonospora cubensis*, can infect corn plants. Symptoms include yellowing of leaves, white or grayish fungal growth on leaf undersides, and reduced plant vigor. *Basiomycetes* spp. are fungi which cause downy mildew in corn. They can lead to leaf discoloration, stunting, and poor crop development (Maitra et al. 2020). These various types of downy mildew in corn are typically managed through cultural and crop rotation practices, the use of resistant corn varieties, and the application of fungicides when necessary. These pathogens thrive in humid and moist conditions and can severely impact maize crops. It affects a wide range of plants, including maize. The symptoms of downy mildew in maize plants can vary, but they often include yellow to pale green lesions on the upper side of leaves, the characteristic downy growth on the undersides of leaves, stunted growth and reduced yield, and premature death of infected plants (Li et al. 2020). Downy mildew pathogens overwinter as resting structures in infected plant debris or soil. The pathogen then grows inside the plant, causing the symptoms mentioned above.

It's essential for farmers and growers to monitor their corn crops regularly to detect early signs of downy mildew and take appropriate measures to prevent its spread. Downy mildews can be a significant concern for maize growers, leading to reduced crop yields and economic losses as shown in Fig. 19.10. Understanding the causes, symptoms, and management strategies for this disease is crucial for effective disease control in maize cultivation.

Tar Spot

Tar spot is a fungal disease that affects maize (corn) plants (Xu et al. 2023). It can be determined by a formation of tiny, brought-up, black specks on maize plant foliage,

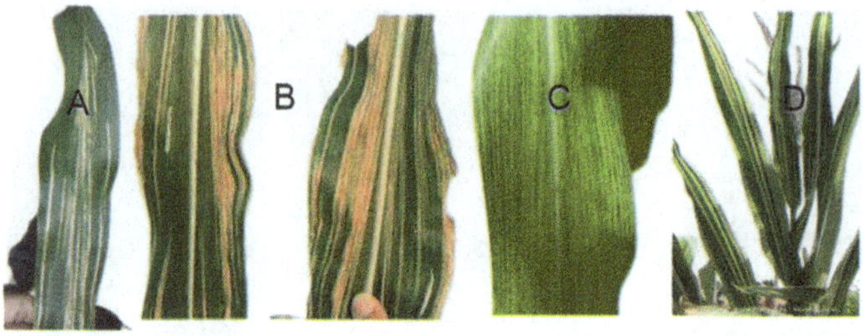

Fig. 19.10 Downy mildew-infected corn leaf

Fig. 19.11 Tar spot-infected corn leaf

husks, and, in rare cases, stems. These black spots often cluster together and can vary in size. The name "tar spot" comes from the resemblance of these black lesions to spots of tar (Santos et al. 2022). The common symptoms of tar spot disease in maize plants are tar-like spots, clustered spots, and yellow halos. The disease can affect the photosynthetic capacity of the plant by reducing the effective leaf area, which can ultimately impact crop yield. While tar spot disease can vary in severity depending on environmental conditions, it can be managed through cultural practices like crop rotation, tillage, and the removal of infected plant debris. In severe cases, fungicides may also be used to control the disease (Zhao et al. 2020). The disease has a life cycle that involves overwintering infected maize debris from the previous growing season, spore production, infection of healthy plants, and the continued production and dispersal of spores during the growing season. Proper management practices are crucial to minimize its impact on maize crop production.

Maize Rusts

Maize rust disease, also known as corn rust, is a plant disease that affects corn (*Zea mays*) plants as shown in Fig. 19.11. Various kinds of fungus from the *Puccinia* genus cause this illness, with *Puccinia sorghi* being one of the most prevalent offenders (Schunck et al. 2021). The hallmark symptom of maize rust is the appearance of small, circular elongated pustules on the leaves, stems, and sometimes even the ears of corn plants. These pustules range in color from orange to reddish-brown, resembling rust, and hence the name "corn rust." The major causes of maize rust disease in corn plants are fungal pathogens, overwintering (alternate hosts, infected crop debris), weather conditions (moisture, temperature, and wind), and susceptible hosts. Maize rust disease is a significant threat to corn crops and can lead to yield, quality reductions, and economic losses for farmers.

Common Rust

Common rust disease, caused by the fungus *Puccinia sorghi* (Miao et al. 2021), is a widespread and economically significant fungal infection that affects various plants, especially those in the Poaceae family, such as corn (maize) (*Zea mays*), and does not infect other plant species. The most noticeable symptom of common rust is the appearance of raised, reddish-brown to orange pustules on the leaves, stems, and sometimes even the husks and ears of infected plants (Qiu et al. 2022). These pustules give the plant a rusty or blistered appearance, hence the name "common rust." The life cycle of the common rust fungus involves both sexual and asexual reproduction. Urediniospores: these are the asexual spores produced in the rust pustules on infected plants. They are transported due to wind and rain. Urediniospores land on healthy corn plants and germinate to form specialized structures called appressoria. These structures penetrate the plant's epidermal cells, initiating infection. The fungus grows within the plant, forming pustules that eventually rupture, releasing more urediniospores. As the season progresses, the fungus undergoes sexual reproduction, producing teliospores within the pustules. Teliospores serve as the overwintering stage, allowing the fungus to survive the winter on infected crop residues or alternate hosts. Common rust can lead to reduced photosynthetic capacity in infected plants, resulting in decreased yields. Severe infections can cause premature leaf senescence, further reducing yield and crop quality. Early detection and management are key to preventing the spread and impact of common rust. Crop rotation, planting resistant varieties, and practicing good agricultural hygiene are important preventative measures as shown in Fig. 19.12.

Polysora Rust

This disease is also known as oxidation in the southern hemisphere. Fungal infection caused by *Polysora* spp. (Qiu et al. 2022) that affects maize (corn) primarily occurs in tropical and subtropical regions. The symptoms include rust-colored pustules.

Fig. 19.12 Common rust-infected corn leaf

Fig. 19.13 Polysora rust (southern rust)-infected corn leaf

Similar to common rust, polysora rust also produces rust-colored pustules on the leaves, stems, and sometimes husks and ears of infected maize plants as shown in Fig. 19.13. Small yellow spots: early symptoms may include small, yellowish spots.

To manage fungal diseases in maize, farmers often use a combination of cultural practices, such as crop rotation, choosing resistant maize varieties, and applying fungicides when necessary. Proper irrigation and drainage management can also help reduce the risk of fungal infections by minimizing moisture stress.

Bacterial Disease in Maize

Bacterial diseases in maize can significantly impact crop yield and quality. One of the most common bacterial diseases affecting maize is bacterial leaf streak (BLS).

BLS in Maize Infectious leaf scar in maize occurs predominantly by the bacteria *Xanthomonas vasicola* pv. *Zeae*. This bacterium is a pathogenic microorganism responsible for the development of BLS (Oehme et al. 2022).

Symptoms BLS symptoms typically appear on the leaves of maize plants. They may include long, narrow, yellow to tan streaks on the leaves, which often parallel the leaf veins. These streaks may expand and turn brown or necrotic, leading to a "burned" appearance. Lesions may coalesce, covering large areas of the leaf surface. In severe cases, the leaves may become tattered and dry.

Disease Progression Bacterial leaf streak can spread rapidly, especially under favorable conditions like high humidity and rainfall. The bacteria enter through natural openings or wounds on the leaves. As the disease progresses, it can lead to defoliation, reducing the plant's photosynthetic capacity and ultimately affecting crop yield.

Favorable Conditions BLS is more likely to occur when environmental conditions favor bacterial growth and infection. These conditions include high humidity and moisture, which promote bacterial growth and the spread of the disease. Warm temperatures, typically between 68 and 86 °F (20 and 30 °C), are conducive to bacterial multiplication. Rainfall or irrigation can provide the moisture necessary for infection.

Management Strategies for BLS in Maize Include:

Crop Rotation: Avoid planting maize in the same field year after year to reduce the buildup of the pathogen.

Resistant Varieties: Plant maize varieties that are resistant or tolerant to BLS.

Fungicide Application: In severe cases, fungicides may be used to control the disease, although this is not always economically viable.

Sanitation: Remove and destroy infected crop residues to reduce the inoculum source.

Proper Water Management: Avoid excessive irrigation, as it can create conditions favorable for the disease.

Bacterial leaf streak is a significant concern for maize growers as it can lead to yield losses and reduce the quality of harvested maize. Understanding the causal agent, symptoms, disease progression, and management strategies for bacterial leaf streak in maize is crucial for effective disease control and maintaining healthy maize crops.

19.2.2 Maize Virus and Mollicute Diseases

Maize lethal necrosis disease (MLND) is a nasty disease that affects maize (corn) plants. It is a relatively recent and destructive disease that has caused significant economic losses in maize production, primarily in regions of East Africa, although cases have been reported in other parts of the world as well (Klupczyńska and Pawłowski 2021). MCMV and SCMV can cause this disease. These two viruses work together synergistically to produce the devastating effects of MLND. Joint infection of twins is caused by oxidation in the southern hemisphere MCMV and SCMV. These viruses belong to different virus families. MLND is primarily transmitted through vectors, particularly insect vectors. These vectors acquire viruses when feeding on infected plants and then transmit them to healthy maize plants while feeding. MLND-infected maize plants display a range of characteristic symptoms, which typically become apparent during the growing season. The leaves of infected plants show chlorosis or yellowing, particularly along the leaf margins. Infected plants often exhibit stunted growth and reduced overall vigor.

Mollicute diseases in maize Mollicutes are a group of bacteria that lack a cell wall and are responsible for causing various plant diseases (Farmers Weekly 2017), including those in maize. One notable mollicute disease in maize is "corn stunt," caused by a bacterium known as *Spiroplasma kunkelii*.

19.2.2.1 Corn Stunt

Corn stunt is caused by *Spiroplasma kunkelii*, a mollicute bacterium. Infected maize plants exhibit stunting, reduced growth, and a characteristic "broom-like" appearance of the leaves. The disease can also lead to reduced yields. Leafhoppers, which act as vectors for the bacteria, are the main source of corn stunt transmission. When they feed on infected plants, they acquire the pathogen and transmit it to healthy plants. In both cases, prevention and control measures often involve planting disease-resistant maize varieties, doing good weed control to reduce potential hosts for vectors, and implementing integrated pest management strategies. It's important to note that particular symptoms and management practices could differ contingent on the maize virus disease and the geographical region in which they occur. Regular monitoring and early detection are key to managing these diseases effectively.

19.3 Conclusion

Maize crop diseases pose significant challenges to farmers worldwide, impacting crop yield, quality, and overall agricultural productivity. Effective identification and management of these diseases are crucial for sustainable maize production. Early detection is essential for effective disease management. Farmers should regularly scout their maize fields to identify symptoms of diseases, including leaf discoloration, wilting, necrosis, and unusual growth patterns. Integrated pest management (IPM) approaches are recommended, combining various strategies to minimize disease impact. Artistic practice, crop rotation, proper spacing, and sanitation help reduce disease pressure. Planting disease-resistant maize varieties is a proactive measure to prevent infections. Fungicides and pesticides should be used judiciously and in accordance with recommended guidelines to minimize environmental impact. Biological control methods, like beneficial microorganisms, can be employed to suppress disease-causing pathogens. Farmers should be educated about common maize diseases, their symptoms, and management options. Extension services and agricultural institutions play a crucial role in disseminating knowledge and best practices to farmers. By combining early detection, integrated management strategies, education, and ongoing research, farmers can minimize the impact of diseases and ensure sustainable maize production.

References

Afzaal M, Hameed S, Rasheed R, Din Khan WU (2022) Microalgal biofuels: a sustainable pathway for renewable energy. In: Algal biotechnology. Elsevier, pp 187–222

Doebley JF, Gaut BS, Smith BD (2006) The molecular genetics of crop domestication. Cell 127(7): 1309–1321

Dotasara SK, Choudhary CS (2023) Fall armyworm in maize crop: challenges and management in India. International Year of Millets

FAOSTAT (2023) https://www.fao.org/faostat/en/#data/QCL

Farmers Weekly (2017) https://www.farmersweekly.co.za/crops/field-crops/maize-production-managing-critical-plant-growth-stages/

Forde BG, Clarkson DT (1999) Nitrate and ammonium nutrition of plants: physiological and molecular perspectives. In: Advances in botanical research, vol 30. Academic Press, pp 1–90

Gao Z-F, Yang X, Mei Y, Zhang J, Chao Q, Wang B-C (2023) A dynamic phosphoproteomic analysis provides insight into the C4 plant maize (*Zea mays* L.) response to natural diurnal changes. Plant J 113(2):291–307

Hallauer AR, Carena MJ (2009) Maize. Springer

Klupczyńska EA, Pawłowski TA (2021) Regulation of seed dormancy and germination mechanisms in a changing environment. Int J Mol Sci 22:1–18

Kominko H, Gorazda K, Wzorek Z (2021) Formulation and evaluation of organo-mineral fertilizers based on sewage sludge optimized for maize and sunflower crops. Waste Manag 136:57–66

Kumar B, Rakshit S, Kumar S, Singh BK, Lahkar C, Jha AK, Kumar K et al (2022) Genetic diversity, population structure and linkage disequilibrium analyses in tropical maize using genotyping by sequencing. Plants Theory 11(6):799

Li Z, Guo R, Li M, Chen Y, Li G (2020) A review of computer vision technologies for plant phenotyping. Comput Electron Agric 176:105672

Maitra S, Shankar T, Banerjee P (2020) Potential and advantages of maize-legume intercropping system. In: Maize-production and use, pp 1–14

McHale LK, Haun WJ, Xu WW, Bhaskar PB, Anderson JE, Hyten DL, Gerhardt DJ, Jeddeloh JA, Stupar RM (2012a) Structural variants in the soybean genome localize to clusters of biotic stress-response genes. Plant Physiol 159(4):1295–1308

McHale LK, Haun WJ, Xu WW, Stinchcombe JR (2012b) Genomic features and evolution of corn plants. Plant Cell 24(11):4815–4832

Miao T, Zhu C, Tongyu X, Yang T, Li N, Zhou Y, Deng H (2021) Automatic stem-leaf segmentation of maize shoots using three-dimensional point cloud. Comput Electron Agric 187:106310

Monterrubio-Solís C, Barreau A, Ibarra JT (2023) Narrating changes, recalling memory: accumulation by dispossession in food systems of indigenous communities at the extremes of Latin America. Ecol Soc 28(1):3

Naseem M, Singh V, Ahmed K, Mahroof M, Ahamad G, Abbasi E (2022a) Architecture of automatic irrigation system in hilly area using wireless sensor network: a review. In: In 2022 2nd International conference on emerging frontiers in electrical and electronic technologies (ICEFEET). IEEE, pp 1–6

Naseem M, Alam M, Ahmad K, Singh V, Mahroof M, Ahamad G (2022b) Machine learning approaches for automatic irrigation system in hilly areas using wireless sensor networks

Oehme LH, Reineke A-J, Weiß TM, Würschum T, He X, Müller J (2022) Remote sensing of maize plant height at different growth stages using UAV-based digital surface models (DSM). Agronomy 12(4):958

Qiu R, Zhang M, He Y (2022) Field estimation of maize plant height at jointing stage using an RGB-D camera. Crop J 10(5):1274–1283

Ranum P, Peña-Rosas JP, Garcia-Casal MN (2014) Global maize production, utilization, and consumption. Ann N Y Acad Sci 1312(1):105–112

Sadras V, Calderini D (2009) Crop physiology: applications for genetic improvement and agronomy. Academic Press

Santos D, Caio L, Abendroth LJ, Coulter JA, Nafziger ED, Suyker A, Jianming Y, Schnable PS, Archontoulis SV (2022) Maize leaf appearance rates: a synthesis from the United States corn belt. Front Plant Sci 13:872738

Schunck D, Magistri F, Rosu RA, Corneließen A, Chebrolu N, Paulus S, Léon J et al (2021) Pheno4D: a spatio-temporal dataset of maize and tomato plant point clouds for phenotyping and advanced plant analysis. PLoS One 16(8):e0256340

Smith AM, Zeeman SC (2006) Quantification of starch in plant tissues. Nat Protoc 1(3):1342–1345

Subedi B, Poudel A, Aryal S (2023) The impact of climate change on insect pest biology and ecology: implications for pest management strategies, crop production, and food security. J Agric Food Res 14:100733

ur Rehman F, Adnan M, Kalsoom M, Naz N, Husnain MG, Ilahi H, Ilyas MA, Yousaf G, Tahir R, Ahmad U (2021) Seed-borne fungal diseases of maize (*Zea mays* L.): a review. Agrinula Jurnal Agroteknologi dan Perkebunan 4(1):43–60

World Agricultural Production (2023) http://www.worldagriculturalproduction.com/crops/corn.aspx

Xu H, Ming B, Wang K, Xue J, Hou P, Li S, Xie R (2023) Quantitative analysis of maize leaf collar appearance rates. Plant Physiol Biochem 196:454–462

Zhang F, Cui Z, Chen X, Ju X, Shen J, Chen Q, Liu X et al (2012) Integrated nutrient management for food security and environmental quality in China. Adv Agron 116:1–40

Zhao M, Tang S, Zhang H, He M, Liu J, Zhi H, Sui Y et al (2020) DROOPY LEAF1 controls leaf architecture by orchestrating early brassinosteroid signaling. Proc Natl Acad Sci 117(35): 21766–21774

Chapter 20
Comprehensive Analysis of Deep Learning Models for Plant Disease Prediction

Narendra Pal Singh Rathor, Praveen Kumar Bhanodia, and Aditya Khamparia

Abstract The major crop across the world is wheat. The growth of the wheat crop is significantly affected by various types of plant diseases. Technology advancements are instrumental in the recognition and prediction of plant diseases by diagnosing the health of plant leaves. To cut losses and achieve intelligent, healthy farming, the use of computer vision and pattern recognition to identify disease has been researched. The rapid and precise automatic detection of diseases is now possible with image recognition techniques. This work focuses on developing methods for wheat plant disease identification using deep learning models. There are many deep learning models proposed by researchers, but the majority provide poor testing results if some variation (rotation, tiling, and other abnormal image orientations) is there in the images; moreover the models do not store relative spatial relationships among the features captured. Thus the intent of this work is to implement the Whe-C-Net hybrid model, which combines features of VGG16 and CapsNet. The VGG16 model is initially employed for feature extraction. After that, misalignment issues with the current deep learning models are dealt with using CapsNet layers. Later, dropouts, sigmoid activation functions, and fully connected layers are employed. To avoid overfitting and create a Whe-C-Net model that is more broadly applicable, dropouts are used. On the dataset of wheat plant photos, the effectiveness of Whe-C-Net is confirmed. Compared to competing models like pre-trained MobileNet, it obtains a better validation accuracy of 98%, which is noteworthy. The accuracy rates for Xception, ResNetAQ1, MobileNet and VGG16 were 96%, 96%, 65%, and 93%, respectively.

Keywords Deep learning · Wheat · Plant disease · CNN · Whe-C-Net · Capsule net

---

N. P. S. Rathor · P. K. Bhanodia (✉)
Computer Science Engineering, Acropolis Institute of Technology and Research, Bhopal, Madhya Pradesh, India
e-mail: narendrasingh@acropolis.in

A. Khamparia
Department of Computer Science, Babasaheb Bhimrao Ambedkar University, Satellite Centre, Amethi, Uttar Pradesh, India

A. Khamparia et al. (eds.), *Microbial Data Intelligence and Computational Techniques for Sustainable Computing*, Microorganisms for Sustainability 47,
https://doi.org/10.1007/978-981-99-9621-6_20

20.1 Introduction

India is primarily an agriculturally oriented nation, with agriculture employing 70% of the people (http://www.fao.org/india/fao-in-india/india-at-a-glance/en/ n.d.). Agriculture has grown far more significant in recent years than it was a few years ago, when plants were just used to feed people and animals. This is because plants are being employed to generate electricity and other forms of energy in order to improve mankind's living conditions. However, there are other plant diseases that can cause significant economic and social impacts. It may possibly result in significant ecological damage. To avoid such losses, it is therefore preferable to detect diseases precisely and promptly. Plant diseases can be detected using a variety of methods, including both manual and computerized technologies. The majority of plant illnesses manifest themselves as spots on the leaves that are more obvious to the naked eye. On the other hand, some diseases do not appear on the leaves at all, while others appear later after they have already caused significant damage to the plants. In such cases, it is suggested that computerized systems would be the only option for obtaining actual, timely exploitation of some pretty advanced algorithms and analytical tools, ideally through the use of powerful microscopes and other instruments (Mohanty et al. 2016). If the system was easy to use and accessible, it may be a valuable tool for farmers in areas of the world where there is a lack of infrastructure for providing agronomic and phytopathological guidance. Furthermore, on a large scale, the technology might be paired with autonomous agricultural tractors to correctly and quickly find phytopathological concerns throughout the cultivation field, utilizing continuous image capture. All of this is true, of course, if the system can attain high levels of performance in detecting and diagnosing specific diseases in real-world situations (Johannes et al. 2017). Machine learning-related artificial intelligence applications have grown exponentially in recent years, thanks to the development of computational systems, particularly Graphical Processing Unit (GPU) embedded processors, resulting in the development of novel methodologies and models, which have now formed a new category, deep learning (DL) (LeCun et al. 2015). A plant pathologist must have good observation skills in recognizing distinctive symptoms in order to diagnose plant diseases accurately (Riley et al. 2002). Because amateur gardeners and hobbyists may have more difficulty diagnosing defected plants than trained plant pathologists, variations in symptoms suggested by diseased plants may lead to an incorrect diagnosis. An automated system that can identify plant illnesses based on the look and visual symptoms of the plant might be extremely useful to both amateur gardeners and skilled professionals as a disease diagnosis verification process (Sladojevic et al. 2016). Nowadays deep learning provides a very efficient way to extract features from images, through which we can easily identify diseases in plant leaves. Here, the work demonstrated various deep learning models to identify the plant leaf identification and detection. In this work four different deep learning models are applied, namely, VGG16, XceptionNet, MobileNet, and ResNet, using transfer learning for the detection of plant leaves. All four model performances on the wheat leaf dataset in terms of accuracy and loss are compared. After that a hybrid model is used to improve the performance to detect the plant leaf disease. In this work a dataset of wheat leaves has been used, and all deep learning models are compared on the same dataset. Our work mainly concentrated on a wheat leaf disease called leaf rust. After applying all models, we have achieved the highest accuracy, i.e., 98.93%, through our hybrid model. The details of the work carried out are mentioned in the coming chapters.

20.2 History

Several researchers have tried on agricultural applications that use image processing, pattern recognition techniques, machine learning, and deep learning throughout the last decade. The disease can affect any part of the plant, including the roots, stem, fruit, or leaves. When technology has updated in the last 5–10 years, everybody is moving to small and handy devices. Everybody wants to process their results quickly and efficiently. Earlier for disease detection in plant leaves we were dependent only on physical scanning using our naked eyes, but now so many technologies are available, using which we can easily and quickly detect diseases and provide information about the diseases. Neural network training proves significant changes in this area. Table 20.1 shows the summary of the work proposed in this domain.

We have used deep learning to recognize wheat plant diseases in our research, which is motivated by the advancement of deep learning techniques and their practical use. Researchers studied deep learning approaches for plant disease recognition using leaf images, but no evidence of this was found in a search of the state-of-the-art literature.

20.3 Research Methodology

DL enhances classical ML by having "depth" to the model and updating the data with different functions that permit the representation of data hierarchically, across a number of levels of abstraction (Patrick et al. 2019). Feature learning, or the automated extraction of features from raw data, is a significant benefit of DL, with higher-level features being generated by the composition of lower-level features (Sarayloo and Asemani 2015b). Because of the more sophisticated models utilized, which allow huge parallelization, DL can handle more complex problems especially well and quickly (Lee et al. 2015). Models used in deep learning (DL) can improve classification accuracy or minimize regression error when substantial datasets defining the problem are readily available. While there are many distinct types of deep learning (e.g., unsupervised pre-trained networks, convolutional networks, recursive networks, and recurrent networks), every DL uses some combination of convolutions and fully connected layers as well as gates and memory cells. Classification and

Table 20.1 A comprehensive summary of noteworthy contributions in the plant disease analysis domain

Author	Steps	Attribute	Plants	Diseases	Analysis
Zahra Sarayloo and Davud Asemani (2015a)	1: Segmentation 2: Extract feature 3: Select features 4: Disease classification	Color, shape, and texture	Wheat	Stem rust, leaf rust, tan spot, stripe rust, pink snow mold, septoria, and powdery mildew	No of images not specified
Sue Han Lee, Chee Seng Chan, Paul Wilkin, Paolo Remagnino (2015)	Convolutional neural network 1: Layer/multilayer perceptron 2: Support vector machine, radial basis function	–	Forty-four plants but specification not mentioned by the author	–	Plant names missing and on which attributed work has performed also missing
Yogesh Dandawate, Radha Kokare (2015)	Image acquisition of soybean leaves. Extraction of soybean Leaves from complex background Statistical analysis and Disease classification	Color transformation, background subtraction, shape analysis	Soybean	–	On which soybean disease has been carried out not mentioned. It just shows healthy vs. infected images. Also worked on only 120 images
Srdjan Sladojevic, Marko Arsenovic, Andras Anderla, Dubravko Culibrk and Darko Stefanovic (2016)	Image preprocessing and labeling, 2. Augmentation process. Neural network training. Performed tests. Fine-tuning. Equipment.		Apple, grapevine, pair, peach, pear	Powdery mildew, downy mildew, apple rust, apple powdery mildew, apple, pear, *Erwinia amylovora*	Presented their work on fruits, leaves, and different diseases
Alvaro Fuentes, Dong Sun Park, Sook Yoon, Hong Youngki, Yujeong Lee (Fuentes et al. 2016)	Dataset, labeling, disease correlation		Tomato	Leaf mold, gray mold, canker, plague	Presented their work on tomato leaf disease detection but not clearly mentioned how many

					images and dataset have been used for the work
Kadir Sabanci, Mustafa Akkaya (2016)	Database (Waikato environment for knowledge analysis), multilayer perceptron algorithm, K-nearest neighbors algorithm, J48 decision tree algorithm, naive Bayes algorithm		Wheat	–	Not properly mentioned on which wheat disease work has been carried out. Work actually carried on wheat seed
Alvaro Fuentes, Sook Yoon, Sang Cheol Kim and Dong Sun Park (2017)	Data collection, data annotation, data augmentation, disease and pest detection		Tomato	Leaf mold, gray mold, canker, plague, miner, powdery mildew, nutritional excess	Work carried out on 5000 tomato images and used faster R-CNN
Yang Lu, Shujuan Yi, Nianyin Zeng, Yurong Liud, Yong Zhang (2017b)	Convolutional layer, stochastic pooling layer, Softmax regression, training algorithm		Rice	Rice false smut (RFS), rice blast (RB), rice brown spot (RBS), rice seedling blight (RSEB), Rice Sheath blight (RSHB), rice sheath rot (RSR), rice bacterial leaf blight (RBLB), rice bacterial sheath rot (RBSR), and rice bacterial wilt (RBW)	Work carried out on 500 images of rice and presented a simple CNN to detect the disease in rice leaf
Bin Liu, Yun Zhang, DongJian He, and Yuxiang Li (2018)	Apple leaf pathological image acquisition, image processing and generating pathological images		Apple leaf	Mosaic, rust, brown spot, and Alternaria leaf spot	Work carried out on 13,689 images of apple leaf using AlexNet, GoogLeNet, and Softmax regression has been used

(continued)

Table 20.1 (continued)

Author	Steps	Attribute	Plants	Diseases	Analysis
Mohammed Brahimi, Kamel Boukhalfa & Abdelouahab Moussaoui (2017)	Preprocessing, feature extraction, classification,		Tomato	Tomato yellow leaf curl virus, tomato mosaic virus, spider mites, Septoria spot, leaf mold, late blight, bacterial spot	Work carried out on 14,828 images, and machine learning algorithms K-nearest neighbors algorithm (KNN) and support vector machine (SVM) have been used
Jayme G.A. Barbedo (2018)	–		Olive tree, corn, apple, tomato, potato, rice	Anthracnose, tropical rust, southern corn rust, scab, southern corn leaf blight, Phaeosphaeria leaf spot, Diplodia leaf streak, Physoderma brown spot, northern leaf blight	Simple CNN has been applied for the work. Details of the steps performed are also missing
Juncheng Maa, Keming Dua,Feixiang Zhenga, Lingxian Zhangb, Zhihong Gongc, Zhongfu Suna (2018)	Image datasets, CNN, evaluation of the DCNN		Cucumber	Anthracnose, downy mildew, powdery mildew, target leaf spots	Work carried out on cucumber leaves and also used KNN and SVM as classifiers
Artzai Picon, Aitor Alvarez-Gila, Maximiliam Seitz, Amaia Ortiz-Barredo, Jone Echazarra,Alexander Johannes (2019)	Deep convolutional neural networks (CNNs)		Wheat	Tan spot, rust, Septoria	ResNet-50 has been used on wheat leaves for disease detection
G. Prem Rishi Kranth, M. Hema Lalitha, Laharika Basava, Anjali Mathur (2018)	Decision tree, K-means clustering, naive Bayes, random forest, artificial neural network		Apple, strawberry	Wilting, spot, powdery mildew, galls, dryness	Machine learning algorithm used for the detection of apple and strawberry diseases

Md Mehedi Hasan, Joshua P. Chopin, Hamid Laga and Stanley J. Miklavcic (2018a)	CNN		Wheat		Worked on wheat images and presented simple CNN and performed work to identify SPIKE in wheat
Konstantinos P. Ferentinos (2018)	VGG, AlexNet		Apple, banana, blueberry, cabbage, cassava, celery, cherry, corn, cucumber, eggplant, gourd, grape, onion, orange, peach, soybean	Apple scab, black rot, black Sigatoka, banana speckle, powdery mildew, downy mildew, esca, leaf blight, Huanglongbing, early blight, cercospora	Pre-trained model architecture have been used to identify the plant leaves disease detection
M. Akila, P. Deepan (2018)	Data collection, image annotation and augmentation, image analysis, feature extraction		Sugarcane, cotton, potato, carrot, chilly, brinjal, rice, wheat, banana and guava	–	Faster region-based convolutional neural network (faster R-CNN) and region-based fully convolutional network (R-FCN) are mainly used, but the disease name on which work has been carried out is not mentioned specifically
Satwinder Kaur, Garima Joshi, Renu Vig (2019)	–	–	Apple, blueberry, corn, cherry, grapes, orange, peach, pepper, potato, soybean, squash, tomato, strawberry	Apple scab, apple cedar rust, cherry powdery mildew, corn leaf spot, corn common rust, grape black Measles, peach bacterial spot, potato early blight, squash powdery mildew, tomato leaf mold, tomato mosaic virus, tomato leaf	GoogLeNet architecture has been used to identify disease

(continued)

Table 20.1 (continued)

Author	Steps	Attribute	Plants	Diseases	Analysis
				curl Virus	
Abirami Devaraj, Karunya Rathan, Sarvepalli Jaahnavi and K Indira (2019)	Image acquisition, image preprocessing, image segmentation, classification		Rice	Alternaria alternata, bacterial blight	K-means and random forest machine learning algorithm have been used. Not clearly mentioned in the dataset details
Chowdhury Rafeed et al. (2021)			Tea, apple, tomato, grapevine, peach, and pear	False smut, sheath blight, sheath rot, bacterial leaf blight (BLB), neck blast and brown spot	VGG16, ResNet-50, inception-v3, Xception, inception-ResNet-v2 pre-trained model architecture have been used and mainly focused on rice leaves
Justine Boulent, Samuel Foucher, Jerome Theau1, and Pierre-Luc St-Charles (2019)	–	–	–	–	AlexNet, DenseNet, inception, VGG, ResNet model details have been mentioned but not clearly mentioned on which plant and diseases work has been carried out
X.E. Pantazi, D. Moshou, A.A. Tamouridou (2019)	Local binary patterns (LBPs) for feature extraction and one-class classification for classification		*Vitis*, *Cucurbita pepo*, *Cucumis sativus*	Powdery mildew, black rot, downy Mildew	Support vector machine learning algorithms have been used

prediction are especially effectively performed by DL models due to their highly hierarchical structure, enormous learning capacity, and adaptability to many extremely difficult tasks (from a data analysis standpoint) (Lee et al. 2015). Deep learning is very useful nowadays in the agriculture domain. In deep learning various models are proposed by authors for wheat leaf detection. This research covered simple convolution neural networks, XceptionNet, MobileNet, ResNet, and VGG16. All these models are available and trained at Keras. After deep learning models are studied, move on to the wheat leaf diseases dataset. Dataset plays a key role in deep learning to train the model. The dataset collection visited a number of recognized and reputed dataset repositories like PlantVillage, Mendeley dataset, and Kaggle. Mohammed, Assem (Mohammed 2020) presented a Wheat dataset which is included in our research. The detailed description of the dataset is covered in the next chapter. After that research moved on to transfer learning studies and implementation of deep learning models. Transfer learning (Lumini and Nanni 2019) is a technique which provides a way through which we can reuse an already developed model with our weights and trained model according to our problem. Covered ResNet, VGG, MobileNet, and XceptionNet in this research use transfer learning. Implemented a model which was trained with cutomized weights on Wheat dataset and listed out accuracy, loss, Precision, Recall, and F-Score, to provide a complete view of system's performance (Goutte and Gaussier 2005). Then our research moved to developing a new hybrid model which is better in terms of the performance of existing implemented transfer learning models as shown in Fig. 20.1. Sabour et al. (2017) presented a new architecture, i.e., Capsule Networks. Deep learning models that use Capsule Networks as their fundamental building blocks have grown in popularity since they were first introduced in 2017. A routing technique known as "routing by agreement" is used in the most common version of CapsNets. The vector output of this method substitutes the scalar output of CNNs and the pooling technique of CNNs. Due to the fact that CNNs require large amounts of training data and are unable to identify object position and deformation, the development of Capsule Networks was made possible. Known as Capsule Networks, they are the latest craze in deep learning. They have lived up to this expectation since their performance in issues such as image recognition, natural language processing, object identification, object segmentation, and language translation has consistently outperformed that of convolutional neural networks (Patrick et al. 2019).

20.4 Deep Learning for Wheat Leaf Disease Detection

Deep learning methods are increasingly being used to solve machine vision challenges. Various researchers have previously investigated plant and leaf identification using various methodologies. Initially, color information was employed to identify the plant from the soil in these difficulties (Woebbecke et al. 1995; Gerhards and Christensen 2003). The morphology of veins was used in some investigations (Sack et al. 2008; Scoffoni et al. 2011). The leaf veins include a variety of texture and form features that can help with plant identification using eyesight. The shape information was also utilized in certain leaf-based research (Agarwal et al. 2006; Neto et al.

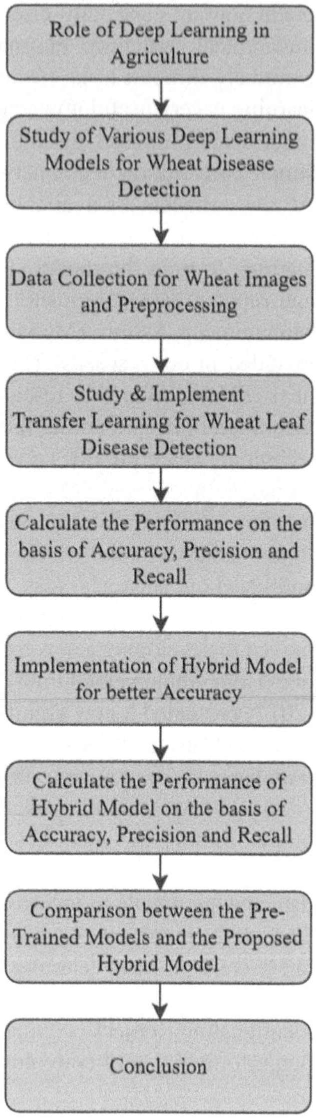

Fig. 20.1 Systematic steps involved in plant disease analysis and prediction

2006). Several researchers (Husin et al. 2012) integrated the shape and texture information retrieved from the leaf, while others used color and texture information (Pydipati et al. 2006). Larese et al. (2014a) offer a method for extracting the morphological leaf vein information using computer vision techniques. Machine learning methods are then utilized to predict the three different plant species using the extracted data. Later, in a new study endeavor, these researchers enhanced this strategy (Larese et al. 2014b). Plant genre classification uses spectroscopic approaches as well. The feature extraction in these approaches is based on the reflectance of the infrared, multispectral, and visible bands (Mattila et al. 2013; Wang et al. 2007; Tyystjärvi et al. 2011). In their research, Muthevi and Uppu

(2017) used Local Binary Patterns (LBP). For feature extraction, different types of LBP extraction methods were used, including the signed component of CLBP (SCLBP). Murat et al. (2017) used a combination of form descriptors to classify the leaves of tropical shrub species. Flavia and Swedish leaf datasets were used to test the suggested technique. View rotation invariant characteristics taken from the fast Fourier transform and discrete wavelet transform were employed by Yousefi et al. (2017). Yu et al. (2016) devised a strategy based on leaf venation and contour. A method for invariant leaf detection was proposed by Horaisová and Kukal (2016).

Deep learning is a relatively recent approach to plant and leaf identification. To work with leaf veins, Grinblat et al. (2016) presented a CNN model. Leaf veins are retrieved as a binary mask and used to train an end-to-end CNN model. To create a weed segmentation system, dos Santos Ferreira et al. (2017) integrated the superpixel segmentation algorithm with CNN. On images segmented with the superpixel method, CNNs are used in weed detection. For leaf identification, Barré et al. (2017) developed a CNN architecture. For the classification, they used the Foliage, LeafSnap, and Flavia datasets and developed a CNN model. In their work, they developed an architecture that was similar to well-known CNN models like AlexNet and CifarNet. Transfer learning from pre-trained CNN models was employed by Jeon and Rhee (2017). For feature extraction, a pre-trained GoogLeNet was employed. GoogLeNet was given leaf photos in various scales as input, and the activation values of different layers were saved as features. Lu and Hu (Jiang et al. 2017a) proposed an automatic wheat leaf disease detection system and proposed a CNN model to detect the disease. Hasan et al. (2018b) has also worked on wheat spikes detection and analysis. They have worked on the ResNet-50 deep learning model in their research.

20.5 Transfer Learning for Wheat Leaf Disease Detection

Transfer learning is typically a machine learning method, wherein the CNN models trained on one task have been used as the base model for some other task (Lumini and Nanni 2019). We can use a pre-trained network on big, labeled datasets, like public picture datasets, to establish the weights instead of starting the training from scratch by randomly initializing the weights. In the same fashion (Chen et al. 2020) Junde Chen used pre-trained models on the ImageNet large dataset, which they applied to the specific task established by the objective dataset. The main processes of the transfer learning approach (Chen et al. 2020) are described as follows.

20.5.1 Identify Transfer Learning Base Model

Using the pre-trained CNN model, determine the transfer learning base networks and give the network weights (W1, W2,. . . .W n) to the base networks. It is possible to obtain the weights of bottom layers from a well-trained CNN (https://ker-as.io/applications/) via the Ker-as API.

20.5.2 *Create a New Neural Network*

The network structure can be changed based on the bottom layers, which can include altering layers, adding layers, and removing layers from networks, among other things. It is possible to generate a new network structure in this way.

20.5.3 *Perform Fine-Tuning on Created New Neural Networks*

In this step we are able to modify the layer parameters. Edna et al. (Too et al. 2019) have performed fine-tuning on VGG16, done by truncating the original Softmax layer and replacing it with their own.

Kathiresan et al. (2021) have also proposed the idea of transfer learning for the detection of rice leaf diseases as shown in Fig. 20.2. Goyal et al. (2021) have worked on wheat leaf and spike detection using transfer learning. They have included the VGG16 and ResNet-50 base models for transfer learning.

20.6 Hybrid Approach

In the hybrid approach we apply pre-trained models using the transfer learning method on the wheat leaf dataset.

The pre-trained models used for the classification and detection of plant diseases are as follows.

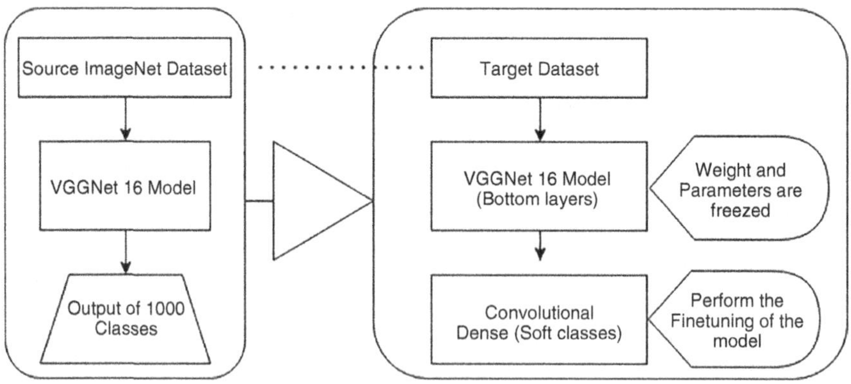

Fig. 20.2 Process involved in transfer learning

20.6.1 MobileNet

This is the first TensorFlow model designed specifically for mobile apps, and it uses significantly separable convolutions. When compared to MobileNet's use of conventional convolutions on networks with the same depth, the number of parameters is significantly reduced. It further leads to deep networks with low weight (Howard et al. 2017).

20.6.2 XceptionNet

For the deep neural network it has 71 deep layers proposed by Francois Chollet based on smart and thorough separation. In comparison to traditional convergence depth this separable convolution model is more efficient and effective (Chollet 2017).

20.6.3 VGG16

The VGG16's pre-training architecture consists of 13 convolutional layers, 5 additional layers, and 3 dense pool layers. A global average pooling layer and two dense layers with the activation algorithms ReLu and Softmax are among the changes made to the VGG16. The drop-out rate for both thick layers is 0.5 (Simonyan and Zisserman 2014).

20.6.4 ResNet

Kaiming He et al. in 2015 introduced ResNet referred to as the Residual Network, with this model the problem of vanishing gradient specifically in deep systems by permitting the alternative for flow across gradient (He et al. 2015).

In this study the hybrid model is implemented using Google Colab, Keras, and TensorFlow. Pre-trained VGG model is used for transfer learning (Hasan et al. 2018b), with dense and global average pooling layers; the activation function used here is ReLu. The VGG model is followed by a capsule network with three added routings. From the results obtained it is evident that Whe-C-Net superseded ResNet, XceptionNet, MobileNet, and VGG model. Refer to Fig. 20.3 for the proposed architecture of the Whe-C-Net model architecture.

The Whe-C-Net architecture has three distinct layers:

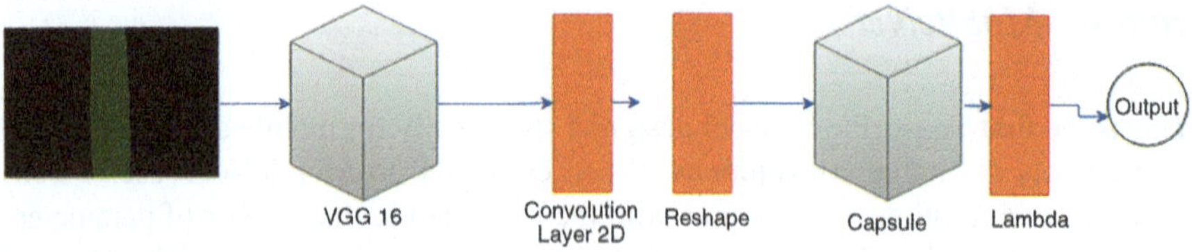

Fig. 20.3 The proposed Whe-C-Net model architecture

20.6.5 Extraction of Features Using VGG and Capsule Network Layers

The VGG16 is a pre-trained CNN model consisting of 13 convolutional, 5 max pooling, and 3 dense layers. In total out of 23, only 14 are weighted layers.

20.6.6 Classification Layers

In the output of capsule and VGG16 networks the first layer is known as the flattened layer followed by the next layer, i.e., the dense layer, with ReLu activation function and 0.5 dropout. The third and the last is also termed the dense layer, with the Softmax activation function used for deduction in depth.

20.6.7 Spatial Transformer Layers

In the spatial transformer Lambda λ is used Default [−0.5: 0.5] to convert the wheat-leaf image characteristics to a normal value of 0.0. For faster processing divide the input into mini batches and apply batch normalization.

20.6.8 Implementation Steps Involved

Image dataset loaded in RAM for processing like the previous attempt when wheat leaf image format pictures were saved in RGB. Defining the implementation architecture is shown in Fig. 20.4. The next step is to implement the metrics involved, which includes threshold accuracy with β threshold and Fβ score. Thereafter the model is employed for training and validation. The last step is validation of the model. All steps are implemented in Figs. 20.5, 20.6.

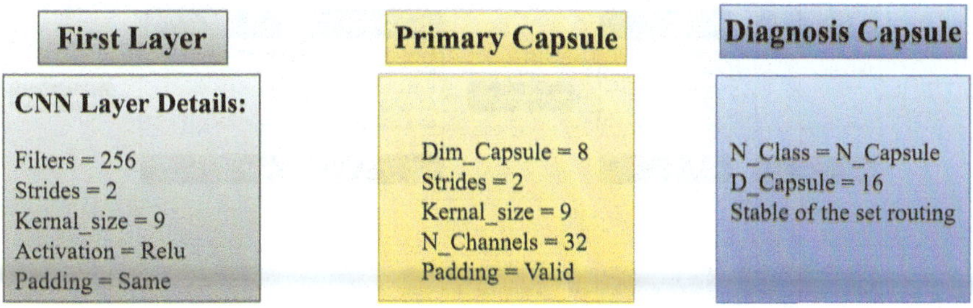

Fig. 20.4 Capsule model summary

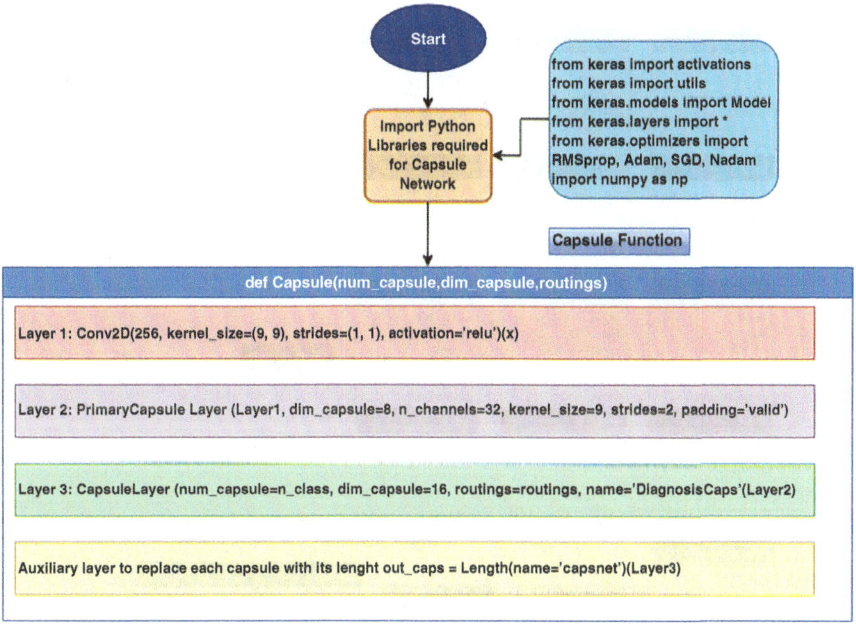

Fig. 20.5 Capsule layer configuration

Table 20.2 shows that Whe-C-Net performed better for detecting plant disease by diagnosing the wheat leaves with 98% accuracy. However, further training with more epochs can enhance the proposed model performance.

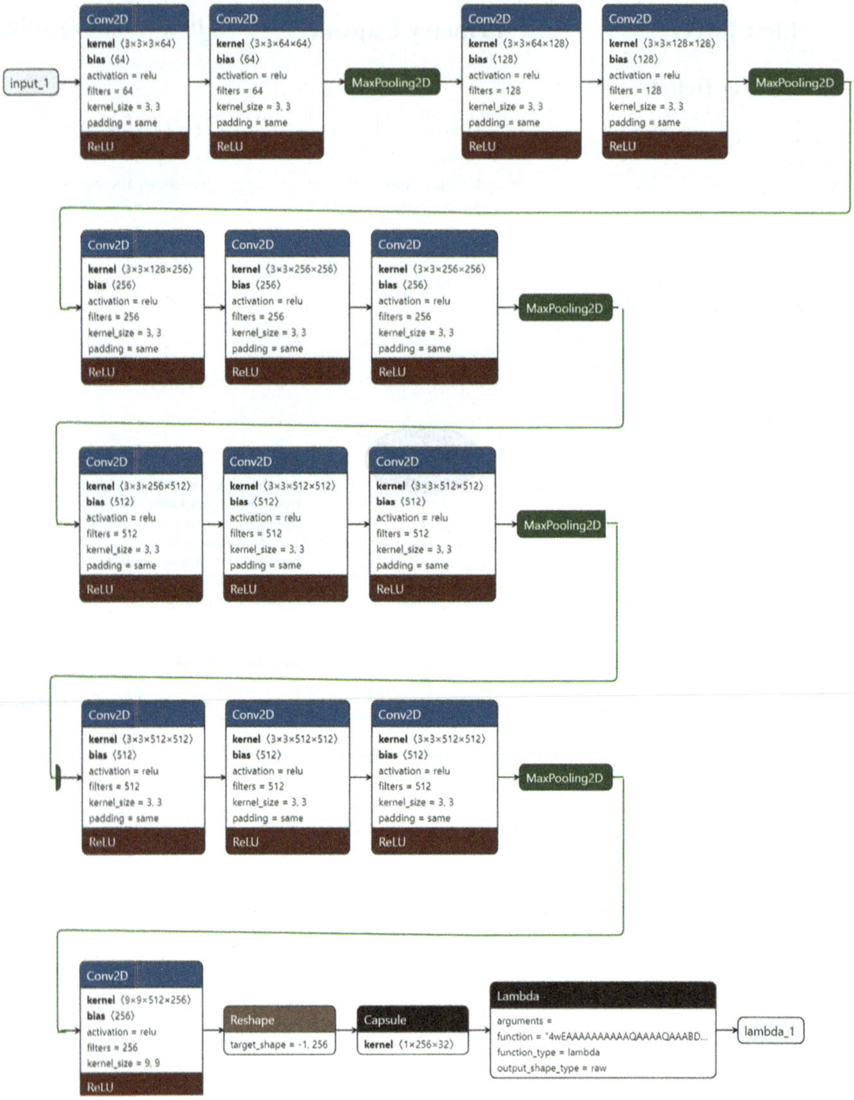

Fig. 20.6 Hybrid model detailed architecture

20.7 Result and Discussion

The goal of the study is to use the deep learning models discussed to provide wheat leaf disease detection. We used photos of normal leaves and leaves with rust to train the models. To begin with, the ResNet-50 models have been used, which Artzai Picon et al. (2019) also used to detect wheat leaf diseases. The validation accuracy for the ResNet-50 model was 65% after training. Precision, recall, and f1-score are

Table 20.2 Model performance analysis and parameters

Dataset	Design (model)	Recall	Precision	f1-score	Validation accuracy	No of parameters
Mendeley dataset (Kaur et al. 2019)	ResNet-50	0.58	0.57	0.57	65%	Parameter: 23,788,418 Trainable: 200,706
	MobileNet	0.44	0.43	0.43	81%	Parameter: 3,379,395 Trainable: 150,531
	XceptionNet13	0.12	0.67	0.20	52%	Parameter: 21,271,082 Trainable: 409,602
	VGG	0.87	0.98	0.92	93%	Parameter: 16,824,130 Trainable: 2,109,442
	Whe-C-net	0.99	0.98	0.98	98%	Parameter: 25,339,968 Trainable: 17,704,704

additional parameters that the author has not addressed but eventually are crucial to analyzing the classifier. Thereafter, the MobileNet model was used to forecast the diseases, with an accuracy rate of 81%. Despite the fact that we have achieved better accuracy, it is not a reliable indicator of model performance.

Further in comparison to ResNet-50, the MobileNet model has higher accuracy but worse precision, recall, and f1-score. Later we began training using a new model, XceptionNet13, for the detection and prediction of possible disease across wheat crops but, regrettably, did not attain the desired outcome. The results obtained were lower in precision, recall, and f1-score. The VGG was the next model trained on the dataset, and the results obtained in terms of accuracy, precision, recall, and f1-score were anticipated. Finally, we used a hybrid model that offered greater precision, recall, accuracy, and f1-score compared to the rest of the models trained and tested. With the hybrid model, we can forecast the diseases that affect wheat leaves with a sizable degree of accuracy, precision, recall, and f1-score. The results are shown in Fig. 20.7.

20.8 Conclusion

The aim of this work is to propose Whe-C-Net a hybrid deep learning model for the detection of diseases in wheat leaves. For the experimental study the technique is applied to a standard dataset of wheat plants downloaded from Mendeley. The model

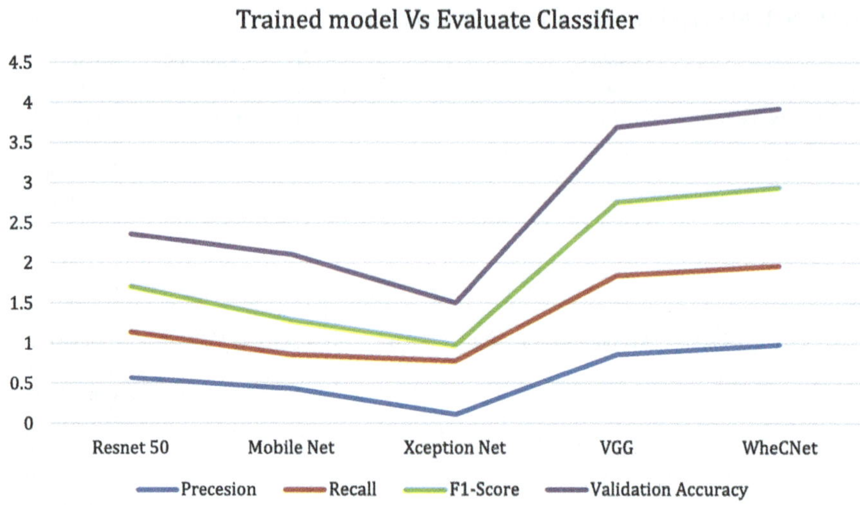

Fig. 20.7 Evaluative comparative analysis of different CNN-based classification models

Whe-C-Net has performed fairly significantly and generated an optimum validation accuracy of 98%, compared to ResNet-50 (65%), Xception13 (52%), MobileNet (81%), and VGG (93%). A further prerequisite is the gathering of actual images from the fields in order to increase the effectiveness of the Whe-C-Net model and utility in the area of agriculture for farmers to detect diseases in wheat plants using leaves. Thus this hybrid model is used to boost accuracy without lengthening training time because the simple CNN typically performs poorly in data augmentation. The outcome of the suggested hybrid model Whe-C-Net describes the deep learning model that can be used to enhance the diagnosis of plant disease using the leaves in comparison to the conventional manual approaches.

References

Agarwal G, Belhumeur P, Feiner S, Jacobs D, Kress WJ, Ramamoorthi R, Bourg A, Dixit N, Ling H, Mahajan D et al (2006) First steps toward an electronic field guide for plants. Taxon 55(3):597–610

Akila M, Deepan P (2018) Detection and classification of plant leaf diseases by using deep learning algorithm. Int J Eng Res Technol 6(07)

Barbedo JG (2018) Factors influencing the use of deep learning for plant disease recognition. Biosyst Eng 172:84–91

Barré P, Stöver BC, Müller KF, Steinhage V (2017) Leafnet: a computer vision system for automatic plant species identification. Ecol Informat 40:50–56

Boulent J, Foucher S, Théau J, St-Charles PL (2019) Convolutional neural networks for the automatic identification of plant diseases. Front Plant Sci 10:941

Brahimi M, Boukhalfa K, Moussaoui A (2017) Deep learning for tomato diseases: classification and symptoms visualization. Appl Artif Intell 31(4):299–315

Chaudhari SB, Wagaskar V, Shaikh M, Shelke O, Shirsath V (2021) Plant disease detection implementation using tensorflow. Int Res J Mod Eng Technol Sci 3(6)
Chen J, Chen J, Zhang D, Sun Y, Nanehkaran YA (2020) Using deep transfer learning for image-based plant disease identification. Comput Electron Agric 173:105393., ISSN 0168-1699. https://doi.org/10.1016/j.compag.2020.105393
Chollet F (2017) Xception: deep learning with depthwise separable convolutions. In: Proceedings of the IEEE conference on computer vision and pattern recognition, pp 1251–1258
Dandawate, Y., & Kokare, R. (2015). An automated approach for classification of plant diseases towards development of futuristic Decision Support System in Indian perspective. In 2015 International conference on advances in computing, communications and informatics (ICACCI) (pp. 794–799). IEEE
Devaraj, A., Rathan, K., Jaahnavi, S., & Indira, K. (2019). Identification of plant disease using image processing technique. In 2019 International Conference on Communication and Signal Processing (ICCSP) (pp. 0749–0753). IEEE
Dos Santos Ferreira A, Freitas DM, da Silva GG, Pistori H, Folhes MT (2017) Weed detection in soybean crops using convnets. Comput Electron Agric 143:314–324
FAO (n.d.). http://www.fao.org/india/fao-in-india/india-at-a-glance/en/
Ferentinos KP (2018) Deep learning models for plant disease detection and diagnosis. Comput Electron Agric 145:311–318
Fuentes A, Lee Y, Hong Y, Yoon S, Park D (2016) Characteristics of Tomato Plant Diseases—A study for tomato plant disease identification
Fuentes A, Yoon S, Kim SC, Park DS (2017) A robust deep-learning-based detector for real-time tomato plant diseases and pests recognition. Sensors 17(9):2022
Gerhards R, Christensen S (2003) Real-time weed detection, decision making and patch spraying in maize, sugarbeet, winter wheat and winter barley. Weed Res 43(6):385–392
Goutte C, Gaussier E (2005) A probabilistic interpretation of precision, recall and F-score, with implication for evaluation. Lect Notes Comput Sci 3408:345–359. https://doi.org/10.1007/978-3-540-31865-1_25
Goyal L, Sharma CM, Singh A, Singh PK (2021) Leaf and spike wheat disease detection & classification using an improved deep convolutional architecture. Inform Med Unlocked 25: 100642
Grinblat GL, Uzal LC, Larese MG, Granitto PM (2016) Deep learning for plant identification using vein morphological patterns. Computer Electron Agric 127:418–424
Hasan MM, Chopin JP, Laga H, Miklavcic SJ (2018a) Detection and analysis of wheat spikes using convolutional neural networks. Plant Methods 14(1):1–13
Hasan MM, Chopin JP, Laga H, Miklavcic SJ (2018b) Detection and analysis of wheat spikes using convolutional neural networks. Plant Methods (14):100. https://doi.org/10.1186/s13007-018-0366-8. Erratum in: Plant Methods 2019 Mar 20;15:27. PMID: 30459822; PMCID: PMC6236889
He K, Zhang X, Ren S, Sun J (2015) Spatial pyramid pooling in deep convolutional networks for visual recognition. IEEE Trans Pattern Anal Mach Intell 37(9):1904–1916
Horaisová K, Kukal J (2016) Leaf classification from binary image via artificial intelligence. Biosyst Eng 142:83–100
Howard AG et al (2017) Mobilenets: efficient convolutional neural networks for mobile vision applications. arXiv:1704.04861
Husin Z, Shakaff A, Aziz A, Farook R, Jaafar M, Hashim U, Harun A (2012) Embedded portable device for herb leaves recognition using image processing techniques and neural network algorithm. Comput Electron Agric 89:18–29
Jeon W-S, Rhee S-Y (2017) Plant leaf recognition using a convolution neural network. Int J Fuzzy Logic Intell Syst 17(1):26–34
Johannes A, Picon A, Alvarez-Gila A, Echazarra J, Rodriguez-Vaamonde S, Navajas AD, Ortiz-Barredo A (2017) Automatic plant disease diagnosis using mobile capture devices, applied on a wheat use case. Comput Electron Agric 138:200–209

Kathiresan G, Anirudh M, Nagharjun M, Karthik R (2021) Disease detection in rice leaves using transfer learning techniques. J Phys Conf Ser 1911(1):012004. IOP Publishing
Kaur S, Joshi G, Vig R (2019) Plant disease classification using deep learning Google net model. Int J Innovat Technol Explor Eng 8(9):319–322
Kranth GPR, Lalitha MH, Basava L, Mathur A (2018) Plant disease prediction using machine learning algorithms. Int J Comput App:1–7
Larese MG, Namías R, Craviotto RM, Arango MR, Gallo C, Granitto PM (2014a) Automatic classification of legumes using leaf vein image features. Pattern Recogn 47(1):158–168
Larese MG, Baya AE, Craviotto RM, Arango MR, Gallo C, Granitto PM (2014b) Multiscale recognition of legume varieties based on leaf venation images. Expert Syst Appl 41(10): 4638–4647
LeCun Y, Bengio Y, Hinton G (2015) Deep learning. Nature 521:436–444. https://doi.org/10.1038/nature14539
Lee, S. H., Chan, C. S., Wilkin, P., & Remagnino, P. (2015). Deep-plant: plant identification with convolutional neural networks. In 2015 IEEE International Conference on Image Processing (ICIP) (pp. 452–456). IEEE
Liu B, Zhang Y, He D, Li Y (2018) Identification of apple leaf diseases based on deep convolutional neural networks. Symmetry 10(1):11
Lu J, Hu J, Zhao G, Mei F, Zhang C (2017a) An in-field automatic wheat disease diagnosis system. Comput Electron Agric 142:369379
Lu Y, Yi S, Zeng N, Liu Y, Zhang Y (2017b) Identification of rice diseases using deep convolutional neural networks. Neurocomputing 267:378–384
Lumini A, Nanni L (2019) Deep learning and transfer learning features for plankton classification. Ecol Inform 51:33–43
Ma J, Du K, Zheng F, Zhang L, Gong Z, Sun Z (2018) A recognition method for cucumber diseases using leaf symptom images based on deep convolutional neural network. Comput Electron Agric 154:18–24
Mattila H, Valli P, Pahikkala T, Teuhola J, Nevalainen OS, Tyystjärvi E (2013) Comparison of chlorophyll fluorescence curves and texture analysis for automatic plant identification. Precision Agric. 14(6):621–636
Mohammed A (2020) Wheat rust images for diseases map. V1. https://doi.org/10.17632/25g6cm8vhb.1
Mohanty SP, Hughes DP, Salathé M (2016) Using deep learning for image-based plant disease detection. Front Plant Sci. https://doi.org/10.3389/fpls.2016.01419. Article: 1419, 7
Murat M, Chang S-W, Abu A, Yap HJ, Yong K-T (2017) Automated classification of tropical shrub species: a hybrid of leaf shape and machine learning approach. PeerJ 5:e3792
Muthevi, A., Uppu, R.B., 2017. Leaf classification using completed local binary pattern of textures. In: 2017 IEEE 7th International Advance Computing Conference (IACC).IEEE, pp. 870–874
Neto JC, Meyer GE, Jones DD, Samal AK (2006) Plant species identification using elliptic Fourier leaf shape analysis. Comput Electron Agric 50(2):121–134
Pantazi XE, Moshou D, Tamouridou AA (2019) Automated leaf disease detection in different crop species through image features analysis and one class classifiers. Comput Electron Agric 156: 96–104
Patrick MK, Adekoya AF, Mighty AA, Edward BY (2019) Capsule networks—a survey. J King Saud Univ Comput Inf Sci
Picon A, Alvarez-Gila A, Seitz M, Ortiz-Barredo A, Echazarra J, Johannes A (2019) Deep convolutional neural networks for mobile capture device-based crop disease classification in the wild. Comput Electron Agric 161:280–290
Pydipati R, Burks T, Lee W (2006) Identification of citrus disease using color texture features and discriminant analysis. Comput Electron Agric 52(1–2):49–59
M. B. Riley, M. R. Williamson, and O. Maloy, "Plant disease diagnosis. The Plant Health Instructor," 2002

Sabanci K, Akkaya M (2016) Classification of different wheat varieties by using data mining algorithms. Int J Intell Syst App Eng 4(2):40–44
S. Sabour, N. Frosst, G.E. Hinton, Dynamic routing between capsules, 31st Conference on Neural Information Processing Systems (2017)
Sack L, Dietrich EM, Streeter CM, Sánchez-Gómez D, Holbrook NM (2008) Leaf palmate venation and vascular redundancy confer tolerance of hydraulic disruption. Proc Nat Acad Sci 105(5): 1567–1572
Sarayloo, Z., & Asemani, D. (2015a). Designing a classifier for automatic detection of fungal diseases in wheat plant: by pattern recognition techniques. In 2015 23rd Iranian Conference on Electrical Engineering (pp. 1193–1197). IEEE
Sarayloo Z, & Asemani, D. (2015b). Designing a classifier for automatic detection of fungal diseases in wheat plant: By pattern recognition techniques. In 2015 23rd Iranian Conference on Electrical Engineering (pp. 1193–1197). IEEE
Scoffoni C, Rawls M, McKown A, Cochard H, Sack L (2011) Decline of leaf hydraulic conductance with dehydration: relationship to leaf size and venation architecture. Plant Physiol 156(2): 832–843
Simonyan, K, and Zisserman A. Very deep convolutional networks for large-scale image recognition arXiv:1409.1556 (2014)
Sladojevic S, Arsenovic M, Anderla A, Culibrk D, Stefanovic D (2016) Deep neural networks based recognition of plant diseases by leaf image classification. Comput Intell Neurosci 2016
Too EC, Yujian L, Njuki S, Yingchun L (2019) A comparative study of fine-tuning deep learning models for plant disease identification. Comput Electron Agric 161:272–279. https://doi.org/10.1016/j.compag.2018.03.032. ISSN 0168-1699; (https://www.sciencedirect.com/science/article/pii/S0168169917313303)
Tyystjärvi E, Nørremark M, Mattila H, Keränen M, Hakala-Yatkin M, Ottosen C-O, Rosenqvist E (2011) Automatic identification of crop and weed species with chlorophyll fluorescence induction curves. Precision Agric 12(4):546–563
Wang N, Zhang N, Wei J, Stoll Q, Peterson D (2007) A real-time, embedded, weed detection system for use in wheat fields. Biosyst Eng 98(3):276–285
Woebbecke DM, Meyer GE, Von Bargen K, Mortensen D (1995) Color indices for weed identification under various soil, residue, and lighting conditions. Trans ASAE 38(1):259–269
Yousefi E, Baleghi Y, Sakhaei SM (2017) Rotation invariant wavelet descriptors, a new set of features to enhance plant leaves classification. Comput Electron Agric 140:70–76
Yu, X., Xiong, S., Gao, Y., Zhao, Y., Yuan, X., 2016. Multiscale crossing representation using combined feature of contour and venation for leaf image identification. In: 2016 International Conference on Digital Image Computing: Techniques and Applications (DICTA). IEEE, pp. 1–6

Chapter 21
Enhancing Single-Cell Trajectory Inference and Microbial Data Intelligence

Bhargavi Posinasetty, Mukesh Soni, Sagar Dhanraj Pande, Krishnendu Adhikary, and Dhirendra Kumar Tripathi

Abstract The utilization of single-cell trajectory inference methods to deduce cell differentiation trajectories from single-cell transcriptomic or proteomic data holds significant importance in comprehending the developmental mechanisms of healthy tissues and offering valuable insights into pathological conditions. Nonetheless, the enhancement of accuracy and resilience in existing algorithms for inferring single-cell trajectories presents a persistent obstacle, mostly attributable to the interference caused by the identification of unrelated genes during single-cell sequencing. In order to effectively tackle this matter and expand the suitability of these methods to a wider array of biological data, we introduce iterTIPD, a trajectory inference method that utilizes iterative feature selection. The Iterative Topological Feature Selection (iterTIPD) algorithm is a widely employed approach in the field of genomics for the purpose of detecting differentially expressed genes. It is specifically designed to be applied repeatedly on linear or branching single-cell RNA sequencing data. The described iterative approach involves the selection of a subset of genes that make a significant contribution to the construction of the differentiation trajectory. This selection leads to enhanced precision and robustness in the ordering of cell pseudo-time. Furthermore, iterTIPD exhibits its efficacy not only in conventional single-cell data analysis but also in the realm of microbial data intelligence. The

---

B. Posinasetty
Department of Masters in Public Health, The University of Southern Mississippi, Hattiesburg, MS, USA

M. Soni (✉)
Department of CSE, University Centre for Research & Development Chandigarh University, Mohali, Punjab, India

S. D. Pande
School of Computer Science and Engineering, VIT-AP University, Amaravati, Andhra Pradesh, India

K. Adhikary
Centurion University of Technology and Management, Bhubaneswar, Odisha, India

D. K. Tripathi
Sri Satya Sai University of Technology and Medical Sciences, Sehore, MP, India

A. Khamparia et al. (eds.), *Microbial Data Intelligence and Computational Techniques for Sustainable Computing*, Microorganisms for Sustainability 47,
https://doi.org/10.1007/978-981-99-9621-6_21

experimental findings obtained from the study of four scRNA-seq data sets demonstrate that iterTIPD significantly improves the accuracy and reliability of single-cell trajectory inference techniques. This enhancement renders iterTIPD a valuable resource for researchers across many fields, including the analysis of microbial data. Additionally, iterTIPD not only improves the efficiency of trajectory inference methods but also has robust generalization abilities. The iterTIPD method effectively reconstructs the differentiation track of neural stem cells, demonstrating a strong alignment with established brain progenitor cell differentiation pathways. Moreover, the present research demonstrates that Top2a and Gjal may serve as promising novel markers for characterizing activated neural progenitor cell subgroups. This finding underscores the algorithm's capacity to uncover biologically significant information.

Keywords Single-cell trajectory inference · iterTIPD · Microbial data intelligence · Gene selection · Pseudo-time estimation · Neural stem cells

21.1 Introduction

Single-cell RNA sequencing (scRNA-seq) has become a powerful tool for studying cellular heterogeneity (Karbalayghareh et al. 2019; Gan et al. 2022) and early phases of embryonic development (Ni et al. 2022; Chen et al. 2020), thanks to its ability to analyse the transcriptome at the level of individual cells. The method shown here is well suited to answering important questions in biology. Quality control (Ni et al. 2022; Chen et al. 2020), batch effect correction (Wu and Ma 2023; Lee et al. 2015), data standardization (Lee et al. 2015; Zeng et al. 2021), feature selection (Lee et al. 2015; Zeng et al. 2021), dimensionality reduction (Zhang et al. 2023; Lu et al. 2021), clustering (Ma et al. 2020; Seabolt et al. 2022), identifying cell subpopulations (Ma et al. 2020; Wassan et al. 2019), and trajectory inference (Ma et al. 2020; Cawley et al. 2006) are just a few of the many methods available for analysing single-cell RNA sequencing (scRNA-seq) data. Inferring a cell's specific path of development is one of the many active research areas nowadays. Introducing genetic markers into cells and then following their offspring's development has always been the method of choice for tracing lineages (Li et al. 2021). However, the existence of only a few genetic markers may mask the diversity that actually occurs within cellular populations. Single-cell transcriptomic data or protein expression profiles are analysed to infer the course of cellular differentiation.

Reconstruct the lineage relationship (differentiation trajectory) of cells automatically using omics data, categorize cells depending on their position in the differentiation trajectory, and use this information to probe the dynamic expression of genes. Pseudo-times are assigned to each cell using these methods of calculation; sorting cells based on pseudo-time is known as pseudo-time sorting. A single cell's progression through the many transitional stages of differentiation can be described in terms of pseudo-time, which is the sequence of cells that follows a dynamic process of continuous development in biological systems. Therefore, cells at the start and finish points of a trajectory, as well as cells at intermediate points, can all be identified by pseudo-temporal sorting (Zeng et al. 2020).

Trajectory inference methods based on single-cell sequencing data are mainly used in the following two aspects:

1. Identify transition states: In many biological systems, cells exhibit continuous state transitions, characterized by different changes in transcription, morphology, epigenome, and surface markers (Lee and Friderikos 2022). Using trajectory analysis based on single-cell sequencing data can provide a more direct and unbiased method to identify and correctly sequence different transition stages, that is, reconstruct the lineage trajectory of cells based on scRNA-seq data and discover new transition states (Wang et al. 2022).
2. Identify key regulatory factors: In addition to revealing gene expression dynamics between cells, single-cell trajectory inference can also help identify key regulatory factors that trigger state transitions. For example, in the study of human definitive endoderm cell development (Zhang and Zhang 2020), single-cell trajectory inference was used to arrange cells along developmental trajectories, successfully reconstructing known differentiation pathways, and also discovering new candidates.

Essentially, the application of trajectory inference methods to single-cell RNA sequencing (scRNA-seq) data may reveal connections made over the course of cell differentiation, providing important new understandings of the variability of gene expression and cell dynamics. Therefore, it is of great scholarly significance to explore trajectory inference methods using single-cell RNA sequencing (scRNA-seq) data. However, considering the intrinsic characteristics of scRNA-seq data—high noise and heterogeneity (Afshar et al. 2016)—as well as the dropout effect, which occurs when low-expression genes are difficult to find owing to technological limitations (Li et al. 2021), a more accurate and robust method must be developed. Techniques for cell differentiation trajectories remain challenging. Currently, there are more than 70 different methods available in the field of single-cell trajectory inference (Li et al. 2023). To deal with the problem of high-dimensional noise in scRNA-seq data, single-cell trajectory inference frequently uses feature dimensionality reduction and feature selection algorithms.

Monocle (Ma et al. 2020) is considered the inaugural and archetypal single-cell trajectory inference technique within the realm of methods employing feature dimensionality reduction. Nevertheless, due to the advancements in contemporary single-cell RNA sequencing (scRNA-seq) techniques, which enable the simultaneous measurement of several cells, the construction of a minimal spanning tree (MST) connecting a substantial number of cells has become intricate and challenging to comprehend. Consequently, Monocle, a computational tool, lacks the capability to accurately forecast such complexities. In 2017, the proposal of Monocle2 (Wassan et al. 2019) introduced a method that generated a lineage tree within a low-dimensional space. This approach effectively preserved a greater amount of information regarding the differentiation trajectory present in the original data, hence addressing the challenge of forecasting branch trajectories. The Monocle and

Monocle2 algorithms have been employed to deduce diverse forms of cellular differentiation trajectories. These include the inference of neuronal lineage trajectories (Muller et al. 1989; Dai et al. 2018), the differentiation of progenitor and stem cells within the haematopoietic system (Wassan et al. 2019), as well as investigations into the development of the placenta (Iriya et al. 2020) and other related studies. The SCUBA (single-cell clustering using bifurcation analysis) method, as described in reference (Geard and Wiles 2005), employs t-distributed stochastic neighbour embedding (t-SNE) (Hu et al. 2016) to effectively reduce the dimensionality of the data. Subsequently, the dimensionally reduced data is subjected to a smoothing process. The utilization of SCUBA has been employed to determine crucial genes involved in the process of stem cell differentiation within the small intestine of mice (Zhang and Zhang 2021). The Slingshot algorithm utilizes the dimensionally reduced data and clustering outcomes as its input to create a minimum spanning tree (MST). Subsequently, it enhances the tree structure by using smooth curve fitting techniques to each branch of the MST. Ultimately, each cell is projected onto the curve that is nearest to it, leading to the emergence of a lineage trajectory that is organized and exhibits branching patterns. The utilization of slingshot has been employed in recent studies to forecast the cellular destiny and branching sites within the lineage trajectories of olfactory stem cells (Deconinck et al. 2021).

Different from feature dimensionality reduction methods, using feature selection methods can reduce the dimensionality of genes and identify the most relevant features, improving signal-to-noise ratio and computational efficiency of downstream analysis (such as clustering or pseudo-time inference). Effective feature selection methods can allow researchers to focus on the analysis of key genes, thereby improving the performance of trajectory inference methods that pseudo-temporally sort cells based on gene expression profile similarity. Some current feature selection methods used in single-cell trajectory inference mainly use the average expression of genes in all cells to screen highly variable genes (HVGs), thereby reducing the dimensionality of the data. However, due to the dropout effect in single-cell data, screening highly variable genes directly from the data may miss some important low-expressed genes carrying effective information. In bulk RNA sequencing (bulk RNA-seq) experiments, screening differentially expressed genes is usually used as a feature selection method, but scRNA-seq data generally does not have information about cell subpopulations (in fact, this is also the goal of analysing scRNA-seq data), so screening of differentially expressed genes is difficult to directly apply to scRNA-seq data. Based on the problems mentioned above, this chapter proposes an iterative feature selection optimization method (iterative trajectory inference based on probability distribution, iterTIPD) based on a trajectory inference method based on probability distribution—TIPD algorithm (Liu et al. 2022). The premise assumptions are:

If a gene plays a role in the final constructed single-cell differentiation trajectory, it should be differentially expressed along the cell's trajectory (Wassan et al. 2019). The work and contributions of this chapter can be summarized as follows:

1. iterTIPD was run on four single-cell data sets, and appropriate iteration parameters were determined experimentally. Moreover, in order to explore the effectiveness of this iterative feature selection method, this chapter compares iterTIPD, TIPD, and two other commonly used feature selection methods in terms of accuracy, robustness, and ability to detect gold standard genes.
2. The differentiation trajectory of neural stem cells (NSCs) was successfully reconstructed based on the optimized iterTIPD algorithm, and through comparison, it was found that the differentiation trajectory was highly consistent with the known differentiation trajectory of neural stem cells.
3. The genes Top2a and Gja1 were found to be differentially expressed in three different states of activated neural stem cell populations, which may be new markers of activated neural stem cell populations.

21.2 iterTIPD

iterTIPD is an iterative feature selection algorithm that can be widely used to screen differentially expressed genes on single-cell RNA sequencing data with linear or branched structures. By filtering out the subset of genes that contribute the most to the constructed differentiation trajectory, the accuracy and robustness of pseudotime sorting of cells are improved, thereby improving the performance of the single-cell trajectory inference algorithm. The following explains the iterTIPD algorithm proposed in this chapter based on the TIPD (Liu et al. 2022) algorithm.

21.2.1 TIPD Algorithms

The TIPD method quantifies the cellular differentiation potential by evaluating the signalling entropy (SR) of individual cells, enabling the assessment of diverse states within the cell population. Single-cell RNA sequencing (scRNA-seq) can be employed to provide a quantitative assessment of the differentiation state of individual cells. A high SR value indicates a heightened capacity for differentiation, whereas a low SR value signifies a diminished aptitude for differentiation. The TIPD method integrates signal entropy and clustering outcomes of gene expression data to characterize the probability distribution of diverse states within a population of cells. The Jensen-Shannon divergence (JSD), also known as the symmetric JS divergence, is employed to quantify the dissimilarity between probability distributions of cell clusters. Subsequently, the minimum spanning tree (MST) is computed on the full network formed by the centres of these cell clusters and their corresponding probability distribution distances. The aforementioned procedures establish the fundamental framework of the cellular development trajectory. Subsequently, all cells are mapped onto the curve-fitted framework, resulting in the derivation of the pseudotime sequence of cells.

The MST thus constructed is calculated based on system-level information in the context of signal entropy, so it can reflect the real biological process globally, making the TIPD algorithm show higher performance than other trajectory inference methods.

21.2.2 Iterative Feature Selection of iterTIPD Algorithm

The output of the single-cell trajectory inference method is sorted according to the pseudo-time of cell differentiation. It can be considered that individual cells are in different states. Therefore, differentially expressed genes can be calculated in pseudo-time order as a feature selection method for scRNA-seq data. The TIPD algorithm is used here to calculate pseudo-time (the part marked by the green dotted box in Fig. 21.1).

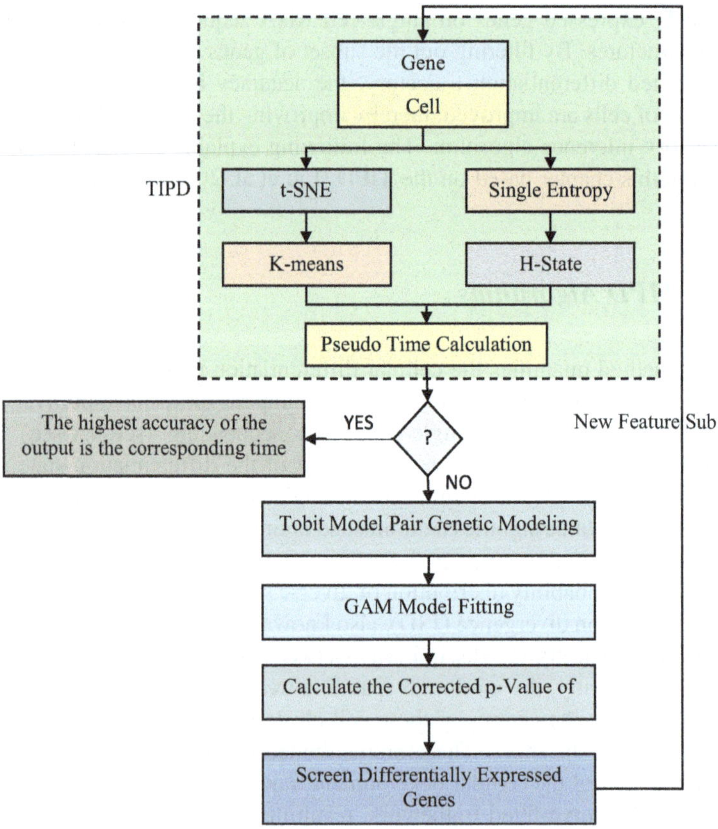

Fig. 21.1 iterTIPD algorithm flow chart

Existing studies have shown that the generalized additive model (GAM) (Ma et al. 2020) is used to calculate differentially expressed genes in pseudo-time and detect gold standard genes. Feature selection can be performed by fitting a GAM to the relationship between a gene's expression and pseudo-time. First, the expression level of each gene in the cell is modelled through the Tobit model (Wei et al. 2021), and then the relationship between the expression of each gene and pseudo-time is fitted using the GAM model, where the expression level of each gene Y depends on the latent variable Y^*, the expression is:

$$Y = \begin{cases} Y^*, Y^* \geqslant \lambda \\ \lambda, Y^* < \lambda \end{cases} \tag{21.1}$$

where λ is the detection threshold, and λ is set to 0.1 by default. The latent variable Y^* depends on the variable x_i in the GAM model, where x_i represents the pseudo-time value corresponding to each cell, $i \in \{1, 2, \ldots, n\}$. Therefore, the expression of the GAM model of differentially expressed genes in pseudo-time is calculated. Thc formula is defined as:

$$E(Y) = s(\psi(b_x, g_i)) + \varepsilon \tag{21.2}$$

Among them, g_i represents a single cell, $i \in \{1, 2, \cdots, n\}$, $\psi_t(b_x, g_i)$ represents the pseudo-time value assigned to cell g_i, and s represents a cubic smooth function with an effective degree of freedom of 3. The error term ε is normally distributed with mean 0. Finally, differentially expressed genes were iteratively screened through eq. (3).

$$G^t = \{a \mid FDRp_a^{t-1} < 0.05, a \in G^{t-1}\} \tag{21.3}$$

Among them, G^t represents the gene set after iterative screening t times; p_a^{t-1} is the p-value of gene a calculated using the approximate χ^2 likelihood ratio test on G^{t-1}; and $FDRp_a^{t-1}$ is the Benjamini-Hochberg test of p_a^{t-1} p-value of gene a obtained after FDR correction (Flexman et al. 2006). Genes with $FDRp_a^{t-1}$ less than 0.05 in each iteration were retained for the next iteration, and these genes were considered to be differentially expressed genes. The implementation of the GAM model and related test functions in the above process is provided by the "VGAM" package (Wei et al. 2019).

The differentially expressed genes in pseudo-time order calculated in the above steps will form a new feature subset for a new round of calculation. The number of iterations is controlled by the specified parameters, and the default is 100, which can be determined according to the accuracy change trend of different data sets during the iteration process (see Sect. 2.3). After the algorithm performs multiple iterations with reference to the iteration parameters, it stops iteration when the accuracy shows an obvious downward trend and outputs the pseudo-time sorting results of cells with the highest accuracy in all calculations and the corresponding gene subsets.

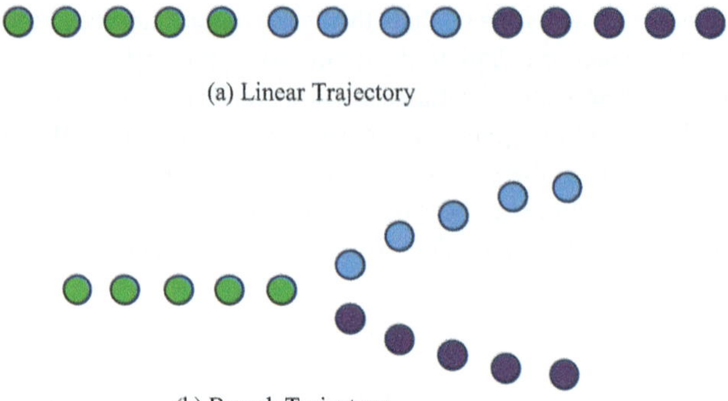

Fig. 21.2 Schematic diagram of pseudo-time sequencing of cells in (**a**) linear trajectories and (**b**) branch trajectories

If the trajectory constructed by the trajectory inference method is a linear trajectory, as shown in Fig. 21.2a, the pseudo-time sequence of differentially expressed genes is directly calculated as a feature subset for screening, if it is a branch trajectory, as shown in Fig. 21.2b, it contains. For the differentiation trajectory of a branch structure, calculate the differentially expressed genes in pseudo-time order for each lineage (such as a-b-c and a-b-d), and finally find the union of these differentially expressed genes as the feature subset for screening. Because the feature subset used in each round is a gene subset calculated based on the gene set of the previous round, as the number of iterations increases, the number of screened genes gradually decreases, and some genes containing effective information may be lost, so the number of iterations should not be too many.

21.3 Experiment

21.3.1 Data Set Description

This study utilizes four single-cell RNA sequencing (scRNA-seq) data sets, which exhibit variations in terms of their sizes and encompass both linear and branched structures. Furthermore, these data sets encompass species such as humans and mice. The individuals in question can be identified as follows:

1. Mouse lung alveolar type 2 cells, often known as AT2 cells. The data set known as AT2 comprises 101 cells of alveolar type 2 in mice. These cells were obtained from the lungs of embryonic mice at four different time points: E14.5, E16.5, E18.5, and adult. This information was taken from a previous study (Dai et al. 2018). The trajectory of differentiation seen in this study exhibits a linear shape.
2. Human skeletal muscle myoblasts (HSMM). The HSMM data set has a total of 271 human skeletal myoblasts, encompassing three distinct cell types:

proliferating cells, developing myoblasts, and interstitial mesenchymal cells (Ma et al. 2020). The trajectory of differentiation observed in this context exhibits a linear shape.

3. Mouse embryonic stem cells (ESC) are a type of pluripotent cells derived from the inner cell mass of mouse embryos. The ESC data set consists of 2717 cells that were collected at four different time intervals (d0, d2, d4, and d7) following the withdrawal of leukaemia inhibitory factor (LIF). The study utilized mouse embryonic stem cells (Schiebinger 2021). The trajectory of differentiation exhibits a linear structure.

Mouse conventional dendritic cells (cDCs) are a type of dendritic cell seen in mice.

The cDC collection is made up of 251 normal dendritic cells (cDCs) from mice. This type of cDC is made up of three different types of cells: pre-dendritic cells (pre-DCs), macrophage dendritic cell progenitors (MDPs), and common dendritic cell progenitors (CDPs) (Geard and Wiles 2005). There are two main types of conventional dendritic cells (cDCs) in mice. These are cDC1 and cDC2 (Rezk et al. 2023; Nguyen et al. 2022). Three different types of cell lines are controlled by different transcriptional programmes. These programmes start when macrophages, monocytes, and dendritic cells (MDPs) differentiate. The next step is the formation of common dendritic cell progenitors (CDPs) and then pre-dendritic cells (pre-DCs). These cells eventually differentiate into cDC1 and cDC2 lines (Chang et al. 2017). There are branches in the process of differentiation.

21.3.2 Evaluation Index

When evaluating the efficacy of algorithms in the field of single-cell differentiation trajectory inference, researchers commonly employ accuracy, robustness, and the ability to detect gold standard genes as complete metrics (Seabolt et al. 2022). This chapter also employs these three variables to assess the efficacy of the trajectory inference technique. This chapter uses pseudo-time sorting of cells to evaluate accuracy. Accuracy is assessed by calculating the Kendall rank correlation coefficient between the externally recorded information in the experimental stage and the data set (Sahayasheela et al. 2022; Chang et al. 2019). The assessment of resilience can be conducted by computing a robustness score for the pseudo-time ordering of both the original and perturbed data (Seabolt et al. 2022). The generation of perturbed data sets involved the random subsampling of 90%, 80%, and 70% of cells from the original data set. This process was repeated 50 times to provide several duplicates. The mathematical expression representing the robustness score is as follows:

$$S_{\pi_1,\pi_2} = \frac{2}{|A||A|-1)} \sum_{i,j \in A; i \neq j} h(\pi_1, \pi_2, i, j) \quad (21.4)$$

Among them, π_1 and π_2 represent the ordering of two pseudo-times, A is the union of all cells in π_1 and π_2, and $|A|$ represents the cardinality of the set A. For a pair of cells i and j, if their order in π_1 is the same as in π_2, then $h(\pi_1, \pi_2, i, j) = 1$, otherwise $h(\pi_1, \pi_2, i, j) = 0$. The higher the robustness score, the more similar the two pseudo-time rankings are.

Each feature selection method's ability to improve the detection of gold standard genes is measured in this study using the mean ranking of gold standard genes (Seabolt et al. 2022). For a given data set of gene expression measurements, the "gold standard" genes are found by carefully combing through the corresponding scholarly literature. Differentiation and development of cells, as well as other biological events, are known to cause these genes to be expressed in unique ways. After applying false discovery rate (FDR) correction, the list of differentially expressed genes is ranked by their p-values (Wei et al. 2019). A measure of the algorithm's performance in detecting known differentially expressed genes is the location of gold standard genes within the set of differentially expressed genes. As the ranking and corresponding numerical value rise, the algorithm's ability to detect known differentially expressed genes improves.

21.3.3 Confirmation of iterTIPD Iteration Parameters

The iterTIPD technique can screen candidate feature genes in a recursive fashion. If the number of iterations through the method is too little, the screened gene subset may contain many false positives. Important feature genes could be lost if the process is repeated too many times. In this piece, we use iterTIPD on the four single-cell data sets described in Sect. 3.1. At the outset, we run 100 iterations. Variations in the correlation between iterations and precision across data sets are tracked. Figure 21.3 displays the obtained data. When the number of iterations is set to 0, the associated accuracy is the accuracy attained by TIPD. The number of iterations where the accuracy improved the most is indicated by the yellow dot. The experimental findings were fitted with a polynomial equation, yielding the red dotted line as the trend line. The accuracy of the four single-cell data sets showed a general trend of first growing and then dropping as the number of iterations rose, as seen in Fig. 21.3, during the iteration process. Because the accuracy is also influenced by the dimensionality reduction phase in the iterTIPD algorithm, it varies. The accuracy trend first rises, then falls. This is because the accuracy of the algorithm is enhanced by increasing the number of iterTIPD iterations when the data contains more redundant features, which in turn reduces the feature subset. When accuracy is the highest, valid information-carrying genes are the first to be filtered out, leading to a decline in accuracy as the number of features is reduced

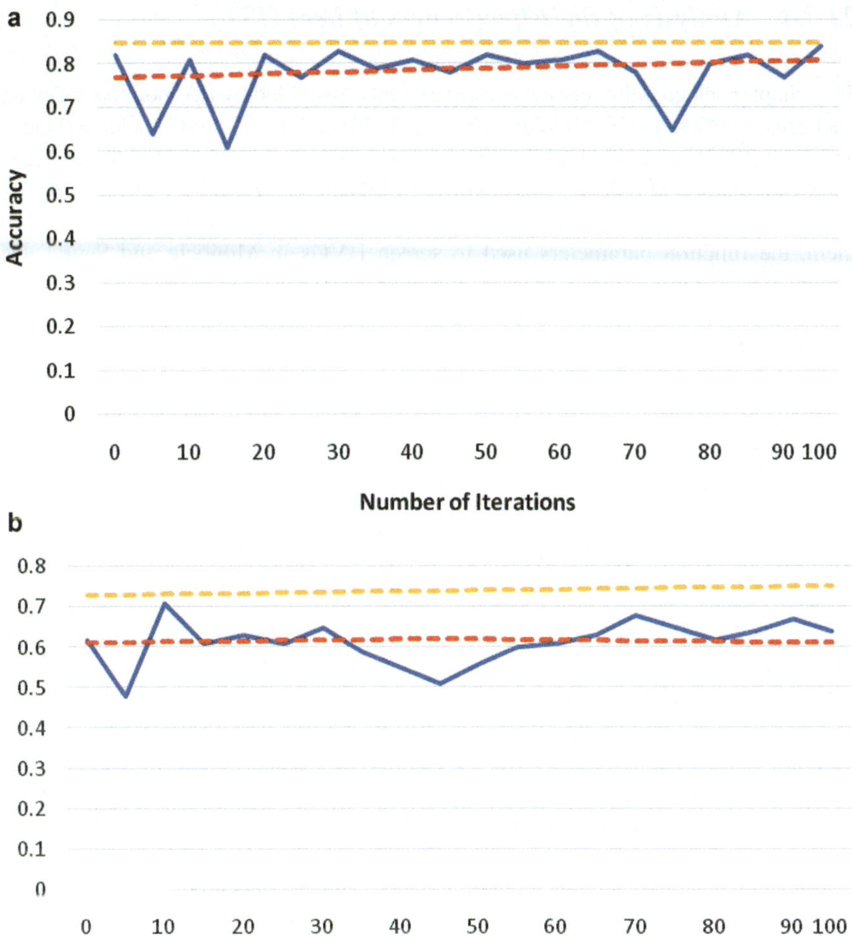

Fig. 21.3 (**a**) Number of iterTIPD Iterations and Accuracy. (**b**) Number of iterTIPD iterations and accuracy

further. The declining accuracy trend for the four single-cell data sets becomes noticeable as the number of repeats approaches 100. For this reason, the initial value for iterTIPD is 100. In the future, researchers can use this parameter as a starting point for inferring cell differentiation trajectories from different data sets using iterTIPD. After the iteration is complete, iterTIPD will return the results of the most accurate pseudo-time sorting of cells throughout the iteration, along with the subsets of genes that were sorted. On the AT2 and cDC data sets, the highest accuracy is reached at the 40th, 56th, 43rd, and 62nd iterations (the number of iterations to reach the greatest accuracy and the associated highest accuracy are marked in brackets) (see Fig. 21.3a, b). Overall, iterTIPD's optimization effect is positive across all four single-cell data sets, but it's most pronounced on the ESC data set, where it improves accuracy by 0.09.

21.3.4 Analysis of the Effectiveness of iterTIPD

This chapter selected the feature selection method provided by two popular software packages, the "Monocle" package (Ma et al. 2020) and the "Seurat" package (Demir and Eren 2022) to screen highly variable genes (HVGs), combined with the TIPD algorithm (named M_HVGs respectively). +TIPD and $S\_HVGs + TIPD$ to compare the optimization effect of iterTIPD's iterative feature selection method. Among them, the function parameters used to screen HVGs in Monocle and Seurat use default parameters, and the process of screening HVGs corresponds to the gene selection step in iterTIPD and TIPD (select the normalized gene expression level to be greater than 1 in at least 30% of the cells genes.); subsequent steps, such as dimensionality reduction, remain consistent. In addition, in order to explore the generalization ability of iterTIPD's feature selection method, this chapter applies it to the gene selection step in Monocle2 (Wassan et al. 2019), and this method is named iter+Monocle2.

21.3.4.1 Accuracy Analysis

Table 21.1 shows that the accuracy of iterTIPD on AT2, HSMM, ESC, and cDC data sets is higher than that of TIPD. The accuracy of the algorithm has been improved by 0.028, 0.062, 0.089, and 0.040, respectively, indicating that the iterTIPD iterative feature selection method has effectively improved TIPD's accuracy degree, and the accuracy is the highest on the other three single-cell data sets, except the ESC data set. The M_HVGs+TIPD method only improves the accuracy of TIPD on the ESC data set, and the S_HVGs+TIPD method only improves the accuracy of TIPD on the ESC and cDC data sets. Comparing the experimental results of Monocle2 and iter +Monocle2, it can be seen that on the AT2, ESC, and cDC data sets, Monocle2 using the iterTIPD feature selection method improved the accuracy of Monocle2 by 0.031, 0.054, and 0.013, respectively. On the HSMM data set, it was only the accuracy difference is 0.018, which shows that the feature selection method of iterTIPD has good generalization performance. The iterTIPD feature selection method improves the accuracy of TIPD and Monocle2 on the ESC data set by 0.089 (the gap between TIPD and iterTIPD) and 0.054 (the gap between Monocle2 and iter+Monocle2), respectively, which is larger than that on the other three data sets, which also shows that on large sample size data sets, excellent feature selection methods can

Table 21.1 Accuracy comparison using different feature selection methods

Method	AT2	HSMM	ESC	cDC
iterTIPD	0.852	0.602	0.658	0.741
TIPD	0.821	0.533	0.570	0.689
M_HVGs_TIPD	0.415	0.478	0.579	0.667
S_HVGs_TIPD	0.698	0.516	0.688	0.735
Monocle2	0.718	0.561	0.419	0.585
Iter+Monocle2	0.752	0.540	0.481	0.595

effectively reduce the noise in the data set and improve the accuracy of the algorithm. Overall, iterTIPD is better than the other two feature selection methods.

21.3.4.2 Robustness Analysis

Theoretically, iterTIPD attains maximum accuracy by eliminating the genes that are deemed irrelevant by progressively decreasing the feature subset throughout the course of iterations. Thus, whether the iterTIPD approach can reliably "target" a subset of genes that yield valuable information more than once indicates how robust the strategy is. In order to determine whether iterTIPD's iterative feature selection may increase TIPD's robustness and whether it is superior to the other two feature selection approaches, this chapter evaluates the robustness of the aforementioned six methods.

All perturbation data sets combined, iterTIPD's average robustness score is higher than TIPD's, and its score fluctuation range is more constrained. All perturbation data sets show consistent robustness scores for the iter+Monocle2 technique. It has been discovered that iterative feature selection techniques improve the algorithm's robustness, as shown by its better performance when compared to Monocle2. However, the feature selection strategies used by Monocle and Seurat do not always improve the robustness of the algorithm. The robustness score of the M_HVGs +TIPD technique is generally lower than that of TIPD. However, the M_HVGs +TIPD technique shows a slightly higher robustness score than TIPD on the perturbation data set that covers 80% of the sample size on the cDC data set.

The S_HVGs+TIPD method demonstrates superior robustness compared to TIPD in certain instances, specifically when applied to perturbed data sets with 90% and 70% sample sizes on the ESC data set, as well as perturbed data with a 70% sample size on the cDC data set. However, in other cases, the robustness of the S_HVGs +TIPD method does not significantly differ from that of TIPD. Hence, the experimental results demonstrate that the iterative feature selection approach employed by iterTIPD significantly enhances the algorithm's resilience, surpassing the feature selection strategies employed by Monocle and Seurat.

21.3.4.3 Analysis of the Ability to Detect Gold Standard Genes

This chapter also conducted comparative experiments to test whether each feature selection method can improve the ability to detect gold standard genes and used the average ranking of gold standard genes as a measurement indicator. The values in Table 21.2 represent the average ranking of gold standard genes in gene rankings obtained by different algorithms.

Table 21.2 shows that the average ranking of gold standard genes on iterTIPD is lower than that of TIPD. The better an algorithm is at finding gold standard genes, the lower the average ranking of gold standard genes. The iter+Monocle2 method also does better in AT2, HSMM, and ESC. It's also lower than Monocle2 on the data

Table 21.2 Average ranking of gold standard genes using different feature selection methods

Method	AT2	HSMM	ESC	cDC
iterTIPD	2151	3342	6044	1592
TIPD	2480	3929	6464	2049
M_HVGs_TIPD	3077	3542	7939	1726
S_HVGs_TIPD	2579	3618	7981	1463
Monocle2	3639	4275	8039	3312
Iter+Monocle2	3105	3983	7220	3731

set, which shows that iterTIPD's method for selecting features over and over again can help the programme find gold standard genes better. The iterTIPD gold standard gene ranks lower than M_HVGs+TIPD and iter+Monocle2 in four sets of data. It also ranks lower than S_HVGs+TIPD in AT2, HSMM, and ESC data sets, with differences of 438, 270, and 1934 points, respectively. On the cDC data set, it ranks a little higher than S_HVGs+TIPD (the difference is only 138). Also, on the HSMM data set and the cDC data set, the average score of gold standard genes for the M_HVGs+TIPD and S_HVGs+TIPD methods is only lower than TIPD. This means that these two feature selection methods can't really make TIPD better at finding gold standard genes. In general, iterTIPD's feature selection method makes TIPD and Monocle2 even better at finding gold standard genes; it is also better than Monocle and Seurat's feature selection methods.

21.3.5 Analysis of Results

In this chapter, an optimization method iterTIPD using iterative feature selection is proposed based on the TIPD algorithm. In order to explore the generalization ability of iterTIPD's feature selection method, this chapter applies it to the gene selection step of Monocle2 and compares it with Monocle2. It also chooses two other classic feature selection methods to combine with TIPD. In terms of accuracy and robustness, comparative experiments were conducted on three indicators including the ability to detect gold standard genes. Experimental results show that iterTIPD's iterative feature selection can improve the accuracy and robustness of the algorithm and its ability to detect gold standard genes. Based on the experimental results on four single-cell data sets, although M_HVGs and S_HVGs use different methods to play the role of feature selection, due to the separation of the cell differentiation trajectory inference algorithm, the effect is usually not optimal. In contrast, iterTIPD combines the feature selection task with the cell differentiation trajectory inference task and performs iteratively. In each iteration, the cell differentiation trajectory inference algorithm is used as the evaluator to screen genes so that the selected feature subset can better meet the needs of cell differentiation trajectory inference, and the feature subset is continuously reduced in multiple iterations so that the inferred cell differentiation trajectories are obtained with optimal accuracy.

In experiments comparing robustness, iterTIPD maintained the highest robustness scores on the four single-cell data sets. This is because iterTIPD continuously narrows down the gene subset and continuously sorts the cell pseudo-time sorting results during the iterative process. Corrections will eventually "lock" on a set of genes that provide useful information to achieve the highest accuracy. The TIPD algorithm has randomness in the dimensionality reduction step, which usually leads to unstable results. iterTIPD's iterative feature selection method can improve the robustness of the algorithm to a certain extent.

However, iterTIPD still has value worthy of further exploration, such as the following:

(1) In the dimensionality reduction step, the values of the retained dimensions have a greater impact on the algorithm results. You can consider flexibly trying other dimensionality reduction algorithms based on the characteristics of the data set; (2) in this chapter, iterTIPD uses 4 public dataset. It has been verified on single-cell data sets, and its generalization ability can be tested on larger-scale data sets in the future; (3) iterTIPD's iterative feature selection method can also be used on other machine learning models by optimizing feature subsets to improve the accuracy of the algorithm.

21.4 Application of iterTIPD in Neural Stem Cell Differentiation

21.4.1 Adult Stem Cell Lineage

The processes of stem cell quiescence and activation are very important for keeping many body systems and tissues healthy, renewing them, and making them adaptable. They are also very important for how quickly we age and get diseases. It is possible for quiescent stem cells to successfully use both external and internal signals to either stay in a dormant state or divide and differentiate to make new cells when they are activated (Flexman et al. 2006). It's important that neural stem cells (NSCs) are present in the adult brain because they are a key source of regenerative cells that can help treat neurodegenerative diseases and damage to neurons. Neurological stem cell (NSC) development processes vary greatly. This is very important to look into if we want to fully understand the functional traits and gene regulatory networks connected to NSC populations. On the other hand, using population-based methods might hide the natural variety that exists within neural stem cell (NSC) lineages. As a result, this could make it harder to find new, unusual cell types or intermediate states and also harder to fully understand how complex transcriptional processes work. The iterTIPD algorithm is used in this study to find adult neural stem cell lineages and look into how NSC populations vary and how genes are expressed in single cells.

Most neural stem cells (NSCs) are in a state called "quiescence," which means they are not actively dividing. This type of NSCs, which are also called quiescent

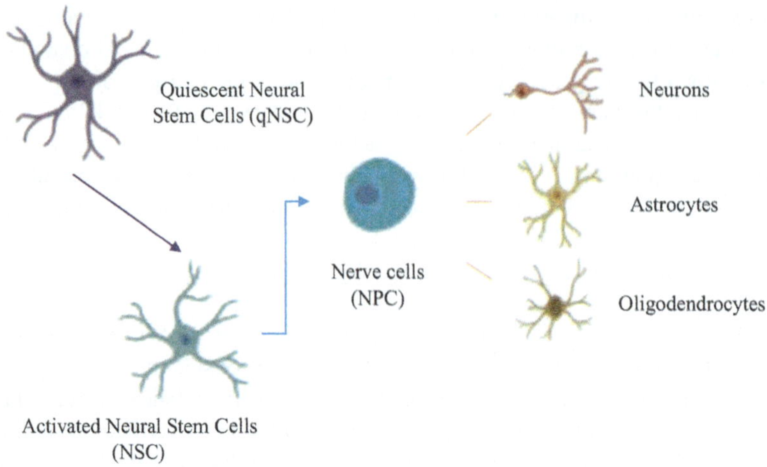

Fig. 21.4 Schematic diagram of neural stem cell differentiation lineage

neural stem cells (qNSCs) and don't divide, can change into activated neural stem cells (aNSCs), which do divide (Mansoor et al. 2015). Neuronal stem cells (aNSCs) in adults can make neural progenitor cells (NPCs). NPCs are a group of cells that divide quickly and show traits of early brain development. Finally, as shown in Fig. 21.4, the neural progenitor cells (NPCs) change into specific cell types such as neurons, astrocytes, and oligodendrocytes. There is still a lot we don't know about the different types of neural stem cells and how their genes change over time, even though new single-cell correlation studies have shed light on the complex makeup of neural stem cell (NSC) populations in certain neurogenic areas of the adult brain (Mikhailov and Sankai 2018; Flexman et al. 2007).

21.4.2 Data Sets and Preprocessing

The neural stem cell differentiation data set used in this chapter is PRJNA324289 (Mikhailov and Sankai 2018), which is in the standardized FPKM format and contains 79 quiescent neural stem cells (qNSC), 152 activated neural stem cells (aNSC), and 64 neural progenitor cells (NPC) and 34 astrocytes.

The preprocessing step refers to the processing method of this chapter (Chang et al. 2017): for cell filtration, contaminated oligodendrocytes and outlier cells (observed in two-dimensional space using PCA (principal component analysis)) are removed. For gene filtering, retain genes expressed more than ten times in at least five samples. Since this chapter hopes to explore the lineage relationship between qNSCs, aNSCs, and NPCs, astrocytes were removed, leaving 250 samples for subsequent analysis.

21.4.3 Neural Stem Cell Differentiation Trajectory Constructed by iterTIPD Analysis

21.4.3.1 iterTIPD Reconstructs the Differentiation Trajectory of Neural Stem Cells

The neural stem cell differentiation data set that had already been cleaned up was fed into the iterTIPD method to figure out the paths of differentiation. iterTIPS prediction of the path of neural stem cell development fits with what other studies have found (Mansoor et al. 2015). First, iterTIPD finds cell states that are not the same. Then, the minimum spanning tree is found by using the JS divergence distance to find the distance between the probability distributions of these different states. Then, iterTIPD picks the qNSC population with the highest average entropy value as the starting point. This lets us figure out that the qNSC-aNSC-NPC differentiate in a straight line. Finally, the simultaneous master curve method (Cawley et al. 2006) was used to make all the cells fit, and each cell's pseudo-time value was found.

21.4.3.2 Analysis to Identify Intermediate States of *aNSC* Populations

This chapter selected relevant marker genes for three different populations of *qNSC*, *aNSC*, and NPC to explore their expression patterns in pseudo-chronological order. Id3 is a key gene related to the quiescent state of neural stem cell differentiation, which was previously reported as a marker of *qNSC* population (Nguyen et al. 2022); *Egfr* is related to the activation process; Cdk4 and *Cdk*1 are related to the cell cycle; *Dlx*2 and *Dcx* are related to the differentiation process of neurons, among which *Dlx*2 is a proneural transcription factor known to promote neural differentiation (Demir and Eren 2022).

The iterTIPD approach calculates the expression of the above marker genes on the pseudo-time sequence. In order to clearly demonstrate the continuous dynamic changes in marker gene expression, the sequential number of the cell pseudo-time is used instead of the pseudo-time value. Id3 is first highly expressed in the *qNSC* population, and then the NSC differentiation process is activated and Egfr is upregulated. Cell cycle-related genes Cdk4 and Cdk1 are highly expressed in *aNSC* populations. *Dlx*2 is highly expressed in a subset of *aNSC* populations, suggesting that a subpopulation of *aNSCs* may exhibit early transcriptomic signatures of neural stem cell differentiation. The NPC population ranks last in pseudo-temporal order and highly expresses the important neurogenesis regulator *Dcx*. The expression pattern of the above marker genes is consistent with existing research (Jiang et al. 2014), which also proves that iterTIPD correctly infers the pseudo-time sequence of single cells.

The dynamic expression of these marker genes along pseudo-time sequence shows successively different differentiation states. This chapter refers to the pseudo-time stage (red) that contains the majority of the *qNSC* population as the *qNSC*-like population and the pseudo-time stage (blue) that contains the majority of the NPC population as the NPC-like population. In aNSC-like populations (green), by observing the specific expression patterns of marker genes such as *Egfr*, Cdk1, Dlx2, and Dcx, this chapter divides aNSC-like populations into three continuous intermediate states and names them *aNSC* early stage, *aNSC* mid-stage, and *aNSC* late stage. Among them, Egfr is low expressed in the qNSC-like population; aNSC has high expression of *Egfr* and low expression of Cdk1 in the early stage; *aNSC* has high expression of *Cdk*1 and low expression of Dlx2 in the middle stage; *aNSC* has high expression of Cdk1 and Dlx2 in the late stage; and the NPC-like population has high expression of *Egfr* and *Dcx*. The expression of genes *Cdk*1 and *Dlx*2 in the *aNSC* population shows different changes, so it is speculated that there may be an intermediate state in the *aNSC* population.

In order to verify the above conjecture, this chapter then calculated the average expression levels of the specifically expressed marker genes *Cdk*1 and *Dlx*2 in the early, middle, and late stages of *aNSC* to observe their trends and patterns. The results are shown in Fig. 21.5a, b. There was a large difference in levels, and there was also a significant difference in the expression of *Dlx*2 on *aNSC* mid-stage and *aNSC* late stage. The specific expression of *Cdk*1 and *Dlx*2 further suggests that novel subpopulations may exist within the *aNSC* population.

In summary, iterTIPD correctly inferred the pseudo-temporal sequence of neural stem cells, which revealed the dynamic characteristics of gene expression. Based on this sequence, we observed the specifically expressed genes *Cdk*1 and *Dlx*2 in the *aNSC* population and calculated that these two genes were differentially expressed in the *aNSC* population, suggesting that new intermediate states of activation and differentiation may exist in the *aNSC* population.

21.4.3.3 Analyse and Discover Markers That Define the Intermediate State of *aNSC*

The dynamic expression of marker genes reveals the possible existence of new subpopulations in *aNSCs*. Based on the specific expression of *Cdk*1 and *Dlx*2 marker genes, this chapter divides the *aNSC*-like population into three different continuous intermediate states: early *aNSC*, mid-stage *aNSC*, and late-stage *aNSC*. In order to find new markers for aNSC populations, this chapter calculated the differentially expressed genes of NSC populations in pseudo-time order and used the Pearson correlation coefficient (PCC) to calculate the similarity between *Cdk*1 and *Dlx*2 and other genes respectively. *aNSC* subpopulation markers were found based on the following criteria: genes with the highest similarity and the smallest corrected *p*-value in differential expression analysis. Two genes were finally identified: one was *Top2a*, which was most similar to *Cdk*1 and had the smallest corrected *p*-value; the other was *Gja*1, which was most similar to *Dlx*2 and had the smallest corrected *p*-value.

The expression of *Top2a* and *Gja*1 in three different states of *aNSC* population is shown in Fig. 21.6a, b. *Top2a* is differentially expressed in early *aNSC* and

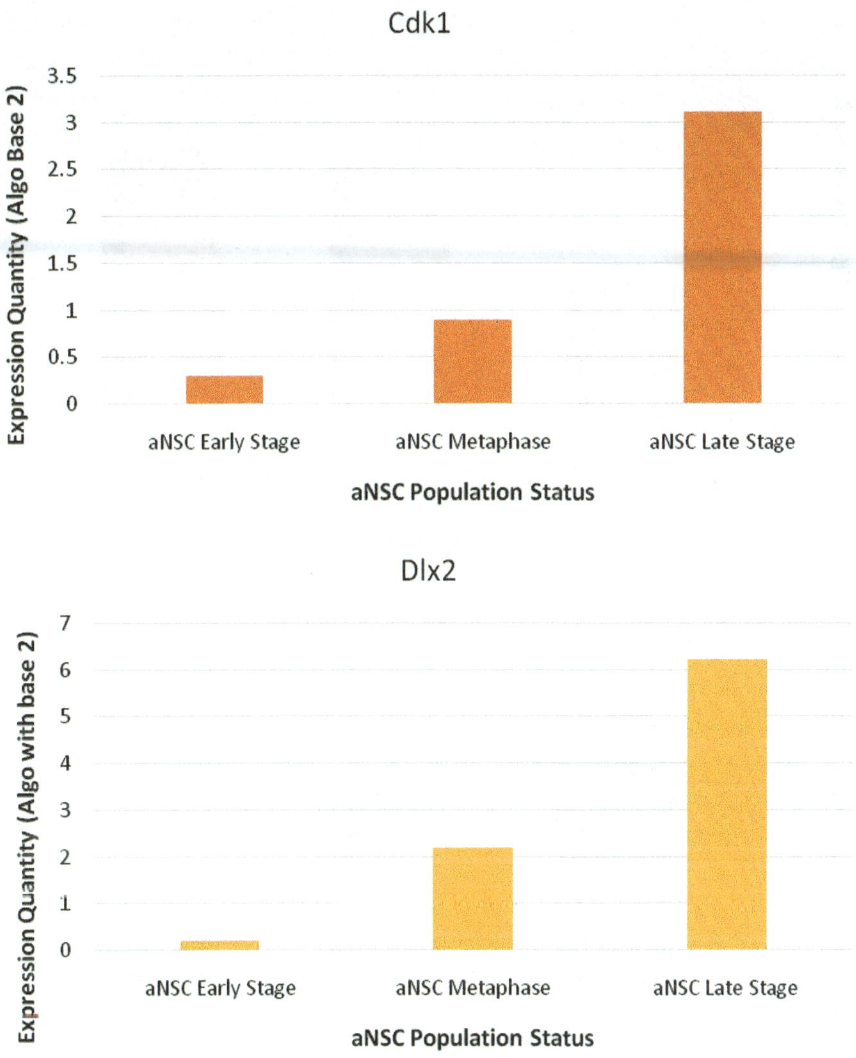

Fig. 21.5 (**a**) Average expression level distribution of Cdk1 aNSC population status. (**b**) Average expression level distribution of Dlx2 aNSC population status

mid-stage *aNSC*, and *Gja*1 is differentially expressed in mid-stage *aNSC* and late *aNSC*, and both show fold changes. Top2a is known to be extremely important for the process of cell division and proliferation, and a recent study showed that Top2a may be a new characteristic molecule expressed by neural stem cells (Mansoor et al. 2015). *Gja*1 is known to be a marker of astrocytes (Mikhailov and Sankai 2018) and is also associated with the process of neural differentiation. Therefore, Top2a and Gja1 are likely to be new markers for aNSC subpopulations.

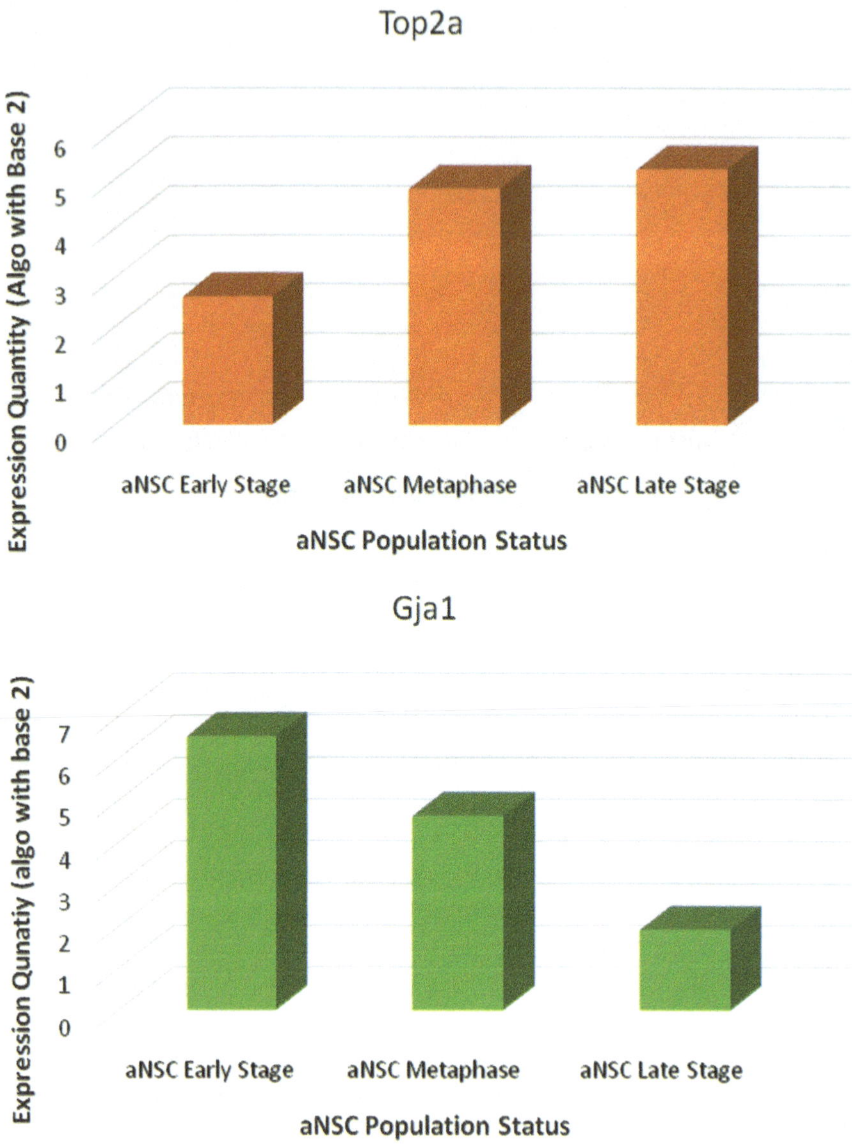

Fig. 21.6 (**a**) Average expression level distribution of Top2a aNSC population status. (**b**) Average expression level distribution of Gja1 aNSC population status

21.5 Conclusion

The majority of genes identified in single-cell RNA sequencing (scRNA-seq) data are not pertinent to investigating the fundamental biological mechanisms. However, by employing feature selection techniques to decrease the dimensionality of genes, the signal-to-noise ratio and computing efficiency of subsequent analyses can be enhanced. This work presents an algorithm called iterTIPD, which is an iterative feature selection optimization algorithm derived from the TIPD algorithm. The novel aspect of iterTIPD consists in its iterative application of a commonly employed feature selection method for bulk RNA-seq data to scRNA-seq data. This approach enhances the accuracy of cell pseudogene identification by effectively filtering out a subset of genes that have the greatest impact on the generated differentiation trajectory. This study performed experiments on four distinct single-cell data sets, where the parameters of iteration were defined. Subsequently, the performance of iterTIPD was compared to two other feature selection approaches, both in combination with TIPD, with respect to accuracy, robustness, and the capacity to identify gold standard genes. A comparative experiment was conducted. This study aims to assess the generalization capability of iterTIPD's feature selection approach by implementing it in the gene selection phase of Monocle2 and conducting a comparative analysis with Monocle2. The experimental findings demonstrate that the iterative feature selection strategy employed by iterTIPD yields enhanced performance for TIPD, exhibits superior generalization capabilities, and outperforms two other conventional feature selection methods.

The present study subsequently employs the iterTIPD algorithm to generate the differentiation pathway of brain stem cells. Through the examination of marker gene expression patterns in a pseudo-chronological sequence, this study has identified the presence of three distinct intermediate states within the population of activated neural stem cells. This finding suggests the existence of additional subpopulations characterized by different activation and differentiation states within the larger population of activated neural stem cells. This study employed a combination of gene correlation analysis and differential expression analysis to identify the differentially expressed genes Top2a and Gja1 in three distinct states of activated neural stem cell populations. Based on a comprehensive review of relevant literature, it has been shown that Top2a exhibits potential as a novel molecular marker expressed by neural stem cells. Conversely, Gja1 has been established as a recognized marker of astrocytes and is additionally implicated in the neural differentiation process. The aforementioned finding has the potential to offer novel insights into the examination of specific subsets of activated neural stem cell populations. Despite the algorithm in this study demonstrating favourable outcomes in accuracy, robustness, and its capacity to identify gold standard genes, there remain certain constraints that necessitate additional investigation.

References

Afshar S, Salimi E, Braasch K, Butler M, Thomson DJ, Bridges GE (2016) Multi-frequency DEP cytometer employing a microwave sensor for dielectric analysis of single cells. IEEE Trans Microw Theory Tech 64(3):991–998. https://doi.org/10.1109/TMTT.2016.2518178

Cawley GC, Talbot NLC, Janacek GJ, Peck MW (2006) Sparse bayesian kernel survival analysis for modeling the growth domain of microbial pathogens. IEEE Trans Neural Netw 17(2): 471–481. https://doi.org/10.1109/TNN.2005.863452

Y. -H. Chang et al., Human induced pluripotent stem cell region recognition in microscopy images using Convolutional Neural Networks, 2017 39th Annual International Conference of the IEEE Engineering in Medicine and Biology Society (EMBC), Jeju, Korea (South), 2017, pp. 4058–4061, doi: https://doi.org/10.1109/EMBC.2017.8037747.

Y. -H. Chang et al., Human induced pluripotent stem cell reprogramming prediction in microscopy images using LSTM based RNN, 2019 41st Annual International Conference of the IEEE Engineering in Medicine and Biology Society (EMBC), Berlin, Germany, 2019, pp. 2416–2419, doi: https://doi.org/10.1109/EMBC.2019.8857568.

Chen Z et al (2020) A branch point on differentiation trajectory is the bifurcating event revealed by dynamical network biomarker analysis of single-cell data. IEEE/ACM Trans Comput Biol Bioinform 17(2):366–375. https://doi.org/10.1109/TCBB.2018.2847690

Dai C et al (2018) Automated non-invasive measurement of single Sperm's motility and morphology. IEEE Trans Med Imaging 37(10):2257–2265. https://doi.org/10.1109/TMI.2018.2840827

Deconinck L, Cannoodt R, Saelens W, Deplancke B, Saeys Y (2021) Recent advances in trajectory inference from single-cell omics data. Curr Opin Syst Biol 27:100344. ISSN 2452-3100

Demir MH, Eren B (2022) Output voltage control of double chambers microbial fuel cell using intelligence-based optimized adaptive neuro fuzzy inference controller. Int J Hydrog Energy 47(45):19837–19849. ISSN 0360-3199.

J. A. Flexman, S. Minoshima, Y. Kim and D. J. Cross, Magneto-Optical Labeling of Fetal Neural Stem Cells for in vivo MRI Tracking, 2006 International Conference of the IEEE Engineering in Medicine and Biology Society, New York, NY, USA, 2006, pp. 5631–5634, doi: https://doi.org/10.1109/IEMBS.2006.259982.

J. A. Flexman, D. J. Cross, Y. Kim and S. Minoshima, Morphological and parametric estimation of fetal neural stem cell migratory capacity in the rat brain. 2007 29th Annual International Conference of the IEEE Engineering in Medicine and Biology Society, Lyon, France, 2007, pp. 4464–4467, doi: https://doi.org/10.1109/IEMBS.2007.4353330.

Gan Y, Li N, Guo C, Zou G, Guan J, Zhou S (2022) TiC2D: trajectory inference from single-cell RNA-Seq data using consensus clustering. IEEE/ACM Trans Comput Biol Bioinform 19(4): 2512–2522. https://doi.org/10.1109/TCBB.2021.3061720

Geard N, Wiles J (2005) A gene network model for developing cell lineages. Artif Life 11(3): 249–267. https://doi.org/10.1162/1064546054407202

Hu Q, Merchante C, Stepanova AN, Alonso JM, Heber S (2016) Genome-wide search for translated upstream open Reading frames in Arabidopsis Thaliana. IEEE Trans Nanobioscience 15(2): 148–157. https://doi.org/10.1109/TNB.2016.2516950

Iriya R et al (2020) Rapid antibiotic susceptibility testing based on bacterial motion patterns with long short-term memory neural networks. IEEE Sensors J 20(9):4940–4950. https://doi.org/10.1109/JSEN.2020.2967058

Jiang X, Xu W, Park EK, Li G (2014) Selecting protein families for environmental features based on manifold regularization. IEEE Trans Nanobioscience 13(2):104–108. https://doi.org/10.1109/TNB.2014.2316744

Karbalayghareh A, Braga-Neto U, Dougherty ER (2019) Classification of single-cell gene expression trajectories from incomplete and Noisy data. IEEE/ACM Trans Comput Biol Bioinform 16(1):193–207. https://doi.org/10.1109/TCBB.2017.2763946

Lee J, Friderikos V (2022) Interference-aware path planning optimization for multiple UAVs in beyond 5G networks. J Commun Netw 24(2):125–138. https://doi.org/10.23919/JCN.2022.000006

Lee KS, Kim TJ, Pratx G (2015) Single-cell tracking with PET using a novel trajectory reconstruction algorithm. IEEE Trans Med Imaging 34(4):994–1003. https://doi.org/10.1109/TMI.2014.2373351

Li M et al (2021) A deep learning-based method for identification of bacteriophage-host interaction. IEEE/ACM Trans Comput Biol Bioinform 18(5):1801–1810. https://doi.org/10.1109/TCBB.2020.3017386

Li X et al (2023) A clustering method unifying cell-type recognition and subtype identification for tumor heterogeneity analysis. IEEE/ACM Trans Comput Biol Bioinform 20(2):822–832. https://doi.org/10.1109/TCBB.2022.3203185

Liu R, Pisco AO, Braun E, Linnarsson S, Zou J (2022) Dynamical systems model of RNA velocity improves inference of single-cell trajectory, pseudo-time and gene regulation. J Mol Biol 434(15):167606. ISSN 0022-2836

Lu F, Lin Y, Yuan C, Zhang X-F, Ou-Yang L (2021) EnTSSR: a weighted ensemble learning method to impute single-cell RNA sequencing data. IEEE/ACM Trans Comput Biol Bioinform 18(6):2781–2787. https://doi.org/10.1109/TCBB.2021.3110850

Ma F, Lian L, Ji P, Yin Y, Chen W (2020) Fault diagnosis scheme based on microbial fuel cell model. IEEE Access 8:224306–224317. https://doi.org/10.1109/ACCESS.2020.3044354

Mansoor A, Patsekin V, Scherl D, Robinson JP, Rajwa B (2015) A statistical modeling approach to computer-aided quantification of dental biofilm. IEEE J Biomed Health Inform 19(1):358–366. https://doi.org/10.1109/JBHI.2014.2310204

A. Mikhailov and Y. Sankai, Donation of neural stem cells? Post mortal viability of spinal cord neuronal cells. 2018 40th Annual International Conference of the IEEE Engineering in Medicine and Biology Society (EMBC), Honolulu, HI, USA, 2018, pp. 5333–5337, doi: https://doi.org/10.1109/EMBC.2018.8513487.

Muller G, Piel H, Roth RW, Aune B, Magne C, Veyssiere A (1989) Field-emission loading of superconducting accelerator cavities at L- and S-band frequencies. IEEE Trans Electr Insul 24(6):1013–1017. https://doi.org/10.1109/14.46329

Nguyen VT, Ta QTH, Nguyen PKT (2022) Artificial intelligence-based modeling and optimization of microbial electrolysis cell-assisted anaerobic digestion fed with alkaline-pretreated waste-activated sludge. Biochem Eng J 187:108670. ISSN 1369-703X

Ni X, Geng B, Zheng H, Shi J, Hu G, Gao J (2022) Accurate estimation of single-cell differentiation potency based on network topology and gene ontology information. IEEE/ACM Trans Comput Biol Bioinform 19(6):3255–3262. https://doi.org/10.1109/TCBB.2021.3112951

Rezk H, Olabi AG, Abdelkareem MA, Sayed ET (2023) Artificial intelligence as a novel tool for enhancing the performance of urine fed microbial fuel cell as an emerging approach for simultaneous power generation and wastewater treatment. J Taiwan Inst Chem Eng 148: 104726. ISSN 1876 1070.

Sahayasheela VJ, Lankadasari MB, Dan VM, Dastager SG, Pandian GN, Sugiyama H (2022) Artificial intelligence in microbial natural product drug discovery: current and emerging role. Nat Prod Rep 39(12):2215–2230. ISSN 0265-0568.

Schiebinger G (2021) Reconstructing developmental landscapes and trajectories from single-cell data. Curr Opin Syst Biol 27:100351. ISSN 2452-3100

Seabolt EE et al (2022) Functional genomics platform, a cloud-based platform for studying microbial life at scale. IEEE/ACM Trans Comput Biol Bioinform 19(2):940–952. https://doi.org/10.1109/TCBB.2020.3021231

Wang H-Y, Zhao J-P, Su Y-S, Zheng C-H (2022) scCDG: a method based on DAE and GCN for scRNA-Seq data analysis. IEEE/ACM Trans Comput Biol Bioinform 19(6):3685–3694. https://doi.org/10.1109/TCBB.2021.3126641

Wassan JT, Wang H, Browne F, Zheng H (2019) Phy-PMRFI: phylogeny-aware prediction of metagenomic functions using random Forest feature importance. IEEE Trans Nanobioscience 18(3):273–282. https://doi.org/10.1109/TNB.2019.2912824

Wei J, Zhou T, Zhang X, Tian T (2019) SCOUT: a new algorithm for the inference of pseudo-time trajectory using single-cell data. Comput Biol Chem 80:111–120. ISSN 1476-9271.

Wei J, Zhou T, Zhang X, Tian T (2021) DTFLOW: inference and visualization of single-cell Pseudotime trajectory using diffusion propagation. Genomics Proteomics Bioinformatics 19(2): 306–318. ISSN 1672-0229.

Wu W, Ma X (2023) Network-based structural learning nonnegative matrix factorization algorithm for clustering of scRNA-Seq data. IEEE/ACM Trans Comput Biol Bioinform 20(1):566–575. https://doi.org/10.1109/TCBB.2022.3161131

Zeng Q, Ma X, Cheng B, Zhou E, Pang W (2020) GANs-based data augmentation for citrus disease severity detection using deep learning. IEEE Access 8:172882–172891. https://doi.org/10.1109/ACCESS.2020.3025196

Zeng S, Zhang H, Di B, Song L (2021) Trajectory optimization and resource allocation for OFDMA UAV relay networks. IEEE Trans Wirel Commun 20(10):6634–6647. https://doi.org/10.1109/TWC.2021.3075594

Zhang L, Zhang S (2020) Comparison of computational methods for imputing single-cell RNA-sequencing data. IEEE/ACM Trans Comput Biol Bioinform 17(2):376–389. https://doi.org/10.1109/TCBB.2018.2848633

Zhang Z, Zhang X (2021) Inference of high-resolution trajectories in single-cell RNA-seq data by using RNA velocity. Cell Rep Methods 1(6):100095. ISSN 2667-2375

Zhang Y et al (2023) Automated dissection of intact single cell from tissue using robotic micro-manipulation system. IEEE Robot Autom Lett 8(8):4705–4712. https://doi.org/10.1109/LRA.2023.3287364

Chapter 22
AI-Assisted Methods for Protein Structure Prediction and Analysis

Divya Goel, Ravi Kumar, and Sudhir Kumar

Abstract Proteins are the workhorses of cells. Their sequence is determined by the genetic code embedded in the DNA, which translates it faithfully into a string of amino acids known as the primary structure of proteins. But for proteins to achieve functional mode, they must be correctly folded into a three-dimensional structure commonly known as their tertiary structure. Determining the tertiary structure of the proteins is often an expensive and time-consuming process. Protein structure prediction has been in play for several decades now. But recent developments in the fields of computational hardware, software, and artificial intelligence have led to the simultaneous development of methods capable of protein secondary and tertiary structure prediction. Machine learning-based methods have recently emerged as aids of choice for the prediction of protein structures. Programs like AlphaFold and AlphaFold2 have revolutionized the structure prediction landscape. This chapter presents a comprehensive account of the basics of AI-associated methods, along with the evolution of protein structure prediction and its subsequent analysis, helping in related applications in diverse fields ranging from drug discovery to enzyme design.

Keywords Artificial Intelligence · Machine Learning · Deep Learning · Protein Structure Prediction · AlphaFold

---

D. Goel
Department of Biotechnology, H.N.B. Garhwal University, Srinagar Garhwal, Uttarakhand, India

R. Kumar
Department of Computer Science Engineering, Lovely Professional University, Jalandhar, Punjab, India

Department of Computer Science Engineering, Jawaharlal Nehru Government Engineering College, Sundernagar, Himachal Pradesh, India

S. Kumar (✉)
Department of Biotechnology, H.N.B. Garhwal University, Srinagar Garhwal, Uttarakhand, India

Special Centre for Molecular Medicine, Jawaharlal Nehru University, New Delhi, India

A. Khamparia et al. (eds.), *Microbial Data Intelligence and Computational Techniques for Sustainable Computing*, Microorganisms for Sustainability 47,
https://doi.org/10.1007/978-981-99-9621-6_22

22.1 Introduction

22.1.1 Overview of Protein Structure Prediction and Analysis

Proteins are fundamental macromolecules that are pivotal in all biological processes. Elucidation of their three-dimensional (3D) structure is essential for comprehending their functions and interactions. However, experimental determination of protein structures, while more accurate, is often expensive and time consuming. As a result, protein structure prediction and analysis via computational methods have emerged as invaluable tools in modern biology. The amino acid sequence and protein structure have always been connected, and they need to be analyzed critically, which would enable the prediction of function from genome sequence data and facilitate the intentional modification of annotated protein functions by crafting amino acid sequences with targeted structures.

Protein structure prediction helps to deduce the three-dimensional protein structure solely from the sequence of amino acids, without experimental data. This field encompasses two main approaches: template-based methods and ab initio methods.

Template-based methods rely on known protein structures with sequence similarity to the target protein. They use these templates to generate a model for the target protein. On the other hand, ab initio methods predict protein structures from scratch without relying on known templates. They utilize principles from physics and optimization algorithms to explore the conformational space of protein folding. The last 2 years have been phenomenal for the development and advancement of AI and AI-assisted techniques, especially in protein structure determination.

22.1.2 Role of AI in Advancing Protein Structure Research

Artificial intelligence (AI) has revolutionized various scientific domains, and protein structure research is no exception (Paul et al. 2021). AI-driven approaches have significantly advanced our understanding of protein structures and their functions. AI-driven approaches, such as homology modeling and ab initio methods, have significantly improved protein structure prediction. Machine learning algorithms can learn from large databases of experimentally determined protein structures to identify patterns and relationships between protein sequences and their corresponding structures. These recognized patterns can then be applied to previse the structures of novel proteins (AlQuraishi 2021). AI-driven classification methods, such as clustering algorithms and neural networks, have facilitated the categorization of proteins into structural families based on their fold and function. These automated approaches can handle large datasets and uncover new relationships between protein structures (Vijayan et al. 2022). AI-based function prediction methods use deep learning models that are trained on vast databases of studied protein functions to devise the functions of novel proteins (Kandathil et al. 2023). AI has also been used to finetune experimentally determined protein structures. By combining experimental data with AI-driven approaches, such as molecular dynamics (MD) simulations and neural network-based models, researchers can obtain more accurate and reliable protein structures (AlQuraishi 2019a). AI-driven protein structure research has opened new opportunities in drug design and personalized medicine. Virtual screening of potential drug compounds against the predicted protein structures has become expeditious, facilitating drug discovery. Moreover, AI algorithms can analyze protein structures that are part of individual patients' profile, which enables researchers to design personalized treatments tailored to their unique genetic makeup (Johnson et al. 2021). This chapter will provide a background of protein structure fundamentals, basics of machine learning, secondary structure prediction, and tertiary structure prediction with a special mention of AlphaFold.

22.2 Fundamentals of Protein Structure

Proteins undergo a process of adopting stable three-dimensional configurations, known as conformations, which are dictated by their specific arrangement of amino acids. This process, known as protein folding, is central to the proper functioning of proteins and is a strictly regulated process (Diaz-Villanueva et al. 2015; Englander et al. 2007). The entire architecture of a protein is generally characterized at four separate levels of intricacy: primary, secondary, tertiary, and quaternary structure (Sun et al. 2004).

At the core of protein structure analysis lies the primary structure, which points toward the sequence of amino acids that constitute a protein. The sequence is encoded by the genetic information in the DNA, and variations in the sequence give rise to the vast diversity of proteins found in nature. The primary structure forms the foundation for higher-order structures and is essential for determining a protein's functional properties (Anfinsen 1973).

The secondary structure is the next level of the structural hierarchy of a protein, which arises from local interactions within the polypeptide chain. These interactions lead to the formation of distinct repetitive patterns, primarily the alpha-helix, beta-sheet, and turns. The alpha-helix resembles a right-handed coil, while the beta-sheet adopts a pleated, sheet-like arrangement. Turns and loops serve as connectors, linking helices and β-sheets together. Hydrogen bonding between the backbone atoms stabilizes these secondary structure elements, influencing the protein's overall shape and properties (Sun et al. 2004).

Tertiary structure represents the three-dimensional arrangement of all secondary structures in terms of their accurate spatial arrangement and location of functional groups in a single polypeptide chain. This folding is guided by long-range interactions between different amino acids that may not be adjacent in the linear sequence. Various forces, such as hydrogen bonding, disulfide bridges, van der Waals interactions, and hydrophobic effect, impart equilibrium and stability to the tertiary structure. The compact, folded conformation is critical for the proper functioning of proteins, as it positions the active sites and functional regions in the correct orientation (Novak 2021). The tertiary structure of the protein can be further subdivided into domains, folds, and motifs. A domain is a structurally independent compact folding unit, usually consisting of 50–300 amino acids that may be conserved in nature (Sun et al. 2004). A specific arrangement of secondary structure elements in space depicts a fold which is consistent with a number of different proteins (Minami et al. 2014). Protein motifs are short, conserved sequences or patterns of amino acids that are found in specific protein families or functional domains (Parras-Moltó et al. 2013).

Proteins can also exhibit quaternary structure, which involves the precise assembly of multiple polypeptide chains (subunits) to form a multimeric functional complex of a protein. The subunits might be similar or distinct, and their interaction is mediated by non-covalent bonds, like hydrogen bonding and hydrophobic interactions (Fig. 22.1) (Alberts et al. 2022).

Any prediction of the protein structure must first accurately identify and predict these structural features (Deng et al. 2018). Learning from the vast data available from databases like PDB, NCBI, and UniProt, AI-assisted techniques have come a long way from predicting the smaller secondary structural elements to modeling full-length proteins with increasing accuracy (Pakhrin et al. 2021).

22.3 Machine Learning Basics for Protein Structure Analysis

Prediction of protein structure has remained a fundamental challenge in biochemistry, particularly because there is a direct relation between protein sequence, its structure, and function (AlQuraishi 2021). A branch of artificial intelligence, i.e., machine learning, has emanated as a formidable aid in protein structure research, enabling researchers to make accurate predictions and gain deeper insights into protein structure-function relationships.

Designing algorithms and drafting models for statistical analysis are important components of machine learning that empower computers to recognize motifs and project the protein structure from data even when it is not programmed emphatically. Machine learning is divided into three main categories: supervised learning, unsupervised learning, and reinforcement learning (RL).

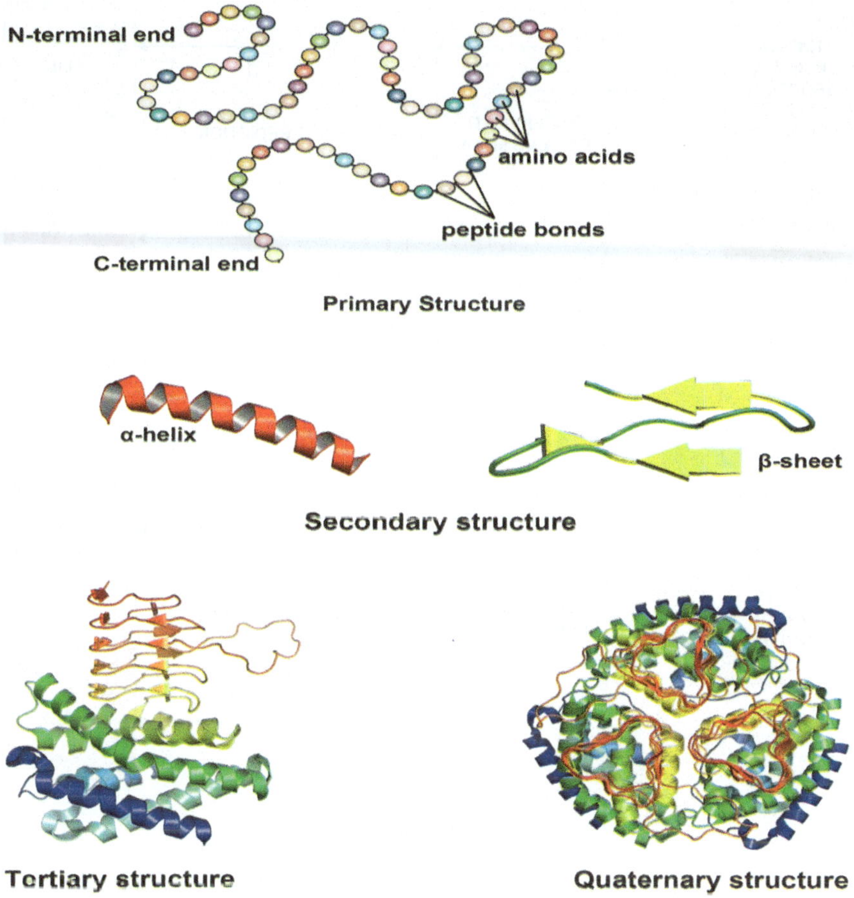

Fig. 22.1 Hierarchical representations of protein structure starting from primary structure (amino acid sequence), secondary structure (α-helix, β-sheets), tertiary structure, and finally quaternary structure. Protein structures are derived from PDB id 3P47 (Kumar et al. 2011), and the images are made using the PyMOL program (Educational version) (Schrodinger, LLC 2015)

22.3.1 Supervised, Unsupervised, and Reinforcement Learning

Supervised Learning

In supervised learning, the algorithm is trained on categorized data, along with the provision of input and compatible output. The objective is to comprehend mapping from corresponding inputs and outputs and estimate new and unseen data, as shown in Fig. 22.2.

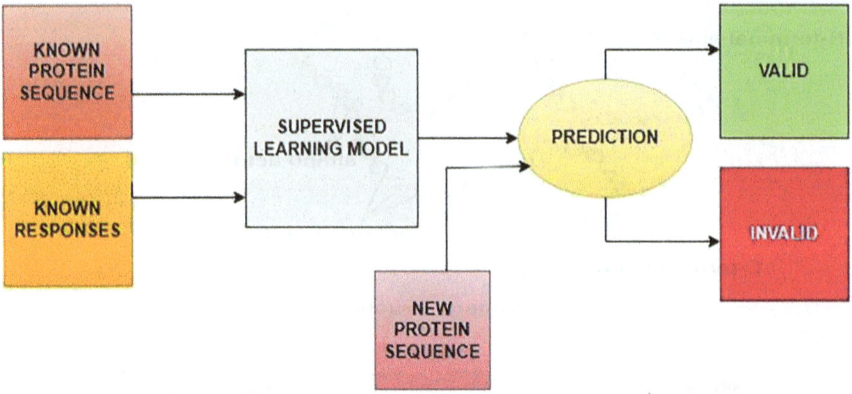

Fig. 22.2 Supervised learning

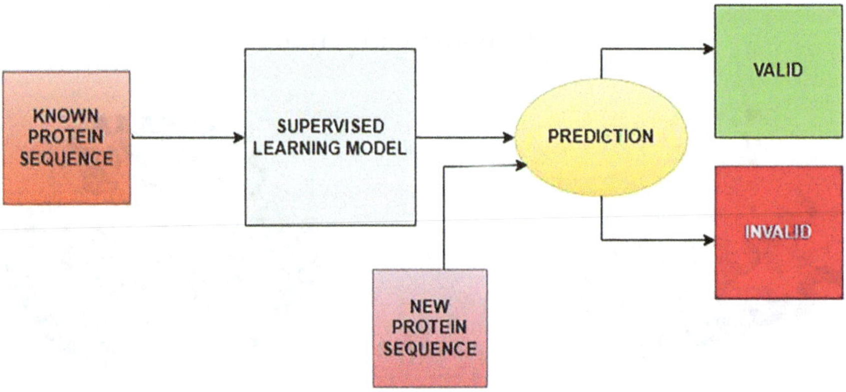

Fig. 22.3 Unsupervised learning

Unsupervised Learning
In unsupervised learning, the algorithm is trained on unlabeled data, and the aim is to identify data patterns or groupings without explicit guidance, as shown in Fig. 22.3.

Reinforcement Learning
Reinforcement learning involves training an agent to work in an environment which reaps supreme results. The agent gathers information from the assessment that is gained in the form of incentives or consequences, as shown in Fig. 22.4.

22.3.1.1 Supervised Learning Algorithms

Supervised learning algorithms constitute a cornerstone of machine learning, leveraging labeled training data to infer correlations between input features and corresponding output labels. When working with protein structure prediction,

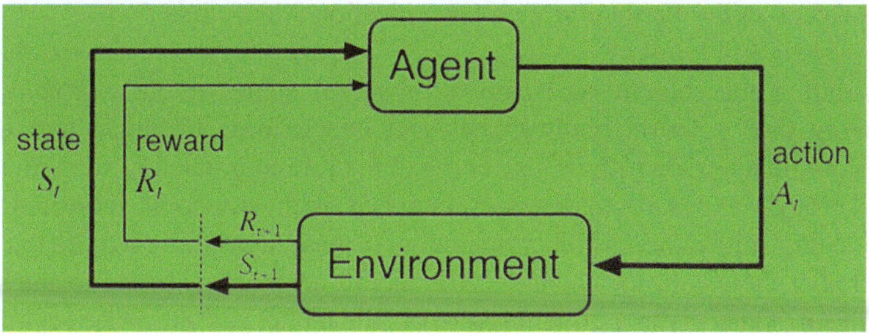

Fig. 22.4 Reinforcement learning

these algorithms learn from known protein structures to predict the structure of a given amino acid sequence. Regression algorithms, such as linear regression and support vector regression, attempt to predict continuous structural properties like the distance between amino acids. Classification algorithms, like random forests and support vector machines, can predict secondary structure elements (helices, strands, coils) or solvent accessibility (Kumar et al. 2011).

Regression Algorithms and Distance Prediction

One of the prime objectives in protein structure prediction involves estimating inter-atomic distances within a protein's structure (Schrodinger, LLC 2015). Regression algorithms, such as linear regression and support vector regression, are instrumental in predicting these distances. By training on datasets containing known protein structures and their corresponding inter-atomic distances, these algorithms learn the intricate relationships between amino acid sequences and spatial arrangements (Zhu et al. 2021). For instance, given a protein sequence, a trained regression model can estimate the distances between key amino acid pairs, elucidating the protein's three-dimensional conformation.

Classification Algorithms and Secondary Structure Prediction

The secondary structure of a protein—comprising alpha-helices, beta strands, and coils—plays a pivotal role in determining its function. Classification algorithms like random forests and support vector machines excel at predicting these secondary structure elements. Utilizing annotated datasets containing experimentally determined secondary structure assignments, these algorithms learn to discriminate sequences based on the propensity of adopting particular secondary structures (Pearce and Zhang 2021). Consequently, when provided with a protein sequence, the model can accurately deduce the distribution of secondary structure elements, facilitating insights into the protein's folding patterns.

Solvent Accessibility Prediction with Classification Algorithms

Understanding the solvent accessibility of amino acids within a protein aid in deciphering their roles in binding sites and interactions (Kuhlman and Bradley 2019). Classification algorithms also come to the fore in predicting solvent accessibility. Through exposure to datasets enriched with structural information and corresponding solvent accessibility annotations, these algorithms acquire the capacity to predict the extent to which specific amino acids are exposed to the solvent environment. Thus, given a protein sequence, the model can anticipate regions with varying degrees of solvent accessibility, contributing to a holistic understanding of the protein's structural landscape.

22.3.1.2 Unsupervised Learning Algorithms

Protein dynamics includes the variations in the three-dimensional configuration of a protein necessary for it to execute its biological role. These variations occur in response to external influences, such as temperature fluctuations and interactions with other molecules, compelling the protein to assume multiple conformations throughout its existence and enabling rapid transitions between these conformations (Chen et al. 2009; Ahmad et al. 2003). Unsupervised learning algorithms are employed to extract meaningful information from vast datasets of protein structures, ultimately aiding in the understanding of protein folding, function, and interactions. The primary technique used by unsupervised learning is clustering.

Clustering

Clustering is an unsupervised classification approach utilized to unveil the inherent structure within data by uncovering concealed patterns. The primary aim of clustering techniques is to autonomously partition a dataset comprising instances into clusters or groups characterized by their similarity while simultaneously identifying clusters that exhibit dissimilarity (Chen et al. 2009). By identifying clusters of similar protein structures, researchers can gain insights into protein families, evolutionary relationships, and functional similarities (Teletin et al. 2018). Clustering can further be of many types:

(a) K-Means Clustering: K-means is an algorithm that partitions protein structures into a predetermined number of clusters. It is one of the most used clustering algorithms. It employs a method of iteratively allotting each structure to the closest cluster centroid and then upgrading the centroids established from the average of the structures within each cluster (Tokuriki and Tawfik 2009).
(b) Hierarchical Clustering: Hierarchical clustering constructs a hierarchical tree-like structure of clusters, thereby enabling researchers to investigate the

inconsistencies of different stages in their data. The hierarchical tree can be agglomerative (bottom-up) or divisive (top-down) (Zaslavsky et al. 2016).

(c) Density-Based Clustering: Density-based clustering methods, such as DBSCAN, recognize clusters based on regions of higher data density. This approach is particularly useful when clusters have irregular shapes and varying densities (Melman and Roshan 2018).

(d) Dimensionality Reduction: Methods like Principal Component Analysis (PCA) and t-Distributed Stochastic Neighbor Embedding (t-SNE) are utilized in conjunction with clustering to reduce the dimensionality of protein structure data, making it more manageable and informative (Jisna and Jayaraj 2021).

(e) Graph-Based Clustering: Graph-based methods represent protein structures as nodes in a graph, with edges denoting structural similarity. Community detection algorithms are then applied to identify clusters (Dang et al. 2021).

Principal Component Analysis

PCA is an elementary unsupervised machine learning technique widely applied for protein structure prediction. It plays a pivotal role in alleviating the multiple dimensionalities of complex protein structure data and retaining critical information at the same time (Zhu et al. 2016). Its principal goal is to modify high-dimensional data into a lower-dimensional representation, known as principal components, while minimizing the loss of information (Russo and Borras 2022). In the context of protein structure prediction, PCA operates by finding linear combinations of the original variables (atom coordinates) that maximize variance, effectively capturing the most significant structural variations in the data (Chen et al. 2009; Jolliffe and Cadima 2016).

Application in Protein Structure Prediction

Dimensionality Reduction: Protein structures are inherently high-dimensional, with each atom's coordinates representing a separate dimension. PCA addresses this challenge by projecting the information onto a reduced set of orthogonal axes, the primary elements, which capture the essential structural features. This transformation significantly simplifies subsequent analysis and visualization.

Noise Reduction: Protein structure data may contain noise or redundant information. PCA identifies and emphasizes the principal components that contribute most significantly to the structural variance, effectively filtering out irrelevant details and noise.

Visualization: PCA enables the visualization of protein structures in a lower-dimensional space, facilitating it to discern patterns, similarities, and differences among structures. Researchers can visualize and explore structural variations along the principal components to gain insights into the protein's conformational flexibility.

Clustering: PCA can be used in conjunction with clustering techniques to group similar protein structures into clusters based on their principal component representations. This facilitates the identification of structural subgroups within a dataset.

22.3.1.3 Reinforced Learning

Reinforcement learning (RL) is a component of machine learning which is implicated in sequential decision-making in environments where an agent (algorithm that learns from trial and error) interacts with the surroundings. The agent is trained to act to intensify a cumulative reward signal over time. In protein structure prediction, RL is employed to optimize the folding by iteratively adjusting the conformation of the protein until it reaches its most stable or biologically relevant state.

Unlike traditional machine learning approaches, in reinforcement learning, desired outcome is not explicitly defined. Instead, features are entered into the environment, and rewards are assigned. These rewards can be either positive or negative, based on the agent's actions and decisions (David and Jacobs 2014).

22.4 AI-Assisted Secondary Structure Prediction

22.4.1 Traditional Methods

Protein secondary structure prediction includes the accurate predictions of secondary structures, including alpha-helices, beta-sheets, and coil regions. Traditional methods for protein secondary structure prediction primarily rely on empirical rules and physicochemical properties of amino acids. Methods such as Chou-Fasman (Kitao 2022), Garnier-Osguthorpe-Robson (GOR) (Arora et al. 2018), and Dictionary of Secondary Structure of Proteins (DSSP) (Ashok Kumar 2013) were used over the years for secondary structure prediction. Chou-Fasman is an early method based on the propensity of amino acids to form secondary structures. It assigns secondary structure elements by analyzing the local sequence properties, such as the presence of specific amino acid residues that favor the formation of alpha-helices or beta-sheets. While simple, Chou-Fasman suffers from limited accuracy due to its reliance on fixed parameters. GOR or Garnier-Osguthorpe-Robson method utilizes statistical analysis of amino acid sequences to predict secondary structure. It calculates probabilities for each amino acid to belong to alpha-helices, beta-sheets, or coil regions based on databases of known protein structures. GOR offers improved accuracy compared to Chou-Fasman but is still hampered by its dependency on predefined datasets. (Dictionary of Secondary Structure of Proteins) DSSP is a commonly employed method for secondary structure assignment in experimentally determined protein structures. It defines secondary structures by assigning helical or strand regions based on hydrogen bonding patterns and geometrical criteria. DSSP is highly accurate but is limited to known protein structures.

While these methods were well established because of their simplicity and transparency, needing only limited computational power, they suffered many limitations, especially when dealing with novel protein sequences.

Combining the simplicity and interpretability of traditional methods with the predictive power of AI can lead to more robust and accurate predictions. Additionally, ongoing advancements in AI and deep learning algorithms, as well as the availability of larger and more diverse protein databases, will continue to enhance the accuracy and applicability of AI-based approaches.

22.4.2 *Deep Learning Models for Secondary Structure Prediction*

Deep learning constitutes a specific domain within machine learning which revolves around artificial neural networks. It places a strong emphasis on employing interconnected layers to convert input data into features that are conducive to predicting corresponding outputs. When provided with a dataset that is sufficiently expansive, pairing inputs with their respective outputs, a training algorithm can be applied to autonomously acquire the ability to map inputs to outputs (Kouza et al. 2017). This is achieved by adjusting a group of specifications at all the levels within the neural network (Kabsch and Sander 1983; Taye 2023). Deep learning models, particularly convolutional neural networks (CNNs) and recurrent neural networks (RNNs), exhibited an exemplary success in protein secondary structure prediction. These models have been specifically drafted to automatically decipher intricate motifs and relationships between amino acid sequences, enabling accurate predictions (Goodfellow et al. 2016).

22.4.2.1 Convolutional Neural Networks (CNNs)

CNNs are structural designs intended for the handling of data that exhibits consistent spatial organization. A CNN layer uses a smart trick by using the same small filters repeatedly across data, which is helpful in two ways: it prevents overfitting because there are fewer things to adjust compared to the input layer, and it works well even if the data shifts around. A CNN module typically has many of these layers stacked together. As you move deeper into these layers, they can understand more complicated features (Taye 2023).

In 2016, Wang and colleagues introduced a convolutional neural network (CNN) approach for protein secondary structure prediction (PSSP), achieving an impressive accuracy rate of up to 84% (Wang et al. 2016a, b). Their method, known as "Deep Convolutional Neural Fields" (DeepCNF), comprised two key components: a conditional random fields (CRF) module, originally developed in their earlier work from 2011, and a deep convolutional neural network (DCNN) module that handled the input data leading up to the CRF, as described in their 2016 publication (Ismi et al. 2022). Both one-dimensional CNN (CNN-1D) and two-dimensional CNN (CNN-2D) were used previously in PSSP models. CNN-1D utilized sequences as inputs and thus was used more often. Nevertheless, some models have incorporated CNN-2D into their architectures to enhance the extraction of time- and space-related features from input sequences. Specifically, feature vectors, such as position-specific scoring matrix (PSSM) and one-hot encoding, representing a fixed-length residue window, were assigned as input in studies for the two-dimensional CNN (Goodfellow et al. 2016; Wang et al. 2016a, b).

CNNs bring two advantages in the prediction of secondary structures of proteins:

(a) Local Sequence Patterns: CNNs are well suited to recognize short-range interactions and local sequence features crucial for secondary structure prediction. They apply a set of convolutional filters covering all the different positions in the amino acid sequence, enabling them to learn and extract relevant features effectively (Guo et al. 2018).
(b) Translation Invariance: CNNs exhibit a remarkable property called translation invariance. This means they can identify patterns even when the data is slightly shifted or moved, making them suitable for the flexible and dynamic nature of protein structures (Liu et al. 2017).

22.4.2.2 Recurrent Neural Networks (RNNs)

Recurrent neural networks (RNNs) design enables them to learn the global patterns from sequential data. They provide a potent means for handling sequential data by virtue of their intrinsic capability to “retain” past inputs within the sequence through hidden states (Gelman et al. 2021). In the process of handling an input sequence, an RNN module employs an internal state vector to condense the knowledge acquired after refining the different components of the sequence. This module encompasses a parameterized sub-component that receives both the previous internal state vector and the present input element of the sequence as inputs, resulting in the generation of the current internal state vector. Ultimately, this final state vector serves as a concise representation, summarizing the entirety of the input sequence (Taye 2023).

One of the key advantages of RNNs in protein structure prediction is their ability to handle variable-length input sequences. Proteins can vary significantly in length, but RNNs can process sequences of different lengths without the need for fixed-size inputs. This flexibility is essential for modeling the diversity of protein sequences encountered in practice (Alzubaidi et al. 2021). Furthermore, RNNs can capture long-range dependencies in protein sequences. Proteins often exhibit complex folding patterns that result from interactions between distant amino acids in the sequence. RNNs, with their recurrent connections, can capture these long-range dependencies, allowing them to learn intricate structural motifs and patterns (Gelman et al. 2021). A very simple RNN-based model known as LocalNet achieved an accuracy of 85.15% in predicting all three types of secondary structures (Helix, sheets, turns) using only local amino acid information (Lee 2023).

In recent years, researchers have made significant strides in applying RNNs to protein structure prediction. The development of specialized RNN variants, such as long short-term memory (LSTM) and gated recurrent unit (GRU), has improved the ability of RNNs to capture long-range dependencies (Taye 2023).

Long short-term memory (LSTM) networks are a specialized type of recurrent neural network (RNN) architecture modulated to deal with some of the shortcomings of traditional RNNs when it comes to detecting long-range dependencies in sequential data. LSTMs were introduced by Hochreiter and Schmidhuber in 1997 and have since become a crucial component in various machine learning applications, including natural language processing, time series analysis, and, as you mentioned, protein structure prediction (Liu 2017). LSTMs are particularly advantageous in the context of protein structural prediction because they can handle long sequences as they have memory cells that can store and retrieve information over long sequences (Yang et al. 2021). They incorporate gating mechanisms, like input gate, forget gate, and output gate, which control the direction of information in and out of the memory cells. It provides the network with the fine-grained control required for modeling complex dependencies in protein sequences and structures (Hochreiter and Schmidhuber 1997). An orchestrated model with dual loss function established for the prediction of protein secondary structure consisting of five sub-models connected via a Bi-LSTM layer achieved 84.3% in Q3 accuracy and 81.9% in segment overlap measure (SOV) score (Yang et al. 2021). A deep bidirectional long short-term memory (DBLSTM) was able to achieve reasonable level of accuracy compared to the state-of-the-art approaches like bidirectional long short-term memory (BLSTM) and DeepCNF (Hu et al. 2019). Another study developed a method called the optimized convolutional and long short-term memory neural network model (OCLSTM) for predicting protein structure using an optimized convolutional neural network to find patterns between nearby amino acids and a bidirectional long short-term memory neural network to discover relationships between distant amino acids. This model showed accuracy ranging from 82.36% to 85.08% on different datasets (Ahmed et al. 2023).

Additionally, the integration of observation techniques has enhanced the model's ability to concentrate on relevant parts of the input sequence. Prediction of protein structure with RNNs has also been achieved via transfer learning. Pretrained models on large protein databases can be refined on particular assignments, thereby decreasing the requirement of large quantity of task-specific data.

Gated recurrent unit (GRU) is another type of recurrent neural network (RNN) architecture, related to long short-term memory (LSTM), modeled to explore the diminishing gradient issue and capture long-range dependencies in sequential data (Liu et al. 2017; Hochreiter and Schmidhuber 1997). Compared to LSTMs, GRUs have fewer parameters because they lack separate memory cells. This makes them computationally more efficient, particularly in scenarios with limited data and computational resources. However, this simplicity can also make them less expressive for assignments that are needed to decipher very complex long-term dependencies. A modified GRU-based method was recently utilized for the recognition of phosphorylation sites in host cells infected with SARS-CoV-2 (Hattori et al. 2017).

Many researchers have used the hybrid RNN methods to achieve better accuracies on protein secondary structure predictions. Hybrid recurrent neural networks (RNNs) direct toward a branch of neural network architectures that combine different types of recurrent units or modules within a single network. Romana Rahman Ema et al. used gated recurrent unit (GRU), long short-term memory (LSTM), bidirectional gated recurrent unit (BGRU), and bidirectional long short-term memory (BLSTM) neural networks and improved the accuracies to a range of >90% (Ema et al. 2022) (Zhao and Liu 2021). A webserver ET-GRU was also developed using the 1D CNN, GRU, and PSSM profiles capable of identifying the electron transport proteins among the protein sequence samples (Zhang et al. 2023).

22.4.3 Case Studies

DeepPrime2Sec, developed by Ehsaneddin Asgari et al. utilizing the CNN-BLSTM (convolutional neural network-bidirectional long short-term memory) network and combining top-k ensemble models, obtained accuracy rates of 69.9% and 70.4% for predicting eight-class protein secondary structures on the CB513 dataset, which is known to be one of the most demanding datasets for this type of prediction (Ehsaneddin et al. 2019).

A reductive deep learning model called MLPRNN, composed of one bidirectional gate current unit (BGRU) and two multi-layer perceptron (MLP) blocks, has been introduced for predicting protein secondary structures, either in three-state or eight-state classifications. The accuracy of MLPRNN's predictions on the widely used CB513 benchmark (70.07%) dataset is on par with the performance of other cutting-edge models like DeepCNF (68–73%) (Le et al. 2019).

Sharma and Srivastava used a bidirectional long short-term memory (BLSTM) model to project the secondary structures of proteins using the local contextual information captured with character n-gram and achieved a Q3 accuracy of 86.69% on the CASP9 dataset (Ehsaneddin et al. 2019).

Chi-Hua Yu et al. also reported the development of a deep learning model which used convolutional and recurrent architectures along with natural language models to predict alpha-helix and beta-sheets in proteins. Their model could decipher the obscured patterns in protein sequences and use this knowledge to predict the secondary structures (Lyu et al. 2021).

SPIDER2 employs three rounds of deep learning neural network iterations to enhance the efficiency of predictions for multiple structural characteristics, like three-state secondary structure, angle-based representation of backbone structure, and solvent accessible surface area, concurrently. When evaluated on an independent test set comprising 1199 proteins, SPIDER2 attains the following performance metrics: 82% accuracy for secondary structure prediction, a correlation coefficient of

0.76 between predicted and actual solvent accessible surface area, mean absolute errors of 19° and 30° for backbone φ and ψ angles, respectively, and mean absolute errors of 8° and 32° for Cα-based θ and τ angles, respectively (Sharma and Srivastava 2021).

Xin Jin et al. proposed CGAN-PSSP, a novel protein secondary structure prediction (PSSP) model built upon a conditional generative adversarial network (CGAN), which can be used for the prediction of eight-state (Q8) and three-state (Q3) protein secondary structures. It was able to achieve 71.3% Q8 and 84.8% Q3 accuracy on CASP11 dataset (Yu et al. 2022).

22.5 Predicting Tertiary Structure with AI

The tertiary structure of proteins represents the highest level of their structural organization, defining the intricate 3D arrangement of atoms within a single polypeptide chain. This spatial arrangement is crucial in determining a protein's function and interactions with other molecules. Tertiary structure results from the complex folding and bonding of secondary structures (alpha-helices, beta-sheets) and additional irregular segments. Forces responsible for maintaining the tertiary structure include hydrogen bonds, disulfide bonds, hydrophobic interactions, and van der Waals forces. The precise formation of unique 3D conformation of a protein is essential for it to perform its functions inside a cell effectively, such as enzyme catalysis, molecular recognition, and structural support within cells and organisms (Alberts et al. 2022). Understanding and predicting protein tertiary structure are vital endeavors in biochemistry and molecular biology, with implications ranging from drug design to the elucidation of fundamental biological processes. Homology-based methods have been a huge favorite among the researchers to project the protein tertiary structure.

22.5.1 Homology Modeling and Comparative Protein Structure Prediction

Homology modeling, or comparative modeling, is the method to construct a protein's three-dimensional structure by utilizing structural insights derived from known configurations of related proteins. This modeling strategy capitalizes on the principle that a protein's structural conformation tends to exhibit greater conservation across evolution compared to its specific sequence of amino acids. Consequently, minor-to-moderate alterations in the amino acid sequence typically correspond to minimal deviations in the 3D structure. A critical prerequisite for the effective application of homology modeling is the presence of substantial similarity, typically exceeding 30%, between the protein sequences under consideration.

22.5.2 Template-Based Structure Prediction

Homology modeling starts with identifying the suitable template and aligning the query and target sequences. It then proceeds to the generation of the model including backbone and side chains. Finally, the model is optimized and validated. The query sequence is subjected to BLAST (Yang et al. 2017) against the PDB database. The best structure with the lowest e value and maximum overall coverage is then selected as the template structure. This structure then helps to determine the structure of the query sequence, which is then refined individually for loops, side chains, and secondary structures, giving an output structure. This method is very popular because of its ease and accuracy, but the latter depends on the quality of the reference structure. The extent to which the query sequence shows similarity or identity with the reference sequence and its coverage is the critical step in this type of modeling and is also its major limitation.

MODELLER is a popular program used for homology modeling (Jin et al. 2022). A simple script is needed for modeller along with a sequence alignment file (Query vs template) and the atomic coordinates of the templates (PDB file). MODELLER then computes a model that includes all non-hydrogen atoms. It can also execute various supplementary tasks, such as assigning folds, aligning two protein sequences, or aligning their profiles (Altschul et al. 1990). There are several other software or web-based programs like SWISS-Model (Webb and Sali 2016) and I-TASSER (Eswar et al. 2006) that can be used for template-based homology modeling.

22.5.3 Threading or Fold Recognition-Based Methods

Threading, also known as fold recognition, is a computational method used for protein tertiary structure prediction when there are no close homologous proteins with known structures available in the database. This approach assumes that even distantly related proteins may share similar structural folds, and it seeks to identify the best structural match or "thread" for a target protein sequence within the archive of known protein folds. Although it follows similar steps like homology modeling, fold identification and fold fitting steps are different. This method compares the query sequence with a collection of structural templates containing the different protein folds and allows the best fitting of the folds as judged by the scoring functions. After generating the initial model, some threading methods may perform model assessment and refinement steps to improve the quality of the predicted structure. This can involve energy minimization, molecular dynamics simulations, or other optimization techniques. Examples include Phyre2 (Waterhouse et al. 2018), RaptorX (Yang and Zhang 2015), and HHpred (Kelley et al. 2015). Threading methods have limitations when dealing with highly novel or extremely divergent protein sequences, as the chances of finding a suitable template decrease.

22.5.4 Ab Initio Modeling

Ab initio protein modeling, also referred to as de novo modeling, is based on the fundamental laws of thermodynamics and physics. By minimizing the free energy of a protein's conformation, it seeks to discover the 3D structure of the molecule. This refinement can achieve its aim of the protein's lowest-energy state or its most stable configuration out of an infinite number of potential conformations.

Ab initio modeling often entails a conformational search procedure directed by a carefully designed energy function. The final models are often selected from a variety of structural decoys, or alternative conformations, that this process typically generates (Kallberg et al. 2012). Energy functions used here are either physics based or knowledge based, depending upon their working. The efficiency of ab initio modeling thus depends on three important factors: the availability of a precise energy function that can identify a most thermodynamically stable protein state, the application of a quick search technique to identify low-energy conformational states, and the quick identification of a near-native structural model (Zimmermann et al. 2018).

In physics-based techniques, parameters related to bond lengths, angles, torsional angles, van der Waals interactions, and electrostatics are combined into a molecular mechanics force field. Herein the protein is considered as a sophisticated atomic model (Lee et al. 2009). Amber which started in 1970 is a suite of biomolecular simulation programs using physics-based force fields and molecular dynamics (MD) simulations to derive ab initio protein structure (Lee et al. 2017). Another program, CHARMM (Chemistry at HARvard Macromolecular Mechanics), developed by Martin Karplus and his group, also uses empirical energy functions for molecular simulations but also has the capacity for new model building (Wang et al. 2004; Case et al. 2005). The high cost of physics-based energy computations limited their use. However, programs like UNRES and ASTRO-FOLD attempted to reduce this problem. UNRES represented the polypeptide chain as a sequence of α-carbon atoms (C_α) linked by virtual bonds of length 3.8 Å and the sidechains represented as a single interaction site, effectively reducing the number of atoms under consideration by a factor of 10 and consequently decreasing the time required for structure prediction of proteins up to 100 residues long (Brooks et al. 2009). ASTRO-FOLD calculates the free energy function from the oligopeptide overlapping and interaction between hydrophobic residues to predict the secondary structures in the proteins. This data is then transformed into a whole sequence, leading to the complete model of the protein (Brooks et al. 1983). Other methods based on these energy functions increased the length of the successfully predicted model to about 150 residues (Zimmermann et al. 2018).

22.5.5 Knowledge-Based Energy Function

The statistical analysis of recognized protein structures yields knowledge-based energy functions, also known as statistical potential functions. These programs estimate the likelihood or favorability of interactions or structural properties in each protein conformation using data from databases of experimentally structured proteins (such as the Protein Data Bank [PDB]). Knowledge-based energy functions have inbuilt propensities for secondary structures, but they may not be enough for local structure modeling. A fragment-based approach utilizes secondary structure fragments taken from aligning sequences for protein structure modeling. Several software programs deploy this method to reduce the entropy of conformational search and achieve accuracy in local structure modeling. Examples include Rosetta and TASSER/I-TASSER programs. Simons et al. developed Rosetta based on the fragment assembly approach given by Bowie and Eisenberg (Bowie and Eisenberg 1994; Simons et al. 1997). To explore the relevant conformational space of the target sequence and rank the resulting models on the basis of their energy function, Rosetta employs Metropolis Monte Carlo sampling approaches combined with knowledge-based energy functions (Bowie and Eisenberg 1994). Rosetta was able to predict a 70-amino-acid-long protein structure to near atomic resolution in CASP6 (Simons et al. 1997). TASSER was a purely knowledge-based program for 3D model construction of proteins. To investigate potential folding patterns, the query sequence is first passed through a selection of typical protein structures. The threaded and aligned sections are then used to extract continuous segments with at least five residues. These segments are then used to reconstruct entire models of the protein, whereas non-aligned sections are built using a lattice-based ab initio modeling strategy (Kallberg et al. 2012; Kaufmann et al. 2010). I-TASSER is a graded template-based application for the prediction of the structure and function of a protein, which was ranked at the top of many CASP events (Eswar et al. 2006). These methods provided a stable benchmark for the AI-associated techniques to evolve.

22.5.6 AI-Driven De Novo Protein Structure Prediction

Before delving into the world of AlphaFold, evolutionary algorithms and Monte Carlo Methods need to be understood. Monte Carlo simulations (MCS) represent a widely used approach for calculating the trajectories and thermodynamic characteristics of proteins. Monte Carlo simulation employs stochastic sampling and statistical modeling to approximate mathematical functions and replicate the behavior of intricate systems (Bradley et al. 2005). In a simulation run, a sequence of random movements is made within the conformational space, each of which introduces alterations to certain molecular degrees of freedom. The acceptance of each step is contingent upon the likelihood associated with the alteration in the energy function's value. An energy landscape is a topographical representation across the space of conformations, illustrating the potential energy associated with every conceivable molecular conformation (Zhang et al. 2002).

The theory of the energy landscape says that although there are n number of possible conformations of the protein structure during folding, it can be effectively represented using only one or a limited set of reaction coordinates without sacrificing a remarkable volume of kinetic information (Harrison 2010). The application of Monte Carlo and conformational space learning techniques enabled the progressive protein structure prediction software to perform better in a limited set of time. Protein segments can be constructed incrementally, adding one amino acid at a time, with each consecutive step tailored to promote the adoption of low-energy conformations (Edwards et al. 2012). Such methods have been developed and used previously for biological macromolecular structure prediction (Travaglini-Allocatelli et al. 2009; Wong et al. 2018; Tang et al. 2014).

22.5.7 Deep Learning Models for Tertiary Structure Prediction

CASP 13 (Critical Assessment of Protein Structure Prediction) revolutionized the AI-associated structure prediction landscape with the introduction of AlphaFold, a deep learning-based template free modeling (FM) program (Wick and Siepmann 2000). This has been described as watershed moment in the field of structural prediction for complex macromolecules like proteins (Zhang et al. 2007). CASP, a biennial event that is organized with the goal of evaluating and assessing the accuracy of protein structure predictions on a global scale, has been held since 1994. CASP13, which was held in December 2018, brought the real force of AI to the fore of protein structure prediction. DeepMind's AlphaFold was ranked first among the template free modeling (FM) programs. It doubled the accuracy achieved during the earlier CASP event and increased the prediction range up to 100 (Senior et al. 2020).

The success of AlphaFold comes from the groundwork prepared by several generations of evolution in protein research. It is a co-evolution-dependent method which, in the first step, generates a large repertoire of multiple sequence alignments (MSAs) of evolutionary varied homologous proteins in the range of 10^5–10^6 sequences. It then extracts the information from the residues that seem to co-evolve, sharing a similar conformational space. AlphaFold takes these binary contact distributions as a protein-specific statistical potential function, which is directly optimized to facilitate protein folding (AlQuraishi 2020). AlphaFold's core concept is that a distribution of distances between protein residues can be transformed into a continuous function, which can then be minimized to predict protein structures accurately. The major portion of the success of AlphaFold was attributed to the use of very advanced neural networks with "dilated convolutions" capable of capturing long-distance interactions in the protein sequence (Senior et al. 2020).

22.5.8 AlphaFold2

In CASP 14 in 2020, Jumper et al. submitted their newer version of the AlphaFold under the name of AlphaFold2 (AF2) while claiming that it is completely different from its previous version (AlQuraishi 2019b). To provide accurate end-to-end structure prediction, AF2 proposes a unique architecture that simultaneously integrates pairwise characteristics and multiple sequence alignments (MSAs), as well as an updated output representation and loss function. An attention mechanism-based transformer (Sippl 1990) was used in AF2, which is a recently developed deep neural network that uses the self-attention process to collect intrinsic properties and exhibits considerable potential for widespread AI applications. This transformer system is made up of two modules—a decoder module and an encoder module—each of which contains several transformer blocks with the identical structure. A feed-forward neural network, a shortcut connection, a multi-head attention layer, and layer normalization are all contained within each transformer block. This is termed as "Evoformer" in AF2, which can model the prediction of protein structure as a three-dimensional graph inference issue, where the edges of the graph are determined by the closeness of residues (AlQuraishi 2019b).

Proteins have been known to conserve their structure during evolution, which means that even though the sequence of a particular protein may undergo changes as much as 80%, its 3D structure may still be the same. If two residues show conservation among themselves, it generally means that there is some sort of interaction among them. This information becomes the basis for AF2. The AF2 architecture consists of three layers: the first is the input module, the second is Evoformer, and the third is the structure module. The input module takes the amino acid sequence, and MSA is fabricated by searching for homologous sequences and their corresponding structures, if any. It then generates a pairwise distance matrix between the amino acids. Evoformer accepts the MSA as input and subjects it to deep neural learning layers which outputs the updated MSA and an updated pair representation. Forty-eight such layers are present in AF2, which are independent of each other and contain two pathways: one for the MSA and the other for the pairwise representation which represents attention as a triangle of residues. Here each triangle means that any two residues can influence the third one. The third module, which is like a decoder, decodes the information generated by the Evoformer and translates it into the 3D atomic coordinates for the query sequence (Jumper et al. 2021).

The way AF2 was trained is a key factor in its success. The creators used a method called self-distillation, which combined data from the Protein Data Bank (PDB) with a new dataset of predicted protein structures. In this training process, 25% of the examples came from well-known PDB structures, while 75% came from the new self-distillation dataset. The goal was to help AF2 replicate challenging protein structure predictions by using various training data enhancement techniques. By using this combination of datasets, including the data predicted by AF2, the model's performance was significantly improved.

22.5.9 *Applications of AlphaFold2*

AlphaFold has revolutionized the protein structure landscape, and the major beneficiary is structural biology community. There is a growing divide between the proteomic data and the experimental protein structures available, as it is both time consuming and costly to determine the structure of a protein through wet lab procedures. However, advanced predictive methods such as AF2 still need experimental validation. Other areas that have potential applications of AF include protein-protein interactions, protein drug-interactions, enzyme mechanisms, drug discovery, and protein design (Jumper et al. 2021). AF has been able to accurately decipher the structure of ubiquitin-specific peptidase 7 (USP7), a complex enzyme with multiple domains showing dynamic behavior. The predictive model was able to match the interactions and movements of various domains as shown by the experimentally derived models of the same protein (Vaswani et al. 2017; Yang et al. 2023). Mehmet Akdel et al. did a comprehensive analysis of the AF2 predictions on a large dataset comprising reference proteomes for 11 species. They compared the results with the PDB data for various parameters, including typical structural elements, missense variants, ligand binding, and interaction analysis. They found out that AF2 can predict about 25% more residues with confidence as compared to template-based homology modeling. It was also successful in predicting the protein disorder more accuratcly and was able to predict structural folds that were rare even for PDB (Kim et al. 2016). AF2 has since then found applications in antibiotic discovery (Perrakis and Sixma 2021), disordered protein structure prediction (Akdel et al. 2022), predicting novel human proteins with knots (Wong et al. 2022), and viral research (Ruff and Pappu 2021), to name a few. It has also been combined with X-ray crystallography, NMR, mass spectrometry, and CryoEM for solving the experimental data-based structures (Perlinska et al. 2023; Gutnik et al. 2023; Allison et al. 2022; Arantes et al. 2022; Laurents 2022).

22.6 Conclusion and Future Scope

Protein structure has always fascinated scientists as it allows the visualization of a biological machinery at its finest. There has been constant effort by the researchers to accurately predict the structure of proteins from their sequences. The collective effort of studying the intrinsic properties of proteins has enabled the scientists to now achieve the accuracy of near-native structures using advanced AI tools. However, it is still far from complete, as there are several limitations to these tools which need to be addressed. Specifically, the effects of epigenetic modifications on proteins, point mutations, post-translationally modified proteins (Yang et al. 2023; Nagaratnam et al. 2022; Paul et al. 2022; Azzaz et al. 2022) (Buel and Walters 2022), etc., are some of the areas where AF2 is still in its infancy. Recently protein language models have been developed which promise better and more accurate protein folding predictions (Bertoline et al. 2023). Whatever the case may be, it is a fast-changing landscape, with new developments coming every day.

References

Ashok Kumar T (2013) CFSSP: Chou and Fasman secondary structure prediction server. Wide Spectr Res J 1:15–19. https://doi.org/10.5281/zenodo.50733. (ISSN 2250-2815)

Ahmad S, Gromiha MM, Sarai A (2003) Real value prediction of solvent accessibility from amino acid sequence. Proteins 50(4):629–635. https://doi.org/10.1002/prot.10328

Ahmed SF, Alam MSB, Hassan M, Rozbu MR, Ishtiak T, Rafa N et al (2023) Deep learning modelling techniques: current progress, applications, advantages, and challenges. Artif Intell Rev 56(11):13521–13617. https://doi.org/10.1007/s10462-023-10466-8

Akdel M, Pires DEV, Pardo EP, Janes J, Zalevsky AO, Meszaros B et al (2022) A structural biology community assessment of AlphaFold2 applications. Nat Struct Mol Biol 29(11):1056–1067. https://doi.org/10.1038/s41594-022-00849-w

Alberts B, Johnson A, Lewis J, Raff M, Roberts K, Walter P (2022) The shape and structure of proteins. In: Alberts B (ed) Molecular biology of the cell, 4th edn. Garland Science, New York

Allison TM, Degiacomi MT, Marklund EG, Jovine L, Elofsson A, Benesch JLP et al (2022) Complementing machine learning-based structure predictions with native mass spectrometry. Protein Sci 31(6):e4333. https://doi.org/10.1002/pro.4333

AlQuraishi M (2019a) End-to-end differentiable learning of protein structure. Cell Systems 8(4): 292–301.e3. https://doi.org/10.1016/j.cels.2019.03.006

AlQuraishi M (2019b) AlphaFold at CASP13. Bioinformatics 35(22):4862–4865. https://doi.org/10.1093/bioinformatics/btz422

AlQuraishi M (2020) A watershed moment for protein structure prediction. Nature 577(7792): 627–628. https://doi.org/10.1038/d41586-019-03951-0

AlQuraishi M (2021) Machine learning in protein structure prediction. Curr Opin Chem Biol 65:1–8. https://doi.org/10.1016/j.cbpa.2021.04.005

Altschul SF, Gish W, Miller W, Myers EW, Lipman DJ (1990) Basic local alignment search tool. J Mol Biol 215(3):403–410. https://doi.org/10.1016/S0022-2836(05)80360-2

Alzubaidi L, Zhang J, Humaidi AJ, Al-Dujaili A, Duan Y, Al-Shamma O et al (2021) Review of deep learning: concepts, CNN architectures, challenges, applications, future directions. J Big Data 8(1):53. https://doi.org/10.1186/s40537-021-00444-8

Anfinsen CB (1973) Principles that govern the folding of protein chains. Science 181(4096): 223–230. https://doi.org/10.1126/science.181.4096.223

Arantes P, Nierzwicki L, Belato H, D'Ordine A, Jogl G, Lisi G et al (2022) Assessing structure and dynamics of AlphaFold2 prediction of GeoCas9. Biophys J 121:45a. https://doi.org/10.1016/j.bpj.2021.11.2474

Arora D, Mishra D, Budhlakoti N, Srivastava S, Singh A, Kumar S (2018) Introduction of reinforcement learning in bioinformatics. Biotech Today 8:25. https://doi.org/10.5958/2322-0996.2018.00019.4

Azzaz F, Yahi N, Chahinian H, Fantini J (2022) The epigenetic dimension of protein structure is an intrinsic weakness of the AlphaFold program. Biomol Ther 12(10). https://doi.org/10.3390/biom12101527

Bertoline LMF, Lima AN, Krieger JE, Teixeira SK (2023) Before and after AlphaFold2: an overview of protein structure prediction. Front Bioinform 3:1120370. https://doi.org/10.3389/fbinf.2023.1120370

Bowie JU, Eisenberg D (1994) An evolutionary approach to folding small alpha-helical proteins that uses sequence information and an empirical guiding fitness function. Proc Natl Acad Sci U S A 91(10):4436–4440. https://doi.org/10.1073/pnas.91.10.4436

Bradley P, Malmström L, Qian B, Schonbrun J, Chivian D, Kim DE et al (2005) Free modeling with Rosetta in CASP6. Proteins 61(S7):128–134. https://doi.org/10.1002/prot.20729

Brooks BR, Bruccoleri RE, Olafson BD, States DJ, Swaminathan S, Karplus M (1983) CHARMM: a program for macromolecular energy, minimization, and dynamics calculations. J Comput Chem 4(2):187–217. https://doi.org/10.1002/jcc.540040211

Brooks BR, Brooks CL 3rd, Mackerell AD Jr, Nilsson L, Petrella RJ, Roux B et al (2009) CHARMM: the biomolecular simulation program. J Comput Chem 30(10):1545–1614. https://doi.org/10.1002/jcc.21287

Buel GR, Walters KJ (2022) Can AlphaFold2 predict the impact of missense mutations on structure? Nat Struct Mol Biol 29(1):1–2. https://doi.org/10.1038/s41594-021-00714-2

Case DA, Cheatham TE 3rd, Darden T, Gohlke H, Luo R, Merz KM Jr et al (2005) The Amber biomolecular simulation programs. J Comput Chem 26(16):1668–1688. https://doi.org/10.1002/jcc.20290

Chen C, Chen L, Zou X, Cai P (2009) Prediction of protein secondary structure content by using the concept of Chous pseudo amino acid composition and support vector machine. Protein Pept Lett 16:27–31. https://doi.org/10.2174/092986609787049420

David CC, Jacobs DJ (2014) Principal component analysis: a method for determining the essential dynamics of proteins. Methods Mol Biol 1084:193–226. https://doi.org/10.1007/978-1-62703-658-0_11

Deng H, Jia Y, Zhang Y (2018) Protein structure prediction. Int J Mod Phys B 32(18):1840009. https://doi.org/10.1142/S021797921840009X

Diaz-Villanueva JF, Diaz-Molina R, Garcia-Gonzalez V (2015) Protein folding and mechanisms of Proteostasis. Int J Mol Sci 16(8):17193–17230. https://doi.org/10.3390/ijms160817193

Edwards SA, Wagner J, Gräter F (2012) Dynamic Prestress in a globular protein. PLoS Comput Biol 8(5):e1002509. https://doi.org/10.1371/journal.pcbi.1002509

Ehsaneddin A, Nina P, Alice CM, Mohammad RKM (2019) DeepPrime2Sec: deep learning for protein secondary structure prediction from the primary sequences. bioRxiv:705426. https://doi.org/10.1101/705426

Ema RR, Khatun A, Hossain MA, Akhond MR, Hossain N, Arafat MY (2022) Protein secondary structure prediction using hybrid recurrent neural networks. J Comput Sci 18(7):599. https://doi.org/10.3844/jcssp.2022.599.611

Englander SW, Mayne L, Krishna MM (2007) Protein folding and misfolding: mechanism and principles. Q Rev Biophys 40(4):287–326. https://doi.org/10.1017/S0033583508004654

Eswar N, Webb B, Marti-Renom MA, Madhusudhan MS, Eramian D, Shen MY et al (2006) Comparative protein structure modeling using Modeller. Curr Protoc Bioinformatics Chapter 5: Unit-5 6. https://doi.org/10.1002/0471250953.bi0506s15

Gelman S, Fahlberg SA, Heinzelman P, Romero PA, Gitter A (2021) Neural networks to learn protein sequence-function relationships from deep mutational scanning data. Proc Natl Acad Sci U S A 118(48). https://doi.org/10.1073/pnas.2104878118

Goodfellow I, Bengio Y, Courville A (2016) Deep learning. MIT Press

Guo Y, Wang B, Li W, Yang B (2018) Protein secondary structure prediction improved by recurrent neural networks integrated with two-dimensional convolutional neural networks. J Bioinforma Comput Biol 16(05):1850021. https://doi.org/10.1142/S021972001850021X

Gutnik D, Evseev P, Miroshnikov K, Shneider M (2023) Using AlphaFold predictions in viral research. Curr Issues Mol Biol 45(4):3705–3732. https://doi.org/10.3390/cimb45040240

Harrison RL (2010) Introduction to Monte Carlo simulation. AIP Conf Proc 1204:17–21. https://doi.org/10.1063/1.3295638

Hattori L, Benítez M, Lopes H. A deep bidirectional long short-term memory approach applied to the protein secondary structure prediction problem. 2017 IEEE Latin American Conference on Computational Intelligence (LA-CCI), 1–6.2017

Hochreiter S, Schmidhuber J (1997) Long short-term memory. Neural Comput 9(8):1735–1780. https://doi.org/10.1162/neco.1997.9.8.1735

Hu H, Li Z, Elofsson A, Xie S (2019) A Bi-LSTM based ensemble algorithm for prediction of protein secondary structure. Appl Sci 9. https://doi.org/10.3390/app9173538

Ismi DP, Pulungan R, Afiahayati. (2022) Deep learning for protein secondary structure prediction: pre and post-AlphaFold. Computational and structural. Biotechnol J 20:6271–6286. https://doi.org/10.1016/j.csbj.2022.11.012

Jin X, Guo L, Jiang Q, Wu N, Yao S (2022) Prediction of protein secondary structure based on an improved channel attention and multiscale convolution module. Front Bioeng Biotechnol 10: 901018

Jisna VA, Jayaraj PB (2021) Protein structure prediction: conventional and deep learning perspectives. Protein J 40(4):522–544. https://doi.org/10.1007/s10930-021-10003-y

Johnson KB, Wei WQ, Weeraratne D, Frisse ME, Misulis K, Rhee K et al (2021) Precision medicine, AI, and the future of personalized health care. Clin Transl Sci 14(1):86–93. https://doi.org/10.1111/cts.12884

Jolliffe IT, Cadima J (2016) Principal component analysis: a review and recent developments. Philos Trans R Soc A Math Phys Eng Sci 374(2065):20150202. https://doi.org/10.1098/rsta.2015.0202

Jumper J, Evans R, Pritzel A, Green T, Figurnov M, Ronneberger O et al (2021) Highly accurate protein structure prediction with AlphaFold. Nature 596(7873):583–589. https://doi.org/10.1038/s41586-021-03819-2

Kabsch W, Sander C (1983) Dictionary of protein secondary structure: pattern recognition of hydrogen-bonded and geometrical features. Biopolymers 22(12):2577–2637. https://doi.org/10.1002/bip.360221211

Kallberg M, Wang H, Wang S, Peng J, Wang Z, Lu H et al (2012) Template-based protein structure modeling using the RaptorX web server. Nat Protoc 7(8):1511–1522. https://doi.org/10.1038/nprot.2012.085

Kandathil SM, Lau AM, Jones DT (2023) Machine learning methods for predicting protein structure from single sequences. Curr Opin Struct Biol 81:102627. https://doi.org/10.1016/j.sbi.2023.102627

Kaufmann KW, Lemmon GH, Deluca SL, Sheehan JH, Meiler J (2010) Practically useful: what the Rosetta protein modeling suite can do for you. Biochemistry 49(14):2987–2998. https://doi.org/10.1021/bi902153g

Kelley LA, Mezulis S, Yates CM, Wass MN, Sternberg MJ (2015) The Phyre2 web portal for protein modeling, prediction and analysis. Nat Protoc 10(6):845–858. https://doi.org/10.1038/nprot.2015.053

Kim RQ, van Dijk WJ, Sixma TK (2016) Structure of USP7 catalytic domain and three Ubl-domains reveals a connector alpha-helix with regulatory role. J Struct Biol 195(1):11–18. https://doi.org/10.1016/j.jsb.2016.05.005

Kitao A (2022) Principal component analysis and related methods for investigating the dynamics of biological macromolecules. J 5:298. https://doi.org/10.3390/j5020021

Kouza M, Faraggi E, Kolinski A, Kloczkowski A (2017) The GOR method of protein secondary structure prediction and its application as a protein aggregation prediction tool. Methods Mol Biol 1484:7–24. https://doi.org/10.1007/978-1-4939-6406-2_2

Kuhlman B, Bradley P (2019) Advances in protein structure prediction and design. Nat Rev Mol Cell Biol 20(11):681–697. https://doi.org/10.1038/s41580-019-0163-x

Kumar S, Raj I, Nagpal I, Subbarao N, Gourinath S (2011) Structural and biochemical studies of serine acetyltransferase reveal why the parasite Entamoeba histolytica cannot form a cysteine synthase complex. J Biol Chem 286(14):12533–12541. https://doi.org/10.1074/jbc.M110.197376

Laurents D (2022) AlphaFold 2 and NMR spectroscopy: partners to understand protein structure, dynamics and function. Front Mol Biosci 9:9. https://doi.org/10.3389/fmolb.2022.906437

Le NQK, Yapp EKY, Yeh H-Y (2019) ET-GRU: using multi-layer gated recurrent units to identify electron transport proteins. BMC Bioinformatics. 20(1):377. https://doi.org/10.1186/s12859-019-2972-5

Lee M (2023) Recent advances in deep learning for protein-protein interaction analysis: a comprehensive review. Molecules 28(13). https://doi.org/10.3390/molecules28135169

Lee J, Wu S, Zhang Y. Ab initio protein structure prediction 2009:3–25. doi: https://doi.org/10.1007/978-1-4020-9058-5_1

Lee J, Freddolino PL, Zhang Y. Ab initio protein structure prediction. 2017:3–35. doi: https://doi.org/10.1007/978-94-024-1069-3_1

Liu X (2017) Deep recurrent neural network for protein function prediction from sequence. arXiv:701.08318. https://doi.org/10.48550/arXiv.1701.08318

Liu Y, Cheng J, Ma Y, Chen Y. Protein secondary structure prediction based on two dimensional deep convolutional neural networks. 2017 3rd IEEE International Conference on Computer and Communications (ICCC), 2017. p. 1995–9

Lyu Z, Wang Z, Luo F, Shuai J, Huang Y (2021) Protein secondary structure prediction with a reductive deep learning method. Front Bioeng Biotechnol 9:687426. https://doi.org/10.3389/fbioe.2021.687426

Melman P, Roshan U. K-means-based feature learning for protein sequence classification. 2018

Minami S, Sawada K, Chikenji G (2014) How a spatial arrangement of secondary structure elements is dispersed in the universe of protein folds. PLoS One 9(9):e107959. https://doi.org/10.1371/journal.pone.0107959

Nagaratnam N, Martin-Garcia JM, Yang JH, Goode MR, Ketawala G, Craciunescu FM et al (2022) Structural and biophysical properties of FopA, a major outer membrane protein of Francisella tularensis. PLoS One 17(8):e0267370. https://doi.org/10.1371/journal.pone.0267370

Novak WRP (2021) Tertiary structure domains, folds, and motifs. In: Bell E (ed) Molecular life sciences: an encyclopedic reference. Springer New York, New York, NY, pp 1–5

Pakhrin SC, Shrestha B, Adhikari B, Kc DB (2021) Deep learning-based advances in protein structure prediction. Int J Mol Sci 22(11). https://doi.org/10.3390/ijms22115553

Parras-Moltó M, Campos-Laborie FJ, García-Diéguez J, Rodríguez-Griñolo MR, Pérez-Pulido AJ (2013) Classification of protein motifs based on subcellular localization uncovers evolutionary relationships at both sequence and functional levels. BMC Bioinformatics. 14(1):229. https://doi.org/10.1186/1471-2105-14-229

Paul D, Sanap G, Shenoy S, Kalyane D, Kalia K, Tekade RK (2021) Artificial intelligence in drug discovery and development. Drug Discov Today 26(1):80–93. https://doi.org/10.1016/j.drudis.2020.10.010

Paul B, Weeratunga S, Tillu VA, Hariri H, Henne WM, Collins BM (2022) Structural predictions of the SNX-RGS proteins suggest they belong to a new class of lipid transfer proteins. Front Cell Dev Biol 10:826688. https://doi.org/10.3389/fcell.2022.826688

Pearce R, Zhang Y (2021) Toward the solution of the protein structure prediction problem. J Biol Chem 297(1):100870. https://doi.org/10.1016/j.jbc.2021.100870

Perlinska AP, Niemyska WH, Gren BA, Bukowicki M, Nowakowski S, Rubach P et al (2023) AlphaFold predicts novel human proteins with knots. Protein Sci 32(5):e4631. https://doi.org/10.1002/pro.4631

Perrakis A, Sixma TK (2021) AI revolutions in biology. EMBO Rep 22(11):e54046. https://doi.org/10.15252/embr.202154046

Ruff KM, Pappu RV (2021) AlphaFold and implications for intrinsically disordered proteins. J Mol Biol 433(20):167208. https://doi.org/10.1016/j.jmb.2021.167208

Russo A, Borras A (2022) Comparison of dimension reduction techniques applied to the analysis of airborne radionuclide activity concentration. J Environ Radioact 244-245:106813. https://doi.org/10.1016/j.jenvrad.2022.106813

Schrodinger, LLC. The PyMOL molecular graphics system, Version 1.8. 2015

Senior AW, Evans R, Jumper J, Kirkpatrick J, Sifre L, Green T et al (2020) Improved protein structure prediction using potentials from deep learning. Nature 577(7792):706–710. https://doi.org/10.1038/s41586-019-1923-7

Sharma KA, Srivastava R (2021) Protein secondary structure prediction using character bi-gram embedding and Bi-LSTM. Curr Bioinforma 16(2):333–338. https://doi.org/10.2174/1574893615999200601122840

Simons KT, Kooperberg C, Huang E, Baker D (1997) Assembly of protein tertiary structures from fragments with similar local sequences using simulated annealing and Bayesian scoring functions. J Mol Biol 268(1):209–225. https://doi.org/10.1006/jmbi.1997.0959

Sippl MJ (1990) Calculation of conformational ensembles from potentials of mean force. An approach to the knowledge-based prediction of local structures in globular proteins. J Mol Biol 213(4):859–883. https://doi.org/10.1016/s0022-2836(05)80269-4

Sun PD, Foster CE, Boyington JC (2004) Overview of protein structural and functional folds. Curr Protoc Protein Sci Chapter 17(1):Unit 17 1. https://doi.org/10.1002/0471140864.ps1701s35

Tang K, Zhang J, Liang J (2014) Fast protein loop sampling and structure prediction using distance-guided sequential chain-growth Monte Carlo method. PLoS Comput Biol 10(4):e1003539. https://doi.org/10.1371/journal.pcbi.1003539

Taye MM (2023) Understanding of machine learning with deep learning: architectures, workflow, applications and future directions. Computers 12. https://doi.org/10.3390/computers12050091

Teletin M, Czibula G, Albert S, Bocicor M-I (2018) Using unsupervised learning methods for enhancing protein structure insight. Procedia Comput Sci 126:19–28. https://doi.org/10.1016/j.procs.2018.07.205

Tokuriki N, Tawfik DS (2009) Protein dynamism and evolvability. Science 19359577(5934): 203–207

Travaglini-Allocatelli C, Ivarsson Y, Jemth P, Gianni S (2009) Folding and stability of globular proteins and implications for function. Curr Opin Struct Biol 19(1):3–7. https://doi.org/10.1016/j.sbi.2008.12.001

Vaswani A, Shazeer N, Parmar N, Uszkoreit J, Jones L, Gomez AN et al (2017) Attention is all you need. In: Guyon I, Luxburg UV, Bengio S, Wallach H, Fergus R, Vishwanathan S et al (eds) 31st Conference on Neural Information Processing Systems (NIPS 2017). NeurIPS Proceedings, Long Beach, CA, USA

Vijayan RSK, Kihlberg J, Cross JB, Poongavanam V (2022) Enhancing preclinical drug discovery with artificial intelligence. Drug Discov Today 27(4):967–984. https://doi.org/10.1016/j.drudis.2021.11.023

Wang J, Wolf RM, Caldwell JW, Kollman PA, Case DA (2004) Development and testing of a general amber force field. J Comput Chem 25(9):1157–1174. https://doi.org/10.1002/jcc.20035

Wang S, Peng J, Ma J, Xu J (2016a) Protein secondary structure prediction using deep convolutional neural fields. Sci Rep 6(1):18962. https://doi.org/10.1038/srep18962

Wang S, Li W, Liu S, Xu J (2016b) RaptorX-property: a web server for protein structure property prediction. Nucleic Acids Res 44(W1):W430–W435. https://doi.org/10.1093/nar/gkw306

Waterhouse A, Bertoni M, Bienert S, Studer G, Tauriello G, Gumienny R et al (2018) SWISS-MODEL: homology modelling of protein structures and complexes. Nucleic Acids Res 46(W1): W296–W303. https://doi.org/10.1093/nar/gky427

Webb B, Sali A (2016) Comparative protein structure modeling using MODELLER. Curr Protoc Bioinformatics 54:5.6.1–5.6.37. https://doi.org/10.1002/cpbi.3

Wick CD, Siepmann JI (2000) Self-adapting fixed-end-point configurational-bias Monte Carlo method for the regrowth of interior segments of chain molecules with strong intramolecular interactions. Macromolecules 33(19):7207–7218. https://doi.org/10.1021/ma000172g

Wong S, Liu J, Kou S (2018) Exploring the conformational space for protein folding with sequential Monte Carlo. Ann Appl Stat 12:1628–1654. https://doi.org/10.1214/17-AOAS1124

Wong F, Krishnan A, Zheng EJ, Sträk H, Manson AL, Earl AM et al (2022) Benchmarking AlphaFold-enabled molecular docking predictions for antibiotic discovery. Mol Syst Biol 18(9):e11081. https://doi.org/10.15252/msb.202211081

Yang J, Zhang Y (2015) I-TASSER server: new development for protein structure and function predictions. Nucleic Acids Res 43(W1):W174–W181. https://doi.org/10.1093/nar/gkv342

Yang Y, Heffernan R, Paliwal K, Lyons J, Dehzangi I, Sharma A et al (2017) SPIDER2: a package to predict secondary structure, accessible surface area, and main-chain torsional angles by deep neural networks. Methods Mol Biol 1484, 55:–63

Yang S, Wang Y, Cruz-Gutierrez K, Wu F, Ding C-F (2021) Localnet: a simple recurrent neural network model for protein secondary structure prediction using local amino acid sequences only. Research Square

Yang Z, Zeng X, Zhao Y, Chen R (2023) AlphaFold2 and its applications in the fields of biology and medicine. Signal Transduct Target Ther 8(1):115. https://doi.org/10.1038/s41392-023-01381-z

Yu C-H, Chen W, Chiang Y-H, Guo K, Martin Moldes Z, Kaplan DL et al (2022) End-to-end deep learning model to predict and design secondary structure content of structural proteins. ACS Biomater Sci Eng 8(3):1156–1165. https://doi.org/10.1021/acsbiomaterials.1c01343

Zaslavsky L, Ciufo S, Fedorov B, Tatusova T (2016) Clustering analysis of proteins from microbial genomes at multiple levels of resolution. BMC Bioinformatics 17(Suppl 8):276. https://doi.org/10.1186/s12859-016-1112-8

Zhang Y, Kihara D, Skolnick J (2002) Local energy landscape flattening: parallel hyperbolic Monte Carlo sampling of protein folding. Proteins 48(2):192–201. https://doi.org/10.1002/prot.10141

Zhang J, Kou SC, Liu JS (2007) Biopolymer structure simulation and optimization via fragment regrowth Monte Carlo. J Chem Phys 126(22):225101. https://doi.org/10.1063/1.2736681

Zhang G, Tang Q, Feng P, Chen W (2023) IPs-GRUAtt: an attention-based bidirectional gated recurrent unit network for predicting phosphorylation sites of SARS-CoV-2 infection. Mol Ther Nucleic Acids 32:28–35. https://doi.org/10.1016/j.omtn.2023.02.027

Zhao Y, Liu Y (2021) OCLSTM: optimized convolutional and long short-term memory neural network model for protein secondary structure prediction. PLoS One 16:e0245982. https://doi.org/10.1371/journal.pone.0245982

Zhu Y, Ting KM, Carman MJ (2016) Density-ratio based clustering for discovering clusters with varying densities. Pattern Recogn 60:983–997. https://doi.org/10.1016/j.patcog.2016.07.007

Zhu L, Davari MD, Li W (2021) Recent advances in the prediction of protein structural classes: feature descriptors and machine learning algorithms. Crystals 11. https://doi.org/10.3390/cryst11040324

Zimmermann L, Stephens A, Nam SZ, Rau D, Kubler J, Lozajic M et al (2018) A completely reimplemented MPI bioinformatics toolkit with a new HHpred server at its core. J Mol Biol 430(15):2237–2243. https://doi.org/10.1016/j.jmb.2017.12.007

SPRINGER NATURE

GPSR Compliance

The European Union's (EU) General Product Safety Regulation (GPSR) is a set of rules that requires consumer products to be safe and our obligations to ensure this.

If you have any concerns about our products, you can contact us on ProductSafety@springernature.com

In case Publisher is established outside the EU, the EU authorized representative is:

Springer Nature Customer Service Center GmbH
Europaplatz 3
69115 Heidelberg, Germany

The manufacturer's authorised representative in the EU is Springer Nature Customer Service Centre GmbH, Europaplatz 3, 69115 Heidelberg, Germany. If you have any concerns regarding our products, please contact ProductSafety@springernature.com

Printed and bound by CPI Group (UK) Ltd, Croydon, CR0 4YY

30/09/2026

02225314-0013